Testing of Communicating Systems
Methods and Applications

IFIP - The International Federation for Information Processing

IFIP was founded in 1960 under the auspices of UNESCO, following the First World Computer Congress held in Paris the previous year. An umbrella organization for societies working in information processing, IFIP's aim is two-fold: to support information processing within its member countries and to encourage technology transfer to developing nations. As its mission statement clearly states,

IFIP's mission is to be the leading, truly international, apolitical organization which encourages and assists in the development, exploitation and application of information technology for the benefit of all people.

IFIP is a non-profitmaking organization, run almost solely by 2500 volunteers. It operates through a number of technical committees, which organize events and publications. IFIP's events range from an international congress to local seminars, but the most important are:

- The IFIP World Computer Congress, held every second year;
- open conferences;
- working conferences.

The flagship event is the IFIP World Computer Congress, at which both invited and contributed papers are presented. Contributed papers are rigorously refereed and the rejection rate is high.

As with the Congress, participation in the open conferences is open to all and papers may be invited or submitted. Again, submitted papers are stringently refereed.

The working conferences are structured differently. They are usually run by a working group and attendance is small and by invitation only. Their purpose is to create an atmosphere conducive to innovation and development. Refereeing is less rigorous and papers are subjected to extensive group discussion.

Publications arising from IFIP events vary. The papers presented at the IFIP World Computer Congress and at open conferences are published as conference proceedings, while the results of the working conferences are often published as collections of selected and edited papers.

Any national society whose primary activity is in information may apply to become a full member of IFIP, although full membership is restricted to one society per country. Full members are entitled to vote at the annual General Assembly, National societies preferring a less committed involvement may apply for associate or corresponding membership. Associate members enjoy the same benefits as full members, but without voting rights. Corresponding members are not represented in IFIP bodies. Affiliated membership is open to non-national societies, and individual and honorary membership schemes are also offered.

Testing of Communicating Systems

Methods and Applications

*IFIP TC6 12th International Workshop
on Testing of Communicating Systems,
September 1-3, 1999, Budapest, Hungary*

Edited by

Gyula Csopaki
Technical University of Budapest

Sarolta Dibuz
Ericsson Hungary

Katalin Tarnay
Nokia Hungary

SPRINGER SCIENCE+BUSINESS MEDIA, LLC

Library of Congress Cataloging-in-Publication Data

IFIP TC6 International Workshop on Testing of Communicating Systems (12th:1999:
 Budapest, Hungary)
  Testing of communicating systems : methods and applications : IFIP TC6 12th
 International Workshop on Testing of Communicating Systems, September 1-3,
 1999, Budapest, Hungary / edited by Gyula Csopaki, Sarolta Dibuz, Katalin Tarnay.

  Includes bibliographical references.

  DOI 10.1007/978-0-387-35567-2
  1. Telecommunications systems—Testing Congresses. I. Csopaki, Gyula. II.
Dibuz, Sarolta. III. Tarnay, Katalin. IV. Title.

 TK5101.A1I48292   1999
 004.6'2'0287—dc21   99-34939
                                                                          CIP

Contents

PREFACE

This volume is the proceedings of the 12th IFIP WG 6.1 International Workshop on Testing Communicating Systems (IWTCS'99), which is being held in Budapest, Hungary, from 1 to 3 September, 1999. This workshop being the twelfth of a series of annual meetings sponsored by IFIP WG 6.1, is the continuation of the IFIP International Workshop on Protocol Test Systems. The previous eleven workshops were held in Vancouver, Canada, 1988; Berlin, Germany, 1989; Mclean, USA, 1990; Leidschendam, the Netherlands, 1991; Montreal, Canada, 1992; Pau, France, 1993; Tokyo, Japan, 1994; Evry, France, 1995; Darmstadt, Germany, 1996; Cheju Island,Korea; 1997, Tomsk, Russia, 1998.

The workshop aims at bringing researchers and practitioners together and promoting the exchange of views and experiences as it has happend in the years before. The papers reflect the real requirements regarding how to use the methods and tools in industry. Twenty-nine papers were submitted to IWTCS'99 and all of them have been reviewed by the Program Committee, as well as and addititonal reviewers. Twenty-one papers have been selected for the conference; this proceedings contains these papers together with three invited papers.

IWTCS'99 was organized under the auspices of IFIP WG 6.1 by Ericsson Hungary, the Technical University of Budapest, Nokia Hungary and Conference Tours. It was financially supported by Ericsson and Nokia.

We would like to thank everyone who has contributed to the success of the conference. In particular, we are grateful to the authors for writing and presenting their papers, the reviewers for assessing and commenting on the submitted papers. We are grateful to Hartmut König and Myungchul Kim, members of the Program Committee, for sharing their experiences in organizing the conference with us, and all those who participate in the local organization.

Gyula Csopaki, Sarolta Dibuz and Katalin Tarnay
Budapest, Hungary, September 1999

CONFERENCE CO-CHAIRS

Gyula Csopaki
Technical University of Budapest, Hungary
csopaki@bme-tel.ttt.bme.hu

Sarolta Dibuz
Ericsson Telecommunication Ltd., Hungary
Sarolta.Dibuz@eth.ericsson.se

Katalin Tarnay
Nokia Hungary
katalin.tarnay@ntc.nokia.com

PROGRAM COMMITTEE

A. Ahtiainen (NOKIA, Finland)
B. Baumgarten (GMD, Germany)
G. v. Bochmann (University of Ottawa, Canada)
M. Boda (ERICSSON Radio Systems, Sweden)
E. Brinksma (Univ. of Twente, Netherlands)
S. Budkowski (INT, France)
A. R. Cavalli (INT, France)
S. T. Chanson (Hong Kong Univ. of S. & T., China)
A. T. Dahbura (DEC, USA)
P. Dembinski (IPI PAN, Poland)
R. Dssouli (Univ. of Montreal, Canada)
A. Ek (Telelogic, Sweden)
R. Groz (CNET, France)
T. Higashino (Osaka Univ., Japan)
D. Hogrefe (Univ. of Luebeck, Germany)
K. Inan (Sabanci University, Turkey)
N. Yevtushenko (Tomsk Univ., Russia)
M. Kim (Information and Communications Univ., Korea)
H. Koenig (Technical Univ. of Cottbus, Germany)
E. Kwast (KPN Research, Netherlands)
Gy. Lajtha (Hungarian PTT, Hungary)
G. Leduc (Univ. of Liege, Belgium)
H. Liu (Bellcore, USA)
J. de Meer (GMD, Germany)

A. Petrenko (CRIM, Canada)
R. Probert (Univ. of Ottawa, Canada)
O. Rafiq (Univ. of Pau, France)
S. Randall (PQM Consulting, UK)
D. Rayner (NPL, UK)
I. Schieferdecker (GMD, Germany)
K. Suzuki (KDD K&D Laboratories, Japan)
K. C. Tai (North Carolina State Univ., USA)
J. Tretmans (Univ. of Twente, Netherlands)
K. Turner (Univ. of Stirling, UK)
H. Ural (Univ. of Ottawa, Canada)
S. T. Vuong (Univ. of British Columbia, Canada)
J. Wu (Tsinghua Univ., China)

FURTHER REVIEWERS

S. Boroday
K. El-Fakih
J. Grabowski
A. Idoue

B. Koch
M. Luukkainen
R. G. de Vries

ORGANIZATION COMMITTEE

Tibor Csöndes, Endre Horváth, Balázs Kotnyek, Péter Krémer
    Ericsson Telecommunication Ltd., Hungary

Zsuzsa Karámos, Eszter Zubovics
    Conference Tours Ltd., Budapest

I

INVITED TALKS

1

DIFFERENT APPROACHES TO PROTOCOL AND SERVICE TESTING

Ana Cavalli
Institut National des Telecommunications
9 rue Charles Fourier
F-91011 Evry Cedex
Ana.Cavalli@int-evry.fr

Abstract   This paper presents an overview of the work we have developed at INT on different approaches to protocol and service testing. It is related to performing tests of communication protocols and telecommunication services based on techniques, such as, active test generation, passive testing, embedded testing and test execution.

Keywords: Conformance Testing, Test Generation, Embedded Testing, Passive Testing, SDL, CORBA.

1.   INTRODUCTION

This paper presents an overview of the work we have developed at INT related to different approaches to performing tests of communication protocols and telecommunication services. It concerns active test generation, passive testing, embedded testing, test execution and their application to real protocols and services.

This work is mainly a work developed by my research group at INT, by myself and my PhD students [24, 9, 26, 27, 28, 33, 19, 25] but it is also the result of collaborations with colleagues from CNET [3, 6], Bell Labs [4, 10] and academic collaborations [5, 36, 35, 14]. It concerns different aspects of testing, that we have studied in the last years. This work covers all phases of the testing activity. We started by mainly studying the application of formal methods to test generation and we have developed a prototype tool, called TESTGEN. This tool allows automatic test generation from LOTOS and SDL specifications. In its first version it was strongly influenced by I/O FSMs test generation methods [5].

To solve the problem of combinatorial explosion of states and to perform embedded testing, we have integrated a new composition algorithm. This algorithm hides internal actions and controls the composition process in order to identify global transitions that reflect the behavior of the component under test [26, 28].

Following this work, we have defined and implemented new algorithms to optimize the resulting test sequences and to perform still embedded testing. We have then integrated two new algorithms: one to optimize the test generation [4] and the other to perform embedded testing using a transition coverage criteria of the embedded component [10]. Finally, we have integrated an algorithm that introduces a new approach to testing, passive testing [33].

While we were working on developing these testing methods, we started to work on open distributed architectures for telecommunications networks. We considered that it was interesting to execute the tests on a distributed environment in order to test systems implementations obtained from formal specifications. We have therefore developed an execution test platform into a CORBA environment [25].

In the following, I will provide a more detailed description of each of these steps of our work and give some previews of our future work.

2.   THE TESTGEN TOOL

2.1   PRELIMINARIES

A program specification is composed of a control part and a data part. This section deals with the control part only. The control portion of a program, which will be referred to as a *program specification*, can be modelled as an Input/Output Finite State Machine (I/O FSM) with a finite set of states $S=\{s_1, s_2,..., s_n\}$, a finite set of inputs $I=\{a_1,..., a_k\}$, and a finite set of outputs $O=\{x_1,x_2,...,x_m\}$. Both sets include the symbol *null* used to represent *null* inputs or outputs. The next state (δ) and output (λ) relation are given by the relations $\delta : S \times I \to S$ and $\lambda : S \times I \to O$.

The I/O FSM is usually represented by a directed graph G=(V,E), where V= $\{v_1,v_2,...,v_n\}$ represents the set of states S, and a directed edge represents a transition from one state to another in the I/O FSM. Each edge in G is labelled by an input a_r and a corresponding output x_q. An edge in E from v_i to v_j which has label a_r/x_q means that the I/O FSM, in state s_i, upon receiving input a_r produces output x_q and moves to state s_j. A triple $(s_i,s_j,a_r/x_q)$ is used to denote a transition.

The graph representation is useful for describing and reasoning about test generation algorithms. Within this context, we give some basic definitions from graph theory. A graph is said to be strongly connected if for any pair of distinct vertices v_i and v_j there is a walk which starts on v_i and ends on v_j. A walk W over a graph is a finite, non-empty sequence of consecutive edges. Head(W) and Tail(W) denote respectively the vertex where the walk W starts and the vertex where it ends. A path is a walk in which each edge of G appears exactly once.

An input/output sequence $U = (a_{r1}/x_{q1}, a_{r2}/x_{q2}, ..., a_{rn}/x_{qn})$ is said to be specified for state s_i in a specification I/O FSM if there exists a walk W with origin s_i in the graph representation of the I/O FSM such that $W = (s_i, s_{j1}, a_{r1}/x_{q1})$, $(s_{j1}, s_{j2}, a_{r2}/x_{q2}), ..., (s_{jn-1}, s_{jn}, a_{rn}/x_{qn})$. An I/O FSM is said to be fully specified if from each state it has a transition for every input symbol; otherwise the I/O FSM is said to be partially specified. If an I/O FSM is partially specified and a non specified transition is applied, under the Completeness Assumption it is obtained that the I/O FSM will either stay in the same state without any output or signal an error. The initial state of an I/O FSM is the state the I/O FSM enters after power-up. An I/O FSM is said to have the reset capability if it can move from any state directly into the initial state with a single transition, denoted "*ri/null*" or simply "*ri*". State s_i is said to be weakly equivalent to state s_j if any specified input/output sequence for s_i is also specified for s_j.

A Unique Input/Ouput sequence (UIO sequence) for state s_i, denoted UIO(i), is an input/output sequence with origin s_i such that there is no $s_j \neq s_i$ for which UIO(i) is a specified sequence for starting state s_j [1].

A state possesses multiple UIO sequences if it possesses more than one UIO sequence. Some I/O FSMs do not possess UIO sequences for every state. The proposed approach to identify a state s_i that does not have an UIO is to use a signature, a sequence that distinguishes s_i from each of the remaining states one at a time. A signature for s_i uses (n-1) input/output sequences IO(i,m), each of which distinguishes s_i from one other state s_m. Before applying each IO sequence the I/O FSM must be put back into state s_i. Suppose that after the subsequence IO(i,m) is applied the specification is in state s_k. The method proposed in [1] is to use a transfer sequence Tr(k,i), which is the shortest path from s_k to s_i, in order to bring the I/O FSM back to state s_i. Therefore, the signature for state s_i is formed by concatenating the sequences $IO(i, m) @ Tr(Tail(IO(i, m)), i)$ for all $m \neq i$, where @ indicates the concatenation of each IO sequence with the corresponding transfer sequence. The test generation method presented in the following sections uses a modified version of the concept of signature, based on the SIO sequences (Set of IO sequences), that we will denoted by Partial Unique Input/Ouput sequence (PUIO).

2.2   THE PROCEDURE FOR TESTING

The procedure for testing an implementation (whose behaviour can be described through an I/O FSM) is essentially divided into two parts [5]:

1 state identification: the implementation is forced to display the response of each state to the Unique Input/Output sequence (UIO) (or to the chosen signature) checking whether an implementation I/O FSM has n states and that each state identifier preserves its feature. This is carried out by repeating the following procedure for each state of the I/O FSM:

- the implementation is put into state p_i, presumably isomorphic to s_i, by applying a reset (assumed reliable) followed by some path (already tested) from the initial state s_0 to s_i;

- UIO (or other chosen signature) is applied to the implementation and the output is checked to verify whether it was expected or not.

2 transition testing: each transition is tested. The procedure for testing a transition t = $(s_i,\ s_j,\ a_r/x_q)$ from state s_i to s_j with input a_r is the following:

- the implementation is put into state p_i, known to be isomorphic to s_i, from the state identification part;

- input a_r is applied and the output is checked to assure that it is x_q as expected;

- the new state of the implementation is then tested to verify if it is isomorphic to state p_j. This is done by applying the UIO (or the chosen signature) as in the state identification part.

Any method based on UIOs, or a similar kind of signature, is valid if the UIOs or other signatures are unique both in the specification and the implementation. This uniqueness of the identification state must be verified. This can be done by putting the implementation into state p_i, presumable isomorphic to state s_i, and then applying UIO(j) to the implementation in order to verify that the resulting output is not what it was expected if UIO(j) was applied to s_j.

2.3   THE TOOL TESTGEN

In the first version of our tool TESTGEN, we have adapted the traditional test methods, based on finite automata [1, 32] to test generation and their application to real protocols. In the previous sections we have provided some basic concepts of these methods. These methods present two main advantages: they allow a totally automatic test generation, and they guarantee full coverage for certain types of faults (transfer and output faults) under certain test hypothesis.

In the following we will present test generation from a SDL specification [20], (for a description of test generation from LOTOS, see [7, 8]). The procedure to generate test sequences from a SDL specification can be decomposed in the following steps:

Step 1. Note that the method is applied to a SDL specification. This means that we start with a system of communicating extended finite state machines. This system will be simulated in order to obtain an accessibility graph. The accessibility graph that includes data and control parts of the original specification is computed using the simulator features of the ObjectGEODE tool [34]. This graph is restricted by fixing the values of the parameters in messages received from the environment by the system under test. An I/O FSM is obtained. The constraint on the arguments values is a very important optimization, since it permits the tool to obtain a graph even in the case of a real protocol;

Step 2. The obtained input/output machine is transformed: transitions labelled by internal actions and intermediate states are eliminated using an algorithm described briefly in the following section. We then obtain a *reference finite state machine* (see next section);

Step 3. Certain cases of non-determinism are eliminated using a classical determinization algorithm [2], while others cases, which may be significant for the test, are conserved. Furthermore, the *reference finite state machine* is transformed in a completely specified machine by the addition, on every state, of self-loops having as input labels the inputs not specified for that state and *null* as outputs labels. Notice that this transformation preserves the semantics of SDL. In fact, in certain cases it is not necessary to make every state complete. It is necessary only in certain states, so a unique identification sequence can be obtained;

Step 4. A minimization algorithm based on trace equivalence is applied. This minimization is important, since it makes it possible to obtain states identification sequences of reasonable size even for real protocols. In this step, we also verify that the finite state machine is strongly connected;

Step 5. UIOs , multiple UIOs (MUIOs) or partial UIOs (PUIOs) are defined for each state. TESTGEN allows the use of these different types of UIO sequences. By default, the type that generates the shortest test sequence is chosen;

Step 6. For each state of the minimized automaton, a preamble to put the implementation in that state is computed. Then, if possible, one or more UIO sequences for the state are computed. As some states may not have a UIO sequence, partial UIOs (PUOs) are computed for them.

Step 7. The test sequence is calculated. Once preambles and UIO (or PUIO) are obtained, the generation of the test sequence proceeds in two steps as follows:

- 7.1. First, it is verified that UIO and/or PUIO sequences are uniquely accepted by the states in the implementation;

- 7.2. Then, every transition in the specification is verified to have a corresponding transition in the implementation. This is done by applying to the implementation first the preamble, in order to put it at the start of the transition, then the transition itself, and finally the UIO sequences to verify that the transition put the implementation in the expected end state;

Step 8. The test generation can operate in two modes, optimized or non-optimized. In the optimized mode, one single test sequence is obtained by using an optimization algorithm (either Rural Chinese Postman Tour [1], or greedy algorithm with overlaps and multiple UIOs [11]). In the non-optimized mode, a set of test independent test cases is obtained. In both cases, the resulting test is translated into TTCN.

2.4   GENERATION OF THE REFERENCE FINITE STATE MACHINE

The global finite state machine (obtained using the ObjectGEODE tool) can have redundant states, as well as transitions labelled by internal or empty (*null*) messages. In order to reduce the size of the finite state machine, two transformations are made:

1. *Treatment of internal and empty signals.* In the global finite state machine, we distinguish actions that interact with the PCOs (Points of Control and Observation, which constitute the interface between the tester and the implementation under test) and actions that are not seen at the PCOs. The latter are called internal actions (denoted by τ). The test architecture defining the PCOs will determine the actions that will be considered as internal. Empty signals (denoted *null*) are used to model the input of spontaneous transitions, or the output of actions that do not produce events.

We apply a reduction algorithm to the automaton in order to eliminate part of the internal actions and empty signals, obtaining a finite state machine with fewer states. The principle of this reduction algorithm is to detect certain chains of transitions that can be grouped, for they are all internal. The simplest case is when two consecutive transitions have the form "in/τ" and "τ/out" where "in" and "out" are actions observable at the PCOs. In this case, these two transitions are replaced by a single transition labelled "in/out"; if there are no other incoming or outgoing transitions, the intermediate state is eliminated. *Null* signals are eliminated using the same method. But it must be noted that the algorithm treats "τ" and "*null*" separately and keeps them distinct in the resulting machine. This way, certain spontaneous actions are removed; for example those that appear after the start state in a SDL specification. On the other hand, there may be transitions of the form "in/τ" (a signal from the environment followed by an internal signal), or "in/null" (a signal from the environment followed by a *null* signal) that cannot be removed. This is due to the fact that those transitions model the situations where the specified system accept an input signal but does not produce an observable answer. Transitions labelled with internal signals that cannot be deleted/hidden indicated a bad specified or incomplete machine. A more detailed description of this algorithm is given in [3, 28]. The experience of using this algorithm has shown that the resulting automata has fewer states than the original one. On the other hand, the number of transitions usually increases. The reasons for this is that when one state is eliminated the number of resulting transitions correspond to the number of incoming transitions multiplied by the number of outgoing transitions on that state. However, if the number of states eliminated is substantial, then the number of transitions also decreases. In the case of the MAP protocol [3], the algorithm reduces an automaton with 13.794 states and 45.893 transitions to an automaton with 6.193 states and 43.445 transitions. The application to the INRES protocol [15] gives the following results: it reduced an automaton of 175 states and 267 transitions to an automaton with 60 states and 154 transitions.

2. Minimization based on trace equivalence. This minimization eliminates redundancies and therefore reduces the number of states and transitions of the automata. In the case of the MAP protocol [3], this minimization produced an automaton with 25 states and 187 transitions, from an automaton with 6.193 states and 43.445 transitions.

3.   EMBEDDING TESTING

Traditional testing methods proposed to test systems as a whole or to test their components in isolation. Testing systems as a whole becomes difficult due to their huge size. On the other hand, the purpose of testing can be to test only some system components in order to avoid the redundant testing of other parts of the system. In this case, we are concerned with embedded testing or testing in context [24, 26, 28, 30, 29, 23, 13].

Embedded testing is to test the component's behaviour in context and to detect if the component's implementation conforms to its specification. It is generally assumed that the tester does not have a direct access to the component interfaces. The access is then made through another component of the system, which acts as a sort of "filter". According to the standard: "if control and observation are applied through one or more implementations which are above the protocol to be tested, the testing methods are called embedded" [16].

Many of these systems can be formally specified as a system of distributed interacting extended finite state machines (EFSMs). EFSMs, which are finite state machines extended with variables, have emerged for the design and analysis of both communication protocols and services. They are also the underlying model of some formal specification languages (as for example, SDL [20]).

Concerning embedded testing we have developed two approaches. The first one is based on a trace keeping algorithm for automaton composition. A detailed presentation of this algorithm is given in [26] and [28]. The generation of test sequences from formal specifications has been traditionally based on the exhaustive simulation of the specification in order to obtain an automaton that represents the global behaviour of the system. Since it is impossible, in most of the cases, to deal with the size of the automaton that represents the complete behaviour of these systems, a reasonable approach is to simulate the execution of the specification by controlling the range of values assigned to each internal variable and each parameter of input messages. The closer this range is to the real one, the more realistic and the larger the test will be. Obviously, there is always a compromise between accuracy (completeness) of the automaton and his size. However, even with an automaton of a "computable" size, the process of test sequence generation may not be able to cope with that automaton in a reasonable period of time.

In the first version of TESTGEN, to generate test sequences, what we have done is to take the "big" automaton (that is, the one which is as close to the specification as possible) and then, though the definition of view points (PCOs), abstract the signals which are irrelevant in the current consideration or view point. Then, we may proceed by minimizing the automaton using an algorithm described briefly in the previous section and presented in [5] and [28], which removes all internal signals (if the choice of the PCOs is well done). We thereby obtain an automaton that corresponds to the "big one", by abstracting details we do not yet want to consider. In general, this automaton has a reasonable size, and therefore it can be used as input for the process of deriving test sequences. However, even with a big automaton generated by simulation, the reduced one is often simpler than we would like.

To solve these problems we have considered the use of a composition algorithm in the following way. The idea is to avoid the initial automaton size explosion by dividing (which is often already done, if we are dealing with modular systems) the specification into smaller, interrelated modules which are then simulated to produce more complete or smaller automata. The simple composition of these automata, without taking into account any kind of abstraction (PCOs), would lead to the big automaton of the traditional case. However, if we use information about our abstraction level we are able to compose them and at the same time avoid the explosion of the model. In other words, we compute the global composition of two automata while removing internal actions.

The implementation of the proposed algorithm also includes mechanisms to tackle the following issues: multiple (simultaneous) outputs (i.e. when one signal from the environment stimulates several simultaneous outputs from the system), deadlock and livelock detection (if components exchange messages indefinitely). A complex strategy for transition marking implements a mechanism of behaviour exploration that guarantees that all possible branches are examined, even in the presence of nondeterminism. This is implemented by allowing processes to warn the builder when there was a non deterministic choice for the last input, so that the builder can send the same signal a second time and a different transition will then be traversed. As a result, non deterministic machines can be produced.

The algorithm also allows the detection of undesired situations that may happen during the composition process and that are reported back by the tools as error or warning messages:

- incompatibility errors due that a process was not expecting a given internal signal from its peer machine at its current state. In this case, the internal signal is simply forwarded to the environment that instantiates a global transition with an error message (for it contains an internal output signal that was not supposed to reach the environment);

- unreachability warnings. During the machines join execution, some transitions of either machine may not be traversed and some states even may not be visited. This means that a part of the machine behavior was not exercised in the joint execution (e.g. one machine was not capable of using all the other machine's functionality). This kind of information can be useful, for instance, for the detection of some cases of feature interaction or to identify the executable part of a component in a given context.

We have also developed another approach to embedded testing that is based in a general procedure for embedded testing of EFSMs. This is a work that we have presented in [10]. This procedure can be applied to EFSMs, to Communicating Finite State Machines (CFSMs) and to Communicating Extended Finite State Machines (CEFSMs). The proposed algorithm is based on a "hit or jump" random walk that attempts to cover all transitions of the component EFSMÕs. This procedure has the advantage of avoiding the construction of the system reachability graph. As a matter of fact, the space required is determined by the user, and it is independent of the systems under consideration. The algorithm has been implemented on the ObjectGEODE tool. It takes advantage of some of the functionalities offered by this tool: for instance the use of the simulator to produce partial graphs that will be used to produce the test scenarios.

The first approach for embedding testing that we have developed has been applied to embedded testing of the GSM-MAP protocol [19]. Both approaches have been applied to the embedded testing of services on a telephone network. This case study corresponds to a real system that has been specified using the SDL language. In addition to the Basic Call Services (BCSs), two other services are included: Originating Call Screening (OCS) and Call Forward Unconditional (CFU). These applications are described in [27, 28, 10] respectively.

4.   PASSIVE TESTING

The objective of conformance testing is to evaluate whether a protocol implementation has the same behaviour as its specification. In active conformance testing, the implementation under test (IUT) is placed in a dedicated testing environment. According to predefined test sequences, one (or more) tester sends inputs to the IUT, observes the corresponding outputs and compares them with the expected outputs. On the other hand, passive testing consists in collecting traces of the messages exchanged between an operating implementation and its environment, and verifying whether these traces actually belong to the language accepted by the specification automaton. Though passive testing is sometimes mentioned as an alternative to active testing [17], only little effort has been devoted to this aspect of testing [31, 12]. However, we feel that passive testing is worth investigating, since (a) under certain circumstances, it may be the only type of test available, e.g. in network management, (b) it is relatively cheap and easy to implement, and (c) active testing is sometimes impractical due to the complexity of systems.

In passive testing [31], contrary to active testing, the tester does not control the implementation under test. The implementation is in operating condition, and the tester only observes the messages exchanged by the IUT and its environment, in order to check if they correspond to a behaviour compliant with the specification. The main difficulty in passive testing is that we have no knowledge of the state in which the implementation is at the beginning of the trace (no assumption is made about the moment when the recording of trace begins, and therefore it is not necessarily the initial state). Each input/output pair of the trace is assumed to represent a transition in the specification, and our objective is to match the transitions of the trace with those of the specification. The passive testing process can be decomposed into two steps: the passive homing sequence, in which the current state is found out, and the fault detection phase, in which the trace is compared with the specification.

We have defined an extension of the existing algorithms to consider the specification as an extended finite state machine (EFSM) that is presented in [33]. We also introduce the principles of passive testing, and an application to a real protocol, the GSM-MAP. The proposed algorithm is an extension of this of [12] in order to consider an extended finite state machine as the system specification. The SDL specification of the GSM-MAP that we used in our experiment has been obtained from ETSI standard; this specification has been completed in our group [19, 3]. For the purpose of conformance testing, the specification is usually considered as an Extended Finite State Machine, the underlying model of SDL.

The first results were the detection of discrepancies between the global SDL specification and the behaviour of the FSM obtained through the decomposition/composition process. These discrepancies were detected within 10 input/output pairs after the end of the passive homing sequence and were reported as faults by the passive testing software. They turned out to be mostly due to some variable values that were inconsistent between the various processes. This allowed us to correct these inconsistencies in the specification. After correction, tests were rerun with the same traces, and no faults were detected, which was the expected outcome. In order to assess our tool, faults were also added manually in some of the traces. Those faults were detected within a few input/output pairs after they occurred.

Traces varied from 1.000 to about 12.500 input/output pairs. Computation time varied between 0.04 to 0.14 seconds on a Sun Ultra1. Transition coverage varied between 5% and 25% of the transitions, and state coverage varied between 56% and 71% of the states. The global coverage, obtained by taking into account all the states reached in the various experiments, and all the transitions, was 86% of the states and 46% of the transitions.

5.   EXECUTION OF TESTS ON A CORBA PLATFORM

Intelligent Networks (IN) represented a remarkable step forward in the domain of telecommunications systems. The ability to add "intelligent" nodes (i.e. computers) into switching networks made it possible to introduce new telephonic services. However the service implementation was still dependent on the hardware and operating system upon which it was built. Nowadays, telecommunication services are migrating to open distributed environments where a layer, called middleware, located between the software application layer and the hardware layer (i.e. physical and low level programs like operating systems), assures transparency, interoperability and cooperation among the several service components in spite of the heterogeneous underlying systems. The Telecommunication Information Networking Architectures (TINA) specify the DPE (Distributed Processing Environment) layer which is responsible for providing facilities to service programmers, so that they do not need to cope with communication details.

There are many different validation methods for telecommunications systems as we have mentioned in the previous sections. All these methods correspond to what we call the Testing Generation Phase, which starts with the system specification and ends with the generation of test suites. However very little has been said about validation of telecommunication services, specifically, regarding distributed environments like ORB platforms. Thus, on the one hand we see a world-wide move to using distributed platforms to implement telecommunications services and on the other hand the need for suitable methods to validate these systems.

Our works in this area aim at implementing and validating services on top of a TINA-CORBA like platform [25]. In the following we will provide a short presentation of the implementation and validation phases that we have developed in our test platform.

For the validation a tester (possibly distributed) must be implemented to send stimuli to the several components under test (which compose the System Under Test - SUT) and to gather information concerning their reaction to these stimuli. Eventually, the tester will give a verdict on the observed behaviour of the system. In doing so , we will be able to come up with a practical view on the following aspects :

- test generation and execution;

- distributed test execution and generation;

- how to structure the test architecture for TINA-CORBA-like system/objects.

The distributed platform chosen was Orbix (TM) [18], which is a full implementation of OMG-CORBA. The key advantage of such platforms is that they make application construction and integration much easier, since software interfaces are defined using a standard language, the Interface Description Language (IDL) which allows objects to be accessed from anywhere in a distributed system.

In this platform, all interactions (i.e. service requests) are performed by means of channels. Each channel is composed of a stub (on the client side) and a skeleton (on the server side) which are connected to the Object Request Broker (ORB). Channels allow a transparent communication between the client and the object implementation. The ORB in turn provides the functionality required to communicate across (possibly heterogeneous) platforms.

A service request consists of the following information : a target object, an operation, a list of parameters and an optional request context. "One way" operations (called announcements in the ODP terminology [22]) allow asynchronous message exchange amongst the various system components. It is important to point out that a client can simultaneously be a server to another object (or vice-versa). Objects must cooperate (or interoperate) in a useful way to achieve a given objective, which is exactly what TINA-CORBA provides.

Following the guidelines of classical test architectures for protocols as described in [16], we suggest that a test architecture for an open distributed object-oriented platform must have two sets of components: tester components and components under test that are possibly distributed over several sites. The tester components receive information about the object configuration (i.e. the collection of objects able to interact at their interfaces) and the test suite which is obtained through some test generation method. Simple systems (with local architecture) are usually implemented using one single tester component and a single component under test.

The results of our work are encouraging, particularly taking into account that the majority of our colleagues working on the implementation of multimedia services on a TINA-CORBA platform do testing manually. Once the module representing a service has been implemented, they manually define a test script. They repeatedly express the need for a test environment where all phases of the testing are automated. The testing environment that we have presented in this paper is an effort in this direction: all phases are automated and the test results obtained are more complete. Another advantage is the possibility of using existing formal description techniques to describe the system (SDL), the test suite obtained (using TTCN) and the test script (using MSC) [21].

6.   CONCLUSIONS AND FUTURE WORK

The main objective of all the work we have performed is the application of formal methods for conformance testing to real protocols and services. Our work also contributes to increase the use of formal description techniques and formal methods for testing in industry. In France, the use of the SDL language in industry has considerably increased in the last years and we realize, through our industrial collaborations, that the use of these techniques and associated tools is an accomplished fact.

In the future, we will continue our work related to the test of services on RI and TINA-CORBA architectures. We are applying our algorithms for embedded testing to the test of components of an audio and multimedia services specified following the requirements of this architecture. We are also working on extensions of the testing platform in order to be able to execute test on services implemented in these environments. We have also started to work in how to provide formal descriptions and validation methods for service administration. Furthermore, we will continue our collaborative work concerning more theoretical aspects of embedding testing, as presented in this conference [36, 35], on optimization techniques for test generation [4, 10] and on the test of timed systems [14].

References

[1] A. Aho, A. Dabhura, D. Lee, and U. Uyar. An optimization technique for protocol conformance test generation based on UIO sequences and rural chinese postman tours. In S. Aggarwal and K. Sabnani, editors, *PSTV'88*, pages 75–86, North Holland, 1988.

[2] A. Aho, R. Sethi, and J. Ulman. *Compilers, Principles, Techniques and Tools*. Addison-Wesley Publishing Company, 1986.

[3] R. Anido, A. Cavalli, L. Paula Lima, M. Clatin, and M. Phalippou. Engendrer des tests pour un vrai protocole grâce à des techniques éprouvées de vérification. In *Proceedings of CFIP'96*, Rabat, Marocco, 14-17 October 1996.

[4] C. Besse, A. Cavalli, and David Lee. Optimization techniques and automatic test generation for TCP/IP protocols. Research Report RR 99.06, INT, France, March 1999.

[5] A. Cavalli, B. M. Chin, and K. Chon. Testing methods for SDL systems. In *Computer Networks and ISDN Systems*, volume 28, pages 1669–1683, 1996.

[6] A. Cavalli, J. P. Favreau, and Marc Phalippou. Standardization of formal methods in conformance testing of communication protocols. In *Computer Networks and ISDN Systems*, volume 29, pages 3–14, December 1996.

[7] A. Cavalli, S. Kim, and P. Maigron. Automated protocol conformance test generation based on formal methods for LOTOS specifications. In *IWPTS'92*, Montreal, Canada, September 1992. Elsevier Science Publishers.

[8] A. Cavalli, S. Kim, and P. Maigron. Improving conformance testing for LOTOS. In *FORTE'93*, Boston, USA, October 1993.

[9] A. Cavalli, B. Lee, and T. Macavei. Test generation for the SSCOP-ATM networks protocol. In *SDL Forum'97*, France, September 1997.

[10] A. Cavalli, David Lee, Christian Rinderknecht, and F. Zaïdi. Hit-or-jump: An Algorithm For Embedded Testing With Applications To IN Services. Research Report RR 99.05, INT, France, March 1999.

[11] M. S. Chen, Y. Choi, and A. Kershenbaum. Approaches utilizing segment overlap to minimize test sequences. In *PSTV'90*, pages 67–84, Canada, June 1990.

[12] David. Lee et al. Passive testing and applications to network management. In *ICNP'97 International Conference on Network Protocols*, Atlanta, Georgia, 28-31 October 1997.

[13] M. A. Fecko, U. Uyar, A. S. Sethi, and P. Amer. Issues in conformance testing: Multiple semicontrollable interfaces. In *Proceedings of FORTE'97*, Paris, France, November 1998.

[14] T. Higashino, Nakata, K. Taniguchi, and A. Cavalli. Generating test cases for a timed I/O automaton model. In *Proceedings of the 12th IWTCS*, Budapest, Hungary, September 1999.

[15] D. Hogrefe. Osi formal specification case study: the inres protocol and service, revised. Technical report, Institut für Informatik Universität Bern, may 1992.

[16] International ISO/IEC multipart standard. *ISO/IEC 9646 Information Technology - OSI - Conformance Testing Methodology and Framework.*, 1994.

[17] International Standard 10746-2/ITU-T Recommendation X.902. *Open distributed Processing - Reference Model - Part 2: Foundations.*

[18] IONA Technologies Ltd. *Orbix 2 Programming Guide*, November 1995. P1.

[19] M. Ionescu and A. Cavalli. Test imbriqué du protocole MAP-GSM. In *Proceedings of CFIP'99*, Nancy, France, 26-29 April 1999.

[20] ITU-T, Geneva. *CCITT, Specification and Description Language, CCITT Z.100, International Consultative Committee on Telegraphy and Telephony*, 1992.

[21] ITU-T Rec. Z.120, Geneva. *Message Sequence Chart (MSC)*, 1996.

[22] ITU-T Recommendation X.903. *ISO/ITU-T - Open Distributed Processing - Reference Model - Part 3: Architecture - International Standard 10746-3*, 1995.

[23] D. Lee, K. Sabnani, D. Kristol, and S. Paul. Conformance testing of protocols specified as communicating finite state machines - a guided random walk based approach. In *IEEE Transactions on Communications*, volume 44, No.5, May 1996.

[24] L. P. Lima and A. Cavalli. An embedded testing approach. In *Proceedings of EUNICE Summer School*, Lausanne, Switzerland, 23-27 June 1996.

[25] L. P. Lima and A. Cavalli. Test execution of telecommunication service using CORBA. In *Proceedings FMOODS'97*, pages 409–422, Canterbury, United Kingdom, 21-23 July 1997.

[26] L. P. Lima and A. R. Cavalli. A pragmatic approach to generating test sequences for embedded systems. In *Proceedings of the 10th IWTCS*, pages 125–140, 1997.

[27] L. P. Lima and A. R. Cavalli. Application of embedded testing methods to service validation. In *2nd IEEE Intern. Conf. On Formal Engineering methods*, Brisbane, Australia, 1998.

[28] Luiz P. Lima. *Une Méthode Pragmatique de Génération de Séquences de Test pour les SystŁmes Imbriqués*. PhD thesis, Université d'Évry, France, November 1998.

[29] A. Petrenko and N. Yevtushenko. Testing faults in embedded components. In *Proceedings of IWTCS'97*, Cheju Island, Korea, September 1997.

[30] A. Petrenko, N. Yevtushenko, and G. V. Bochmann. Fault models for testing in context. In *Proceeding of FORTE*, Kaiserslatern, Germany, October 1996.

[31] Charles L. Seitz. *An Approach to designing checking experiments based on a dynamic model.* Z. Kohavi and A. Paz ED. Academic Press, 1972.

[32] D. P. Sidhu and T. K. Leung. Formal methods for protocol testinf: a detailed study. In *IEEE Transaction on Software Engineering*, volume 15, No.4, April 1989.

[33] M. Tabourier and A. Cavalli. Passive testing and application to the GSM-MAP protocol. *Journal of Information and Sotware Technology, Elsevier Science*, December 1999.

[34] Verilog, France. *Geode Editor - Reference Manual*, 1996.

[35] N. Yevtushenko, A. R. Cavalli, and R. Anido. Test suite minimization for embedded nondeterministic finite state machines. In *Proceedings of the 12th IWTCS*, Budapest, Hungary, September 1999.

[36] N. Yevtushenko, A. R. Cavalli, and L. P. Lima. Test suite minimization for testing in context. In *IWTCS'98*, Tomsk, Russia, August 1998.

2

TOWARDS THE THIRD EDITION OF TTCN

Jens Grabowski and Dieter Hogrefe
Institute for Telematics, University of Lübeck,
Ratzeburger Allee 160, D-23538 Lübeck, Germany
{ jens,hogrefe } @itm.mu-luebeck.de

Abstract   The third edition of TTCN (Tree and Tabular Combined Notation) will be a complete redesign of the entire test specification language. The close relation between graphical and textual representation will be removed, OSI specific language constructs will be cleared away and new concepts will be introduced. The intention of this redesign is to modernize TTCN and to widen its application area beyond pure OSI conformance testing. This paper motivates the need for a new TTCN, explains the design principles and describes the status of the work on the third edition of TTCN.

Keywords: ETSI, ITU-T, CORBA, OSI Conformance Testing, Test Specification, TTCN, Programming Languages

1.   INTRODUCTION

The *Tree and Tabular Combined Notation* (TTCN) is a well established notation for the specification of test cases for OSI protocol conformance testing. TTCN is defined and standardized in Part 3 of the international standard 9646 'OSI Conformance Testing Methodology and Framework' (CTMF) [2].

OSI conformance testing is understood as functional black-box testing, i.e., a *system under test* (SUT) is given as a black-box and its functional behavior is defined in terms of inputs to and corresponding outputs from the SUT. Subsequently, TTCN test cases describe sequences of stimuli to and required responses from the SUT.

CTMF and TTCN have been used for testing OSI protocols and protocols in systems following the OSI layering scheme, e.g., ISDN or ATM. CTMF principles and TTCN have also been applied successfully to other types of functional black-box testing, e.g., ISDN service testing and interoperability testing.

The requirements on testing are changing. The testing of new software architectures, e.g., ODP, CORBA, TINA or DCE, with advanced and time-critical applications, like, multimedia, home-banking, or video-conferencing, requires new testing concepts, new testing architectures and a new and powerful test specification language.

Researchers have already started to extend CTMF and TTCN in order to meet the upcoming testing requirements. The proposed extensions were related to testing architectures [5, 9], real-time testing [7, 8] and performance testing [6]. Some of these ideas will find their way into practice.

The international standardization organizations *International Telecommunication Union* (ITU-T) and *European Telecommunications Standards Institute* (ETSI) have studied the new testing requirements [1] and recognized the urgent need for a modern and powerful test specification language. As a consequence, the *specialists task force* (STF) 133 was set up in October 1998. STF 133 will (1) correct the known defects in the second edition of TTCN and (2) develop the third edition of TTCN (TTCN-3) which is an extension and complete redesign of the previous TTCN editions.

The correction of the known defects has been finished in December 1998 and the revised second edition of TTCN will be published by ETSI in 1999. The development of TTCN-3 is an ongoing task and is expected to be completed in spring 2000. This paper introduces TTCN-3 and describes the current status of the work on TTCN-3.

2.   REQUIREMENTS ON TEST LANGUAGES

A test specification language should fulfil some general requirements which distinguish it from other specification or programming languages. In this section, these requirements are listed and compared with the properties of TTCN.

2.1   GENERAL REQUIREMENTS

A test specification language should

1 provide functionality which is specific to testing and which cannot be easily made available in other programming or specification languages,

2 allow to present the details of the implemented test purpose in a human understandable form,

3 be transferable between different computers,

4 be compilable and afterwards be executable on some test equipment,

5 be versatile and not only be suited for one application area, and

6 be easy to learn and understand.

This list is very general and the requirements may not be equally important. However, it is obvious that the fulfillment of each of these requirements will improve the acceptance of a test specification language.

2.2   NEGATIVE PROPERTIES OF TTCN

TTCN provides a graphical form (TTCN/GR) and a textual machine processible form (TTCN/MP). The graphical form is table oriented, i.e., a TTCN/GR test suite is a collection of different kinds of tables. TTCN/GR and TTCN/MP are closely related. In fact, there is a one-to-one mapping between each row in a TTCN/GR table and a line in a TTCN/MP file.

TTCN is too restrictive.   The first negative property of TTCN which is valid for TTCN/GR and TTCN/MP is that TTCN is too restrictive. The browser structure of a TTCN test suite is built into the syntax and provides an OSI conformance testing oriented view. In addition, TTCN includes several concepts and application-specific static semantic rules which have only a meaning in OSI conformance testing. For example, the distinction between upper and lower tester functions or the distinction between PDUs and ASPs. In addition, a compiler will treat some conformance testing specific information only as comments.

The proforma format of TTCN/GR is also too restrictive. Reordering of information to improve readability is not possible. Omitting rows or columns which are not needed is not possible either. In case of long table entries, the column form is not always suitable. Using TTCN/MP in such cases does not help, because TTCN/MP reflects the table structure and cannot be used like a normal programming language.

As a summary, it can be stated that TTCN violates the requirements on readability (2, 6) and versatility (5). An enhanced TTCN should provide views beyond OSI conformance testing, should provide a human readable (usable) textual form and may support other presentation forms than pure tables, e.g., MSC, SDL or Java-like.

TTCN is too complex.   The second negative property of TTCN is its complexity. The grammar rules of TTCN/MP are too complex and TTCN/GR includes too many different proformas. In total, TTCN/GR distinguishes 47 different types of tables (without compact proformas) where some only differ in the table headings. This makes it difficult to learn and read TTCN/GR test suites.

TTCN includes some redundant functionality, e.g., macros versus structs or compact proformas. Such functionality makes a language clumsy, complex and reduces its readability and usability. Typically, it remains in the language definition due to backwards compatibility.

Some TTCN complexity is caused by language constructs supporting functionality which normally should be provided by a tool. For example, the index of test cases and test steps with page numbers is something which should be provided by TTCN tools but should not be part of the language itself.

The conclusion of this discussion is that TTCN violates the requirements on readability and simplicity (2, 5). A new test specification language should be simple, less complex and should not support functionality which can be provided by a tool.

2.3   POSITIVE PROPERTIES OF TTCN

In spite of all the negative properties, TTCN has a lot of positive properties too. TTCN provides functionalitied and concepts which are specific to testing and cannot be provided easily by other languages (requirement 1). These functionalities and concepts are related to

- test case selection and test case parameterization,

- the concept of *points of control and observation* (PCOs),

- test configurations,

- the link to ASN.1,

- the inclusion of encoding information,

- the concept of constraints,

- matching mechanisms,

- verdict assignment, and

- the specific operational semantics (snapshot-semantics)

Beyond that, TTCN has proven to be transferable between different computers, and to be compilable and executable (requirement 3, 4).

2.4   APPLICATION REQUIREMENTS

TTCN was developed to support the conformance testing procedure of OSI protocol implementations. To meet the testing requirements of new software architectures and advanced applications, the enhanced TTCN has to include new concepts [1]. These are:

- options to specify dynamic test configurations,

- support of additional communication mechanisms, e.g., synchronous communication and broadcast communication,

- an extended timer concept to allow the test of hard real-time requirements

- support for the specification of performance tests.

3.   PRINCIPLES OF TTCN-3

As a consequence of the discussions about the requirements on test specification languages and the properties of TTCN, the ETSI *technical committee* (TC) *methods for testing and specification* (MTS) established STF 133 in its funded work program. Beyond the already mentioned and finished correction of the second edition of TTCN, STF 133 will develop TTCN-3 with the following goals in mind:

- simplification of TTCN,

- harmonization of the latest editions of ASN.1 [3] and TTCN,

- integration of new communication concepts such as synchronous communication and monitoring, and

- support of dynamic test configurations in TTCN.

It was decided to base the work of STF 133 on the existing experience with TTCN and CTMF. Due to restricted resources, concepts for real-time and performance testing will not be included in TTCN-3. The goals listed above will be reached by a complete redesign of the existing TTCN. The work will concentrate on the development of a textual syntax with a look-and-feel similar to a programming language.

TTCN-3 may have several graphical representation formats. The ETSI representation format will be defined by STF 133. It will be table-oriented and based on 15 generic proformas. For example, there will be only one generic proforma for the definition of ASPs, PDUs and coordination messages (CMs).

4.   TTCN-3 - STATE OF WORK

At the time of writing this paper, only a few TTCN-3 language issues have been investigated thoroughly and only a few syntax proposals have been made. To give an impression of TTCN-3, the harmonization of ASN.1 and TTCN-3, the concepts for modularization and some ideas about behavior descriptions in TTCN-3 are discussed.

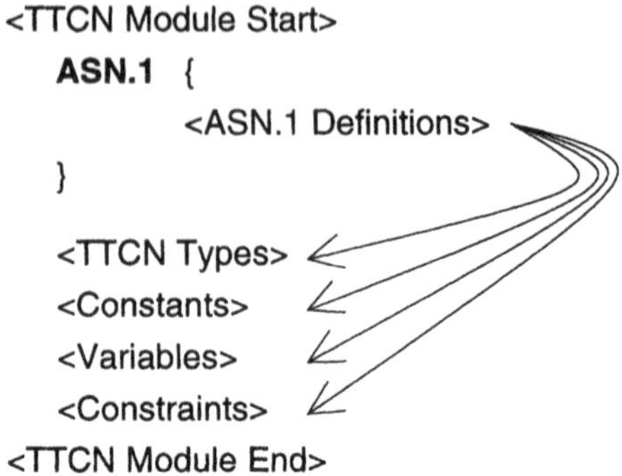

Figure 1   ASN.1 block in TTCN-3

4.1   ASN.1 AND TTCN-3 HARMONIZATION

The harmonization of ASN.1 and TTCN-3 is done to allow the combined use of the latest version of ASN.1 and TTCN-3. It is a difficult task, because ASN.1 and TTCN-3 are distinct languages with different syntax and semantics definitions. In the previous versions, the combined use of ASN.1 and TTCN has led to a mixture of syntax rules. In some cases, it was allowed to use TTCN language constructs in ASN.1 descriptions, e.g., TTCN matching mechanisms in a ASN.1 value notation, and vice versa. A compiler has to decide from the context whether syntax rules of ASN.1, TTCN, or both have to be applied.

The TTCN-3 strategy to avoid such problems is the separation of ASN.1 and TTCN-3 constructs. ASN.1 definitions have to be placed in a ASN.1 block (Figure 1) and all definitions in such a block follow the ASN.1 syntax rules. The definitions can only be used by reference and it is not allowed to use ASN.1 syntax in TTCN-3 constructs.

In some cases, this scheme may lead to minor problems, e.g., the use of matching mechanisms in and the modification of constraints which refer to ASN.1. In this case, the solution is to transform such constraints into the TTCN table notation which can be modified according to the TTCN rules.

The advantages of this harmonization approach seem to be bigger than the recognized problems. In addition, this solution may open a door for TTCN-3 to application areas where other data descriptions are used, e.g., IDL in the CORBA context [4].

4.2   MODULARIZATION

TTCN-3 will only have modules and will not distinguish between test suites and modules. The term *modularization* in TTCN-3 refers to the possibility of reusing the definitions of a module in another module. For this, the definitions have to be imported by the latter one.

TTCN-3 will only have an `import` construct but no `export` construct. All global definitions of a module may be imported by another module.

The names in a module have to be unique. Name conflicts due to the `import` from different modules have to be resolved by implicit and explicit prefixing with the identifier of the module from which the name is imported.

The `import` construct should be easy to use and avoid the writing of long import lists. Additional keywords will help to keep the `import` construct user friendly. Two examples may illustrate this. The statement

```
import typedef MyType from MyModuleC recursively;
```

describes the import of the type definition `MyType` from `MyModuleC`. The keyword `recursively` states that all type definitions which may be used within `MyType` are also imported. The statement

```
import all from MyModuleD exclude all constraints;
```

denotes the import of all definitions except for its constraints from `MyModuleD`.

4.3   BEHAVIOR DESCRIPTIONS

Behavior will be introduced in TTCN-3 on the level of modules, on the level of module operations and on the level of test cases. Behavior on the level of modules allows to specify test control programs. For example, a test case is executed only if another test case has been executed successfully, or a test case is repeated several times. Behavior on the level of module operations corresponds to test suite operations in the second edition of TTCN. Behavior on the level of test cases corresponds to the behavior descriptions of test cases, test steps and defaults in the second edition of TTCN. STF 133 will select the constructs for describing the flow of control (loops, conditions, jumps) in such a way that they can be used on all levels.

On the level of test cases TTCN-3 will distinguish between behavior definitions and the invokation of an instance of a behavior definition. An instance of a behavior definition corresponds to a test case which will be executed when the module is applied to an SUT. A behavior definition may call other behavior definitions for describing more complex behavior and it may be instantiated. This means, it depends on the use of a behavior definition whether it can be interpreted as a test case, test step or default behavior description. Behavior definitions can be imported by other TTCN-3 modules. A behavior definition can be instantiated by reference, i.e., the name of the behavior definition is referenced, or in form of an inline definition, i.e., the behavior definition is provided in the place of its instantiation.

The following example may provide an idea of the TTCN-3 'look-and-feel' in comparison with the second edition of TTCN. In the ETSI abstract test suite for the ISDN supplementary service Multiple Subscriber Number (ETSI EN 300 052 6) the test step definition shown in Figure 2 can be found (for

Test Step Dynamic Behaviour					

Test Step Name : PO49901 (FL : INTEGER)
Group :
Objective : To bring the IUT to the state N0 and terminate the PTC.
Default :
Comments :

Nr	Label	Behaviour Description	Constraints Ref	Verdict	Comments
1		L0!REL START TAC	A_RL3(FL, CREF1, 16)		
2		L0?Rel_COMr CANCEL TAC	A_RC1((FL+1)MOD2, CREF1)		
3		+END_PTC1			
4		?TIMEOUT TAC		(I)	
5		+END_PTC1			
6		L0?OTHERWISE		(I)	
7		+END_PTC1			

Detailed Comments :

Figure 2   A TTCN test step dynamic behavior description

simplicity, the comments and group reference are left out). The corresponding TTCN-3 representation is shown in Figure 3.

```
behaviour PO49901(FL integer)
/* Objective:  To bring the IUT
to the state N0 and
terminate the PTC1.   */
{
  L0!A_RL3(FL,CREF1,16);
  start(TAC);
  alt {
    A1: L0?A_RC1((FL+1) mod 2, CREF1);
        cancel(TAC);
    A2: ?timeout(TAC);
        (inconclusive);
    A3: ?otherwise;
        (inconclusive);
  };
  +END_PTC1;
}
```

Figure 3   TTCN-3 representation of the test step in Figure 2

In Figure 2, ASP or PDU type references can be found on behavior lines, timer commands are related to send and receive events, the verdict assignment is done on behavior lines and the test step END_PTC1 has to be attached three times, i.e., to all three branches of the tree.

In Figure 3, the statements are ordered sequentially. The sequence starts with a send statement, followed by a start timer, followed by an alternative (which for the first alternative includes a sequence of two statements). The behavior description ends with the attachment of END_PTC1. Behavior lines in Figure 2 describing several actions are translated into separate commands (e.g., the first line in Figure 2 L0!REL START TAC is translated into L0!A_RL3(FL,CREF1,16); START TAC;). Verdict assignment is done by special commands which can be used anywhere in the behavior definition. Only constraints are referenced by send and receive events. The ASP or PDU type references found in the second edition of TTCN are superfluous because they are provided within the constraint definitions.

5.   SUMMARY AND OUTLOOK

In this paper, the motivation for the development of TTCN-3 has been discussed and the status of the work has been described. TTCN-3 will be a completely new test specification language which widens the application area of TTCN beyond pure OSI conformance testing as defined in CTMF. This new test specification language will have a look-and-feel like a normal programming language but will keep the testing specific properties of the previous versions of TTCN. TTCN-3 will have a standardized textual syntax, but may have several graphical representation formats. The ETSI representation format will be table-oriented and include 15 different generic proformas. At the time of writing this paper, the development of TTCN-3 was in a state where only a few issues have been studied thoroughly. Nevertheless, it is planned to finish the development of TTCN-3 and the ETSI presentation format in spring 2000.

Acknowledgments

The work presented in this paper has been carried out within STF 133 as part of the funded work program of ETSI. The work on dynamic test configurations within STF 133 is funded by Nortel. One of the authors is a member of STF 133. The authors would like to thank Anthony Wiles, Colin Wilcock and Guy Martin for allowing us to present parts of the STF 133 work.

References

[1] ETSI TC MTS. *Guide for the use of the second edition of TTCN (Revised Version)*. European Guide 202 103, 1998.

[2] International Standardization Organization. *Information Technology — OSI — Conformance Testing Methodology and Framework — Parts 1–7*. ISO, International Standard 9646, 1994 - 1997.

[3] ITU-T Recommendations X.680-683. *Information Technology - Abstract Syntax Notation One (ASN.1)*, 1994.

[4] Object Management Group. *The Common Object Request Broker: Architecture and Specification; Revision 2.0*. Frammingham, Massachusetts, USA, 1995.

[5] I. Schieferdecker, M. Li, A. Hoffmann. *Conformance Testing of TINA Service Components - the TTCN/CORBA Gateway*. In Proceedings of the 5th International Conference on Intelligence in Services and Networks, Antwerp, Belgium, May 1998.

[6] I. Schieferdecker, S. Stepien, A. Rennoch. *PerfTTCN, a TTCN Language Extension for Performance Testing*. In M. Kim, S. Kang, K. Hong, editors, *Testing of Communicating Systems*, volume 10, Chapman & Hall, September 1997.

[7] T. Walter, J. Grabowski. *Real-time TTCN for Testing Real-time and Multimedia Systems*. In M. Kim, S. Kang, K. Hong, editors, *Testing of Communicating Systems*, volume 10, Chapman & Hall, September 1997.

[8] T. Walter, J. Grabowski. *A Framework for the Specification of Test Cases for Real-Time Systems*. Accepted for Publication in: Journal of I nformation and Software Technology, Special Issue on Communications Software Engineering, 1999.

[9] T. Walter, I. Schieferdecker, J. Grabowski. *Test Architectures for Distributed Systems - State of the Art and Beyond* (Invited Paper). In A. Petrenko, N. Yevtushenko, editors, *Testing of Communicating Systems*, volume 11, Chapman & Hall, September 1998.

6.   BIOGRAPHY

Jens Grabowski studied computer science and chemistry at the University of Hamburg, Germany, where he graduated with a diploma degree. From 1990 to October 1995 he was a research scientist at the University of Berne, Switzerland, where he received his Ph.D. degree in 1994. Since October 1995 Jens Grabowski is a researcher and lecturer at the Institute for Telematics. He is a member of the ETSI specialists task force 133 which develops the third edition of TTCN.

Dieter Hogrefe studied computer science and mathematics at the University of Hannover, Germany, where he graduated with a diploma degree and later received his PhD. From 1983 to 1995 he worked at various positions in the industry and at universities. Since 1996 he is director of the Institute for

Telematics and full professor at the Medical University of Lübeck. Dieter Hogrefe is also chairman of ETSI TC MTS.

3

NEW DIRECTIONS IN ASN.1: TOWARDS A FORMAL NOTATION FOR TRANSFER SYNTAX

Colin Willcock

Expert Telecoms Ltd, Mittlerer Pfad 26, 70499 Stuttagrt Germany,

Phone +49 711 1398 8220

C.Willcock@Expert-Telecoms.com

Abstract   At present although the abstract structure of a protocol message can be formally defined, the actual format of the bits transmitted on the line or through the air cannot be formally specified. The small number of standardised encoding rules available are only defined in informal text and in many application domains such as radio interfaces, they do not satisfy today's requirements for very efficient transmission of information. This paper describes a number of the problems associated with the definition of transfer syntax and then discusses a possible solution to these issues involving the development of an advanced encoding control notation.

Keywords: Transfer syntax, abstract syntax, ASN.1, formal description techniques, encoding rules

1.   INTRODUCTION

The advantages of using formal description techniques in the design and development of telecommunications standards has been established beyond doubt [3]. The existing formal description techniques include Abstract Syntax Notation One (ASN.1) [4, 5, 6, 7] for describing data and signal structures, the Specification and Description Language (SDL) [8] for definition and validation of system behaviour and the Tree and Tabular Combined Notation (TTCN) [2] for definition of test suites.

Currently there is a gap in the formal description techniques relative to the definition of encoding rules. This means that the actual transfer syntax of a system cannot be formally specified.

The development of advanced encoding control would allow the completion of the formal description infrastructure, it should also greatly extend the use of ASN.1 in protocol specifications which presently use other less formal notations due to the implied encoding restrictions of the current language.

2.   STATE OF THE ART

ASN.1 is a language for describing structured information, it allows a protocol designer to specify a syntax at a high level without the need to explicitly consider the representation of that syntax in terms of bits and bytes on the line. The actual representation on the line for any message can be derived by applying a set of encoding rules to the associated ASN.1 definition of that message. This combination of abstract syntax and encoding rules which defines the actual bits on the line is known as the transfer syntax.

Relative to encoding rules there are a small number of standardised sets. The first set to be standardised was the Basic Encoding Rules - BER [9]. BER provides a relatively straight-forward and verbose encoding scheme which is still used today in many applications. The relatively poor encoding efficiency of BER lead to the development and standardisation of the packed Encoding Rules - PER [10]. PER is considerably more bandwidth efficient than BER, requiring only between 40 and 60 percent of the encoded size. The encoding efficiency of PER is achieved however at the cost of a considerably more complicated encoding scheme including the use of context specific encoding. In addition to BER and PER, the Canonical Encoding Rules (CER) and the Distinguished Encoding Rules (DER) have been standardised. Both these encoding rule sets are purely restricted forms of BER.

At present the only way to define the transfer syntax for a protocol in Recommendations is to informally associate one of these standardised encoding rules with the ASN.1 abstract syntax definition by means of an ASN.1 comment or simply by stating it in writing in the non-ASN.1 text description.

The TTCN language is typically used in the definition a of test cases to prove that a particular implementation under test conforms to a specified protocol or standard. TTCN supports the use of ASN.1 for type and value definitions and in addition allows the definition of transfer syntax by use of an optional reference mechanism for encoding rule and encoding variant. With this mechanism the test specifier can reference by name the required encoding rule set and required variant of that encoding rule and associate this encoding information with a specific type.

In contrast, the SDL language which is used for specifying system behaviour, although it also supports ASN.1 for type and value definitions, has no concept what so ever for specifying the required transfer syntax. In effect this means that since SDL cannot define the transfer syntax it can never fully specify a real world system that must communicate via cable or air interface.

In addition to the gap in formal description techniques, there are a number of other issues which point towards the need of some form of encoding control. The two areas considered in this paper are tabular based protocols and application specific encoding rules.

2.1   TABULAR BASED PROTOCOLS

Before the general adoption of ASN.1 most telecommunications protocols were defined in the form of tables. The tables define the name, size and order of the constituent fields of the protocol and in addition implicitly define the associated transfer syntax. An example of such a tabular definition is shown in figure 1. Tabular based definitions include many of the worlds most widely used protocols (e.g. Integrated Services Digital Networks (ISDN) basic call [11], Message Transfer Part (MTP) [12], Signalling Connection Control Part (SCCP) [13] and ISDN User Part (ISUP) basic call [14]).

With the existing encoding rules it is not possible to specify the transfer syntax of such protocols using ASN.1. This inability leads to major problems when fields from tabular defined protocols are embedded in ASN.1 protocols. An example of such embedded fields is the Intelligent Network (IN) protocol [15] which has ISUP and ISDN fields embedded inside an ASN.1 octet string as shown in the extract from the recommendation shown below.

CalledPartyNumber ::= OCTET STRING (SIZE (minCalledPartyNumberLength..

$\qquad\qquad\qquad\qquad\qquad\qquad\qquad$ **maxCalledPartyNumberLength))**

– Indicates the Called Party Number. Refer to Recommendation Q.763 for encoding.

CalledPartySubaddress ::= OCTET STRING

– Indicates the Called Party Subaddress. Refer to Recommendation Q.931 for encoding.

2.2   APPLICATION SPECIFIC ENCODING RULES

In many application areas the existing standardised encoding rules are insufficient either in terms of encoding efficiency or simply in terms of ease of data manipulation.

One application area were the requirement for encoding efficiency has lead to the adoption of proprietary non-standardised description techniques is the specification of mobile air interfaces. Within the GSM standardisation community the perceived inadequacy of the current ASN.1 language plus standardised encoding rules has lead to the development of CSN.1 (Concrete Syntax Notation 1) [1]. This notation allows the definition of a non-generic bit-wise transfer

	8	7	6	5	4	3	2	1
1	Odd/even	Nature of address indicator						
2	NI	Numbering plan Ind.			Present. Ind.		Screening	
3	2nd address signal				1st address signal			
⋮								
n	Filler (if necessary)				*n* th address signal			

NOTE - When the address presentation restricted indicator indicates address not available octets 3 to n are omitted.

Figure 1   ISUP Calling Party number parameter field from Q.763

syntax, which is claimed to be considerably more efficient than ASN.1 together with PER encoding.

In addition to pure data efficiency, a number of new encoding rules have been recently proposed purely to enhance the data manipulation or mapping to the application domain. Examples of this class of encoding rules are those proposed for the eXtensible Markup Language (XML) (XML Encoding Rules XER) [17] and intelligent highways (Octet Encoding Rules OER) [16].

XML is a recommendation from the World Wide Web Consortium (W3C) to support client-server applications on the Internet. At present the encoding techniques used in communications are specific to this application and not based on any standardised information description technique. The basic idea of XER is to standardise a set of encoding rules that would allow information defined in ASN.1 to be carried in XML.

OER is based on concepts developed during the National Transportation Communications for ITS (Intelligent Transportation Systems) Protocol. The encoding rules provide a more compact transfer syntax than BER, whilst maintaining full octet alignment for all fields. This allows efficient and fast processing of the messages which is vital for this application.

At present the only solution available to the standardisation bodies relative to these proposed new encoding rules is either to standardise all these encoding rules or incorporate these requirements as alternative forms within existing encoding standards. The former solution could lead to a plethora of new encoding standards which by definition are only intended for a small specific application domain. The latter solution could cause already complex encoding recommendations to become completely unintelligible.

3.   PROPOSED SOLUTION

Clearly it is desirable that any proposed solution for encoding control can at least solve the current problem areas described in clause 2. In addition any solution should try to build on the existing formal description techniques which have already proved their worth. In line with this requirement the proposed solution should provide a common link between ASN.1, SDL and TTCN to allow a universally applicable transfer syntax definition.

In general the encoding control must have the ability to specify for one or more types within an abstract syntax definition the required associated encoding scheme. The encoding control should be general enough to allow definition of entire encoding rules and in addition to allow the direct implementation of the proposed encoding control into associated tool sets the notation must be machine processable.

Considering the motivations and requirements highlighted in section 2 the proposed solution is to develop a new notation for the definition of encoding control modules.

These encoding control modules would provide the ability to define encoding tables which allow the association of particular encoding rules with particular types. The overall structure of such an encoding control module is shown in figure 2. The module consists of a header part which contains a module identifier and optionally a default encoding rule. The module body contains one or more encoder control assignments to form an encoding table.

The encoder control assignments associate various encoding attributes in the form of a property list to a single specified type.

The property list has three components to define encoding attributes associated with tag, length and value as shown below:

<TYPE Name> ::= <Base TYPE>
    [<TAG Encoding Attributes>]
    [<LENGTH Encoding Attributes>]
    [<VALUE Encoding Attributes>]

All three components within the property list are optional, if any component is omitted the property will take the default encoding rule definition for this

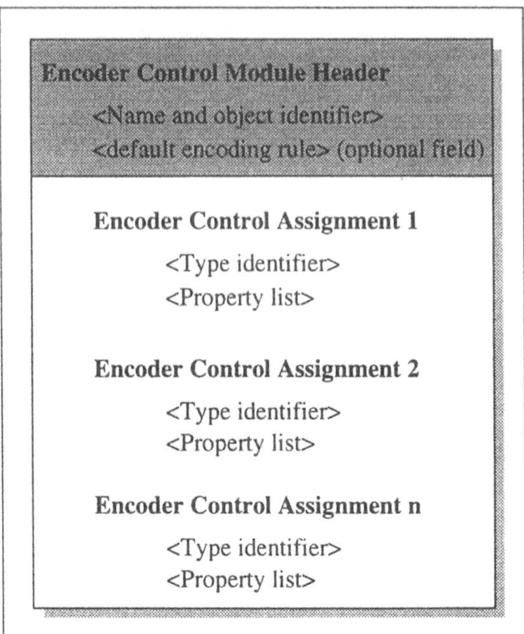

Figure 2   Encoder Control Module Abstract Format

aspect of encoding. In a similar way any type not explicitly included in the encoding table will also be encoded according to the specified default.

A provisional list of encoding attributes for the tag, length and value components is:

```
TAG Attributes -      OMIT, PER, BER, CER,
LENGTH Attributes -   OMIT, PER, BER, CER, ALIGNED LEFT,
                      ALIGNED RIGHT
VALUE Attributes -    PER, BER, CER, ALIGNED LEFT, ALIGNED
                      RIGHT, OCTET ALIGNED,
```

The binding or association of encoding control module to abstract syntax definition could be achieved in one of two ways. Firstly it could be achieved by extension of the existing ASN.1 language to support definition and referencing of encoder control modules. Secondly a new formal notation (Transfer Syntax Notation 1 TSN1) could be defined which includes the current ASN.1 language as a subset , with added functionality for definition of encoder control modules (such a scheme is shown in figure 3).

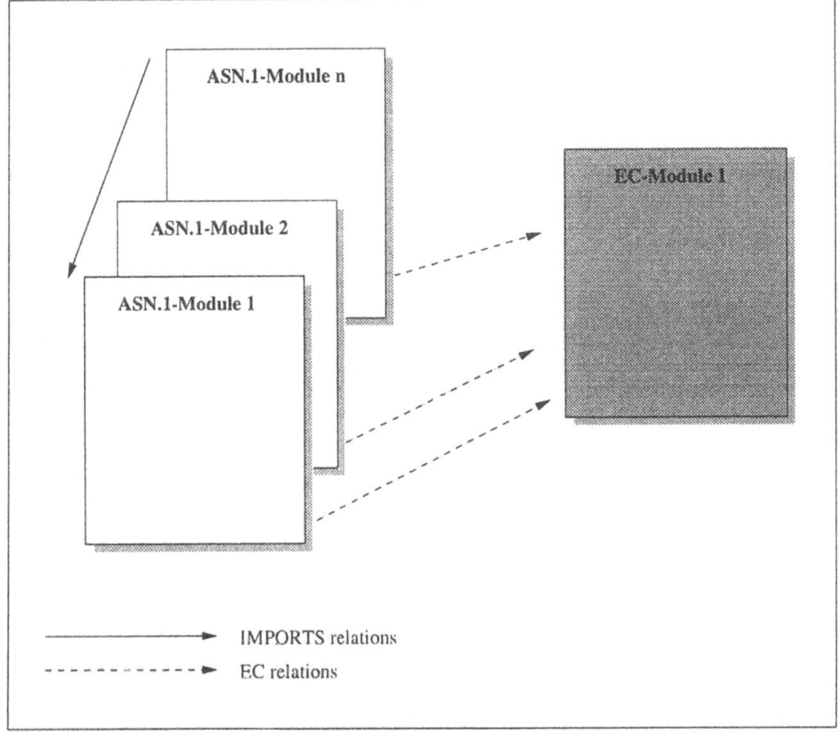

Figure 3   TSN.1 Module Structure

To illustrate the proposed scheme consider the embedded ISUP field shown in figure 1. Disregarding encoding, the field can be described in standard ASN.1 as follows

Digits ::= **SEQUENCE (SIZE(0..8)) OF Digit**

CalledPartyNumber ::= **SEQUENCE {**
    oddEven   **OddEven,**
    nADI   **NADI,**
    nI   **NI,**
    numP   **NumP,**
    presentInd   **PresentInd,**
    screenInd   **ScreenInd,**
    digits   **Digits**
    }

An associated encoder control module for such a type could look like:

```
SignProtocol {itu-t recommendation x 684 example(1)}
ENCODER CONTROL DEFINITIONS
DEFAULT {itu-t recommendation q 690}::=
BEGIN
   CalledPartyNumber {
      oddEven ::=
      TAG             : OMIT,
      LENGTH          : OMIT,
      ALIGNMENT       : OCTET LEFT,
      SIZE-OF         : BIT STRING(SIZE(1));

      ...
```

This example shows only the encoding part for the field oddEven. It specifies that no tag or length should be included in the encoding and that the field should be encoded in one bit with left alignment in the associated octet.

4.   FUTURE WORK

The development of an encoding control notation is still at a preliminary stage. The motivations and possible solutions have been discussed both within the joint ISO/ITU-T ASN.1 experts group and the ETSI (European Telecommunications Standards Institute) Methods of Testing and Specification Technical Committee (MTS-TC). In general this issue has already generated considerable interest leading to the creation of a new work item within the ETSI MTS and the proposal for a ETSI specialist task force (STF) to develop the ideas contained in this paper by defining the required notation.

The next stage towards encoding control is to define a prototype notation which matches the specified requirements. This step requires the definition of syntax and semantics. One of the main issues needing to be resolved is the set of required encoding attributes necessary to provide powerful encoding control but still maintaining the realistic chance of implementing within the associated tool sets.

To validate the proposed notation trial implementations should be made one or more of the stated problem areas. If and when a general consensus is obtained relative to the form and desirability of encoding control modules the relevant standardisation bodies should be approached with a view to either extending ASN.1 or defining the new TSN.1 notation.

5.   CONCLUSION

Without the proposed extensions to ASN.1 (either directly or via TSN.1) it is not possible to formally specify the message structures of any standardized protocol. In addition, the demands of new technologies will cause a proliferation of non-standardized, informal encoding rules which will be specific to a very narrow field of application

The proposed Encoding Control mechanism would provide a possible solution for the current problems and omissions highlighted in this paper. In addition it is intended to provide a generic extendable mechanism which will lend itself to the solution of future encoding problems and lastly an encoding control notation would complete the set of formal description techniques.

References

[1] ETSI GSM 04.07 V7.0.0 (1999-4) Digital cellular telecommunications system (Phase 2+); Mobile radio interface signalling layer 3; General aspects: Annex B Description of CSN.1

[2] ISO/IEC 9646-3:1994, Information technology - Open System Interconnection Conformance testing methodology and framework Part3: The Tree and Tabular Combined Notation (TTCN)

[3] ITU-T Recommendation Z.110 (1993), Formal Description Techniques - Criteria For The Use And Applicability Of Formal Description Techniques

[4] ITU-T Recommendation X.680 (1994) | ISO/IEC 8824-1:1995, Information technology - Abstract Syntax Notation One (ASN.1): Specification of basic notation.

[5] ITU-T Recommendation X.681 (1994) | ISO/IEC 8824-2:1995, Information technology - Abstract Syntax Notation One (ASN.1): Information object specification.

[6] ITU-T Recommendation X.682 (1994) | ISO/IEC 8824-3:1995, Information technology - Abstract Syntax Notation One (ASN.1): Constraint specification.

[7] ITU-T Recommendation X.683 (1994) | ISO/IEC 8824-4:1995, Information technology - Abstract Syntax Notation One (ASN.1): Parameterization of ASN.1 specifications.

[8] ITU-T Recommendation Z.100 (1993), Programming Languages - CCITT Specification And Description Language (SDL)

[9] ITU-T Recommendation X.690 (1994) | ISO/IEC 8825-1:1995, Information technology - ASN.1 encoding rules: Specification of Basic Encoding Rules (BER), Canonical Encoding Rules (CER) and Distinguished Encoding Rules (DER).

[10] ITU-T Recommendation X.691 (1995), Information technology - ASN.1 encoding rules: Specification of Packed Encoding Rules (PER).

[11] ITU-T Recommendation Q.931 (1993), Digital Subscriber Signalling System No. 1 (DSS 1) - ISDN user-network interface layer 3 specification for basic call control

[12] ITU-T Recommendation Q.704 (1993), Signalling System No. 7 - Signalling Network Functions And Messages

[13] ITU-T Recommendation Q.773 (1993), Signalling System No. 7 - SCCP formats and codes.

[14] ITU-T Recommendation Q.763 (1993), Formats and codes of the ISDN User Part of Signalling System No. 7.

[15] ITU-T Recommendation Q.1218 (1995), Interface recommendations for intelligent network CS1.

[16] ITU-T/ISO ASN.1 Experts Meeting Lannion Jan 99 Document 27: ASN.1 Encoding Rules - Octet Encoding Rules

[17] ITU-T/ISO ASN.1 Experts Meeting Lannion Jan 99 Document 28: XER (XML Encoding Rules)

II
TESTABILITY

4

TESTABILITY WITH UNBOUNDED TESTING STRATEGIES

Bernd Baumgarten, Olaf Henniger

GMD – German National Research Center for Information Technology

Rheinstr. 75, 64295 Darmstadt, Germany

{baumgart,henniger}@darmstadt.gmd.de

Abstract   Testability and design for testability are widely discussed practical issues in software engineering, especially in protocol engineering. A number of basic testability qualities were defined formally and independently from any special system model. In this paper we refine these notions on the one hand by a containment order on experiments and on the other hand by a formal distinction between bounded and unbounded experimentation. Containment and bounds will be interpreted mainly temporally, but can also more generally be considered as referring to a "width of view", of which time is only one, though prominent, aspect.

Keywords: Semantic foundations, testing, testability, specification theory

1.   INTRODUCTION

In theory and practice of testing, testability is often emphasized as a desirable property of specifications and systems. There exist various useful informal discussions of testability [5, 7, 8, 9, 10]. In [3], a general semantic framework was used to discuss general qualitative concepts of testability, formalizing what it means to demand that conformance or non-conformance to a specification can be decided by testing. In the present paper, we expand the approach of [3] and explore qualitative testability properties that make sense if unbounded strategies of finite experiments are available. These properties are of interest either if the nature of the tested systems guarantees termination of the strategies, or if it is so important at least to attempt the strategies that the risk that they do not terminate and have to be aborted is accepted.

This paper is part of an effort to introduce a general theory of specification as a basic common language for specification, verification, and testing [4]. It is organized as follows: Section 2 gives a general definition for specification contexts. Section 3 reviews weak/strong non-deterministic/deterministic refutability/validatability as qualitative notions of testability, which were discussed in detail in [3]. Section 4 distinguishes between bounded and unbounded experimentation. Section 5 defines refutability and validatability by strategies and compares them with the corresponding properties defined w.r.t. single experiments. Section 6 gives conclusions and an outlook.

2.   SPECIFICATION CONTEXTS

2.1   DEFINITION

When dealing with specifications, we usually do so within some context, characterized by a population of systems under consideration, the set of their possibly relevant properties, and a set of possible observations of interest.

In [2], a *specification context* is defined as a quintuple $Cont = (Systs,$ $Props,$ $Obs,$ $has\_property,$ $permits)$ where $Systs$ is a set of *systems*, $Props$ is a set of *properties*, Obs is a set of *observations*, $has\_property$ is a relation between systems and properties, and $permits$ is a relation between systems and observations.

Specification contexts [2, 4] form a simple and pervasive paradigm in which notions of specification and test can be formulated, independently of particular formal description techniques. A specification context shall fulfil the following non-degeneracy assumptions:

$$Systs \neq \emptyset \wedge Obs \neq \emptyset. \tag{i}$$

$$\forall sys \in Systs : \exists obs \in Obs : sys\ permits\ obs. \tag{ii}$$

$$\forall obs \in Obs : \exists sys \in Systs : sys\ permits\ obs. \tag{iii}$$

These assumptions are based on practical considerations. A specification context without systems or observations is of little interest. Any relevant system will be observable by at least one observation, and any relevant observation will be possible on at least one system.

What is practically considered as a system, a property, or an observation in a context depends on the chosen part of reality or field of thinking, the chosen level of abstraction, and the envisaged meaningful uses and abuses of the system. In this paper we will not be concerned with $Props$ and $has\_property$.

2.2   SPECIFICATIONS AND CONFORMANCE

The (visible) *behaviour of a system* with respect to a specification context *Cont* consists of all observations that can be made of the system:

$$sys\_beh : \begin{cases} Systs & \to \mathcal{P}(Obs) \\ sys & \mapsto \{obs \in Obs \mid sys \ permits \ obs\}. \end{cases}$$

In general, any subset $Beh \subseteq Obs$ of observations defines a *behaviour*.

Loosely spoken, a *specification* is "anything that defines a behaviour," i.e. a set of permitted observations. This set is usually not listed literally, in particular if it is infinite. Rather, it is syntactically represented by a member of a set *Specs* of specification terms, by way of a pre-determined *observational semantics* $obs\_sem : Specs \to \mathcal{P}(Obs)$, cf. Figure 1.

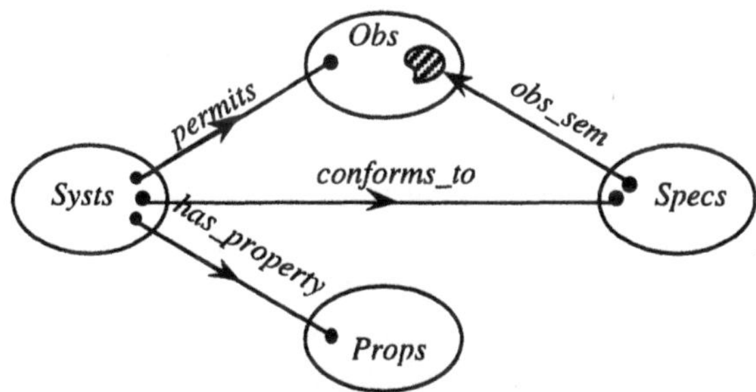

Figure 1   The semantic embedding of specifications into a specification context.

A specification *spec* is called *contradictory* if $obs\_sem(spec) = \emptyset$, and *void* if $obs\_sem(spec) = Obs$, and *non-degenerate* if neither. Non-trivial aspects of voidness have been investigated in [2].

A system *sys conforms to* a specification (term) *spec* if its behaviour stays within the range allowed by *spec*:

$$sys \ conforms\_to \ spec :\Leftrightarrow sys\_beh(sys) \subseteq obs\_sem(spec).$$

As temporary inactivity is observable in practical contexts, one can avoid the well-known disadvantage of untimed trace semantics that total inactivity is always conforming.

The above framework is fairly general and applies not only to digital information processing systems.

2.3   EXPERIMENTAL SPECIFICATION CONTEXTS

To distinguish between the (system-independent) experiments performed on a system and their possible (usually both experiment- and system-dependent) outcomes, the general definition of specification context is refined.

In [3], an *experimental specification context* is defined as a specification context $ExpCont = (Systs, Props, Obs, has\_property, permits)$, fulfilling the Assumptions i, ii, and iii, where Obs is the Cartesian product $Exps \times Outs$ of a set of *experiments* and a set of *outcomes*. Experimental specification contexts are related with the observation frameworks of [6] and even closer with the observation schemes of [1].

The possibility that the experiment exp performed on system sys yields the outcome out is represented by sys $permits$ (exp, out), and is also written as (sys, exp) may_yield out.

$$poss\_outs(sys, exp) := \{out \in Outs \mid (sys, exp)\ may\_yield\ out\}$$

denotes the set of all possible outcomes of exp performed on a system sys. We assume that in an experimental specification context every experiment performed on a system yields at least one outcome,

$$\forall sys \in Systs, exp \in Exps : poss\_outs(sys, exp) \neq \emptyset,$$

such that may_yield is left-total. This means practically that we do not allow eternal experiments, and that we treat test log entries like "nothing happened, so we broke off after one hour" or "test equipment could not be connected to the system under test" as possible outcomes.

The definition of a specification carries over without much change, such that a specification defines which outcomes may be yielded by which experiments. We call an outcome out *allowed* for a specification $spec$ and an experiment exp if $(exp, out) \in obs\_sem(spec)$. Otherwise, we call it *forbidden*. Allowed outcomes form the set

$$allowed(spec, exp) := \{out \in Outs \mid (exp, out) \in obs\_sem(spec)\}.$$

By our definitions, in an $ExpCont$, a system sys conforms to a specification $spec$ if the system may yield only allowed outcomes:

$$sys\ conforms\_to\ spec \Leftrightarrow poss\_outs(sys, exp) \subseteq allowed(spec, exp).$$

We call an outcome out *valid* for a specification $spec$ and an experiment exp if it may have come from a conforming system, i.e. if it is a member of the set

$$valid(spec, exp) := \{out \in Outs \mid$$
$$\exists sys \in Systs : sys\ conforms\_to\ spec \wedge (sys, exp)\ may\_yield\ out\}.$$

Otherwise, we call it *invalid*. Invalid outcomes tell us that the investigated system is non-conforming. They form the set

$$invalid(spec, exp) := Outs \setminus valid(spec, exp).$$

We call an outcome out *validating* for a specification $spec$ and an experiment exp if it definitely comes from a conforming system, i.e. if it is a member of the set

$$validating(spec, exp) := \{out \in Outs \mid$$
$$\forall sys \in Systs : (sys, exp)\ may\_yield\ out \Rightarrow sys\ conforms\_to\ spec\}.$$

In practical contexts, this set will often be empty.

3.   FLAVOURS OF TESTABILITY

3.1   A TAXONOMY OF TESTABILITY

The term "testability" can refer to specifications, systems, and testing environments. In this paper, we deal with qualitative aspects of the *testability of a specification*, asking whether, through testing, conformance to this specification can be

- *refuted* or *validated*

- for arbitrary non-conforming or conforming systems (*strongly*), only for some (*weakly*), or for none of them,

- possibly only with luck (*non-deterministically, ND*) or with a guarantee (*deterministically, D*),

- in bounded or arbitrary finite time.

The first three aspects have been discussed and illustrated by examples and/or counterexamples in [3]. In the present paper we recall the main results in order to refine them in view of the time aspect.

"Through testing" means that information about the behaviour of the system "under test" is procured only by means of observations. Observers do not know in advance the full behaviour of the system "under test", nor will they necessarily know after any combination of observations. For many contexts (in particular black-box contexts for reactive systems), the full behaviour of a system cannot be identified by means of observations, due to non-determinism and infinite cardinality. On the other hand, advance knowledge, or assumptions, about a system may save observation work and may even permit the determination of the full system behaviour.

3.2   REFUTABILITY AND VALIDATABILITY IN EXPERIMENTAL SPECIFICATION CONTEXTS

In *ExpCont*, a non-void specification *spec* is defined to be

- *ND-refutable* if any given non-conforming system can be shown to be non-conforming, though possibly only with luck, i.e. if

$$\forall sys \in Systs : \neg(sys\ conforms\_to\ spec) \Rightarrow$$
$$\exists exp \in Exps, out \in invalid(spec, exp) : (sys, exp)\ may\_yield\ out;$$

- *weakly D-refutable* if some non-conforming system can definitely be shown to be non-conforming, i.e. if

$$\exists sys \in Systs, exp \in Exps : poss\_outs(sys, exp) \subseteq invalid(spec, exp);$$

- *strongly D-refutable* if any non-conforming system can definitely be shown to be non-conforming, i.e. if

$$\exists exp \in Exps : \forall sys \in Systs : \neg(sys\ conforms\_to\ spec) \Rightarrow$$
$$poss\_outs(sys, exp) \subseteq invalid(spec, exp).$$

Fact 1 Every non-void specification *spec* is ND-refutable [3].   □

What is defined as ND-refutability might also be called strong ND-refutability as it allows to reveal any non-conforming system. Due to Fact 1, however, there is no point in differentiating between strong and weak ND-refutability.

In *ExpCont*, a non-contradictory specification *spec* is defined to be

- *weakly ND-validatable* if some conforming system can be shown to be conforming, though possibly only with luck, i.e. if

$$\exists sys \in Systs, exp \in Exps, out \in validating(spec, exp) :$$
$$(sys, exp)\ may\_yield\ out;$$

- *strongly ND-validatable* if any given conforming system can be shown to be conforming, though possibly only with luck, i.e. if

$$\forall sys \in Systs : sys\ conforms\_to\ spec \Rightarrow$$
$$\exists exp \in Exps, out \in validating(spec, exp) :$$
$$(sys, exp)\ may\_yield\ out;$$

- *weakly D-validatable* if some conforming system can definitely be shown to be conforming, i.e. if

$$\exists sys \in Systs, exp \in Exps : poss\_outs(sys, exp) \subseteq validating(spec, exp);$$

- *strongly D-validatable* if any conforming system can definitely be shown to be conforming, i.e. if

$$\exists exp \in Exps : \forall sys \in Systs : sys\ conforms\_to\ spec \Rightarrow$$
$$poss\_outs(sys, exp) \subseteq validating(spec, exp).$$

Practically, validatability occurs only in rather special contexts.

Fact 2 The implications indicated in Figure 2 hold for non-degenerate specifications [3]. $\square$

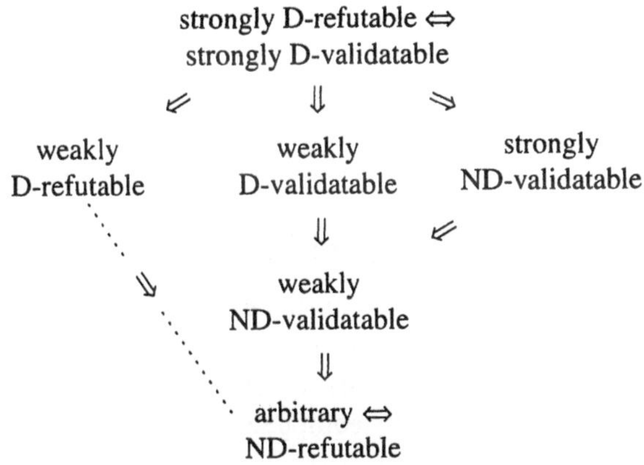

Figure 2   Implications between testability properties.

4.   BOUNDED AND UNBOUNDED EXPERIMENTATION

4.1   OBSERVATIONS ARE FINITE

Up to this point, we have not dealt with questions of duration or complexity of observations, experiments or outcomes.

Nobody has ever performed, or will ever perform, an observation that takes infinitely long. An observation or outcome only achieved after an infinite time span is not achieved in this world. Requirements referring directly or indirectly to infinitely long observations will often be practically void [2], and are then of little interest to all parties economically involved: customers or contractors, implementers, users or testers. Therefore, we consider each observation as obtained in a finite time interval. Similarly, in experimental specification contexts, we consider each outcome as obtained by the performance of an experiment in finite time.

4.2   WHAT ABOUT EXPERIMENTS?

The definitions and assumptions introduced above, imply that each experiment performed on a system yields an outcome, and does so in finite time. A different question is whether experiments are bounded, i.e. whether for every experiment there is a pre-defined time limit within which it leads to an outcome. Of course, an experiment exp can be temporally unbounded, even though it yields for every system some outcome in finite time: let $Systs$ contain infinitely many systems sys_1, sys_2, ..., and let it take n seconds for the outcome to appear if exp is performed on system sys_n.

In commercial testing, usually only a previously fixed period of time is available for the performance of test cases (experiments) and test suites (sets or sequences of experiments). Unbounded testing is thus admittedly of lesser practical importance. It will ordinarily only be employed in high-risk systems. Then it pays to perform permanent testing, probably even in parallel with the operational phase, in order to uncover faults in a controllable way, before they would occur in an uncontrolled manner during regular operation.

4.3   UNBOUNDED EXPERIMENTATION

If we intuitively permitted experiments to be unbounded, then we could now proceed to partition *Exps* into bounded and unbounded experiments. Due to the practical predominance of bounded tests, *we consider* instead *each experiment silently as bounded*, i.e. as previously equipped with an upper bound of "effort" (duration and/or complexity). On this basis, we define in Section 4.5 something new to represent what we intuitively mean by unbounded experimentation, namely an "unbounded strategy." There will also be "bounded strategies," but they will not carry further than single experiments. We will see that unbounded testing can show more than bounded testing with some system populations.

Example A Let $Systs_A$ be the set of all finite binary sequences. We assume that the end of a sequence is recognizable: say, the sequences are read out at one figure per second, and a 2-second gap indicates termination. As one can stop listening to the sequence any time, the set of observations permitted by a system sys is identified with $Pref(sys)$, the set of all initial segments. Thus, $Obs_A = Systs_A$. Consider, with the obvious observational semantics, the simple specification

$spec_{A1}$: the system is a sequence of 0's.

Theoretically, a guaranteed way both to refute and to validate the conformance of a system in finite time is to sit down and listen to the entire sequence. The end of a sequence of all 0's validates conformance to $spec_{A1}$, while the first figure 1 refutes it, both in finite time. In practice, however, one may die of old age (or a listening machine may fall apart due to wear) before a very long sequence is completely read out.   □

In fact, from a practical point of view, when one starts an activity, it does not really matter too much if this activity is guaranteed to terminate within some unknown finite time interval, or if the activity may not terminate at all. A finite duration may be too long for us to wait through it to detect finiteness, and an infinite duration can never be verified to be infinite. These questions are related with a problem of fairness investigated in [2].

How are we going to model unbounded experimentation without giving up our intuitive principle of bounded durations, not only of observations and outcomes, but even of experiments? One possible solution is to *consider unbounded experimentation as an infinite sequence of experiments.*

Example A (continued) In Example A, we can consider unbounded listening until the end of the sequence as a sequence of experiments $(exp_0, exp_1, \ldots)$ where each exp_n amounts to listening to the first $min(n, length(sys))$ members of sys. This strategy can be broken off as soon as we know what we wanted to find out, i.e. after the first figure 1 or at the end of an all-0 sequence.   □

Let us also consider an example where it is, informally spoken, impossible to find out everything about a system in finite time.

Example B Let $Systs_B$ be the set of all infinite binary sequences. We assume that the sequences are read out, on request, at one figure per second. As one can, and by necessity eventually will, stop listening to the sequence after a finite time interval, the set of observations permitted by a system sys is identified with $Pref(sys)$, the set of all (finite) initial segments. Thus, $Obs_B = Systs_A$. Further below we will use, with the obvious observational semantics, three different simple specifications:

$spec_{B1}$: the system is a sequence of 0's.
$spec_{B2}$: the system contains at least one figure 1.
$spec_{B3}$: the system contains infinitely many figures 1.   □

4.4   AN ORDER ON EXPERIMENTS

In Example A, we may consider all the experiments exp_n as started simultaneously, namely as soon as we sit down and listen to the sequence. As soon as we stop listening, say after the n_0-th sequence member, almost all of the experiments ($n > n_0$) will be aborted.

Mark the obvious ordering between experiments of different "strength" (duration, complexity) which also determines an ordering between the outcomes with respect to the amount of information they give us.

Mathematically, this approach prompts the following definitions. An *ordered experimental specification context* is a triple $OrdExpCont = (ExpCont, contained\_in, restr)$ where

- $ExpCont = (Systs, Props, Obs, has\_property, permits)$ is an experimental specification context with experiments in $Exps$ and outcomes in $Outs$,

- $contained\_in$ is a partial order on $Exps$, and

- $restr$ is a function, representing intuitively restriction to "smaller" experiments, that maps outcomes and experiments to outcomes such that

$$exp_1 \ contained\_in \ exp_2 \Rightarrow \forall sys \in Systs : poss\_outs(sys, exp_1) =$$
$$\{restr(out_2, exp_1) \in Outs \mid out_2 \in poss\_outs(sys, exp_2)\}.$$

We assume that *restr* is given effectively, such that what we can find out by an experiment would also be part of what can be found out by any "bigger" experiment, in which it is contained. Note that the application of *restr* is independent of the system observed.

As an illustration consider the scenario of the binary sequences in Example A, setting $exp_i \ contained\_in \ exp_j :\Leftrightarrow i \leq j$. Restriction to exp_i would then mean ignoring all sequence members after the i-th.

Let us consider two examples where *contained_in* is rather connected with "diagnostic breadth" than with duration.

Example C Assume each system $sys \in Systs_C$ is a natural number, and that for each experiment exp_n the only outcome is

- -1, if $sys > n$, and

- the number sys, if $sys \leq n$. □

Example D Assume each $sys \in Systs_D$ is a real number, and that every experiment exp_n consists in finding the interval $[x \cdot 2^{-n}, (x+1) \cdot 2^{-n})$, x an integer, in which sys is contained. We could imagine that the system can be handed over in an instant, e.g. by someone pointing, with a very fine pointer, to a position on a yardstick. In each of the above experiments, an analysis is made to find the interval in which the point is contained, on a grid of some chosen degree of fineness. □

In the Examples C and D, the temporal aspect is not immediately apparent; rather, the "breadth of view" or "level of detail" is what distinguishes the experiments from one another. However, from a practical point of view, temporal aspects can be identified. In Example C, it usually takes longer to read bigger numbers. In Example D, it takes longer to find out the membership to smaller intervals – think of how long it takes to procure more powerful microscopes.

4.5   STRATEGIES OF INCREASINGLY ORDERED EXPERIMENTS

Turning now to "unbounded experiments" in *OrdExpCont*, we define a *strategy* as a growing sequence $strat = (exp_n)_{n=1,2,\ldots}$ of experiments such that

$$\forall m \geq 1 : exp_m \; contained\_in \; exp_{m+1}.$$

Let *Strats* be the set of all strategies.

We call a strategy *strat bounded* if its members are bounded above by some fixed experiment, i.e.

$$\exists exp \in Exps : \forall n \geq 1 : exp_n \; contained\_in \; exp;$$

otherwise *strat* is *unbounded*. With the aid of a bounded strategy, a tester cannot find out more about an unknown system than by the single experiment *exp* dominating the strategy. An unbounded strategy, however, may tell more about a system than any single experiment.

Intuitively, in all our examples, we will find out more and more about the system, the longer we follow a strategy. The examples are, however, distinct in one respect. In Examples A and C, with the obvious strategy (exp_n checks through the first n sequence members, as far as they exist), we are guaranteed to know everything about the system in a finite time, even though – with the investigated system being unknown – we will not know in advance when this will be the case. No single experiment allows this in the two examples. In Examples B and D, no strategy allows us to identify the system completely.

5.   TESTABILITY BY STRATEGIES

5.1   TRUE EXTENSIONS OF THE TAXONOMY OF TESTABILITY

The notions of testability introduced so far (Section 3.2) refer to "experimental" testability, i.e. – according to the boundedness assumption at the beginning of Section 4.3 – testability by experiments within a bounded finite time. Now we try to capture formally the additional power gained by using strategies.

In *OrdExpCont*, a non-void specification *spec* is defined to be *strategically strongly D-refutable* if, with an appropriate sequence of experiments, any given non-conforming system could definitely be shown to be non-conforming, i.e. if

$$\exists (exp_n)_{n=1,2,\ldots} \in Strats : \forall sys \in Systs : \neg(sys \; conforms\_to \; spec) \Rightarrow$$
$$\exists m \in \mathbb{N} : poss\_outs(sys, exp_m) \subseteq invalid(spec, exp_m).$$

In Example A, $spec_{A1}$ is strategically strongly D-refutable but not strongly D-refutable.

In *OrdExpCont*, a non-contradictory specification *spec* is defined to be *strategically strongly D-validatable* if, with an appropriate sequence of experiments, any given conforming system could definitely be shown to be conforming, i.e. if

$$\exists (exp_n)_{n=1,2,\dots} \in Strats : \forall sys \in Systs : sys\ conforms\_to\ spec \Rightarrow$$
$$\exists m \in \mathbb{N} : poss\_outs(sys, exp_m) \subseteq validating(spec, exp_m).$$

In Example A, $spec_{A1}$ is strategically strongly D-validatable but not strongly D-validatable.

Other examples teach us that strategic strong D-refutability and strategic strong D-validatability are independent and not always fulfilled. In Example B,

- $spec_{B1}$ is strategically strongly D-refutable but not strategically strongly D-validatable, because the first figure 1 will show non-conformance, while no observation even of terribly many 0's will ever tell whether the next sequence member is also a 0,

- $spec_{B2}$ is strategically strongly D-validatable but not strategically strongly D-refutable, because the first figure 1 will show conformance, while the absence of any 1 in the figures to come will never be established,

- $spec_{B3}$ is neither strategically strongly D-validatable nor strategically strongly D-refutable; in fact, $spec_{B3}$ is void, cf. [2].

5.2   VAIN EXTENSIONS OF THE TAXONOMY OF TESTABILITY

The testability notions defined in this section turn out to be equivalent to testability notions listed in Section 3.2. However, they help us to prove useful theorems about the true extensions defined in the previous section.

In *OrdExpCont*, a non-void specification *spec* is *strategically weakly D-refutable* if, with an appropriate sequence of experiments, some given non-conforming system could definitely be shown to be non-conforming, i.e. if

$$\exists (exp_n)_{n=1,2,\dots} \in Strats, sys \in Systs, m \in \mathbb{N} :$$
$$poss\_outs(sys, exp_m) \subseteq invalid(spec, exp_m).$$

Lemma 1 Strategic weak D-refutability is equivalent to weak D-refutability. $\square$

Proof of Lemma 1 $\Rightarrow$: Let $sys \in Systs$ be a system and exp_m an experiment in a strategy $(exp_n)_{n=1,2,\dots}$ such that

$$poss\_outs(sys, exp_m) \subseteq invalid(spec, exp_m).$$

Take $exp := exp_m$, then the definition of weak D-refutability is fulfilled.
$\Leftarrow$: Let $sys \in Systs$ be a system and $exp \in Exps$ an experiment such that $poss\_outs(sys, exp) \subseteq invalid(spec, exp)$. Take $\forall n \in \mathbb{N} : exp_n := exp$ as a strategy $(exp_n)_{n=1,2,\ldots}$, then the definition of strategic weak D-refutability is fulfilled. $\square$

In *OrdExpCont*, a non-contradictory specification *spec* is *strategically weakly D-validatable* if, with an appropriate sequence of experiments, some given conforming system could definitely be shown to be conforming, i.e. if

$$\exists(exp_n)_{n=1,2,\ldots} \in Strats, sys \in Systs, m \in \mathbb{N} :$$
$$poss\_outs(sys, exp_m) \subseteq validating(spec, exp_m).$$

Lemma 2 Strategic weak D-validatability is equivalent to weak D-validatability. $\square$

Proof of Lemma 2 ... analogous to that of Lemma 1. $\square$

Strategic weak D-refutability and strategic weak D-validatability do not characterize anything new as they are equivalent to weak D-refutability and weak D-validatability respectively (Lemma 1 and Lemma 2). In other words, if it comes to the certainty of refuting or validating only some systems, then unbounded strategies do not permit more insight than well-chosen simple experiments.

We encounter similar relationships if we attempt to carry over the notions of ordinary "experimental" ND-testability to strategic ND-testability. Strategic ND-refutability amounts to ND-refutability; strategic weak ND-validatability amounts to weak ND-validatability. As we do not need these two notions for proving the theorems about the true extensions of the taxonomy of testability, we omit the full definitions and lemmas for these notions and restrict ourselves to strategic strong ND-validatability.

In *OrdExpCont*, a non-contradictory specification *spec* is *strategically strongly ND-validatable* if, with an appropriate sequence of experiments, any given conforming system could be shown to be conforming, though possibly only with luck, i.e. if

$$\exists(exp_n)_{n=1,2,\ldots} \in Strats : \forall sys \in Systs : sys \; conforms\_to \; spec \Rightarrow$$
$$\exists m \in \mathbb{N}, out \in validating(spec, exp_m) : (sys, exp_m) \; may\_yield \; out.$$

Lemma 3 Strategic strong ND-validatability is equivalent to strong ND-validatability. $\square$

Proof of Lemma 3 ... analogous to that of Lemma 1. $\square$

5.3   EXTENDED TAXONOMY OF TESTABILITY

Theorem 1 Strong D-refutability implies strategic strong D-refutability.   □

Proof of Theorem 1 Let $exp \in Exps$ be an experiment such that

$$\forall sys \in Systs : \neg(sys \; conforms\_to \; spec) \Rightarrow$$
$$poss\_outs(sys, exp) \subseteq invalid(spec, exp).$$

Take $\forall n \in \mathbb{N} : exp_n := exp$ as a strategy $(exp_n)_{n=1,2,...}$, then the definition of strategic strong D-refutability is fulfilled.   □

Theorem 2 Strong D-validatability implies strategic strong D-validatability.   □

Proof of Theorem 2 ... analogous to the proof of Theorem 1.   □

Theorem 3 Strategic strong D-refutability implies weak D-refutability.   □

Proof of Theorem 3 Strategic strong D-refutability obviously implies strategic weak D-refutability, which is equivalent to weak D-refutability (Lemma 1).   □

Theorem 4 Strategic strong D-validatability implies weak D-validatability.   □

Proof of Theorem 4 Strategic strong D-validatability obviously implies strategic weak D validatability, which is equivalent to weak D-validatability (Lemma 2).   □

Theorem 5 Strategic strong D-validatability implies strong ND-validatability.   □

Proof of Theorem 5 Strategic strong D-validatability obviously implies strategic strong ND-validatability, which is equivalent to strong ND-validatability (Lemma 3).   □

Summarizing the implications between experimental and strategic testability properties, we can extend the taxonomy of testability as depicted in Figure 3.

6.   CONCLUSION AND OUTLOOK

As the history of testability notions shows, testability has even more flavours than those offered in the present and its companion paper [3]. It seems to be advisable to define clearly the various testability notions and to distinguish carefully among them. Otherwise, general remarks about testability are prone to ambiguity or meaningless.

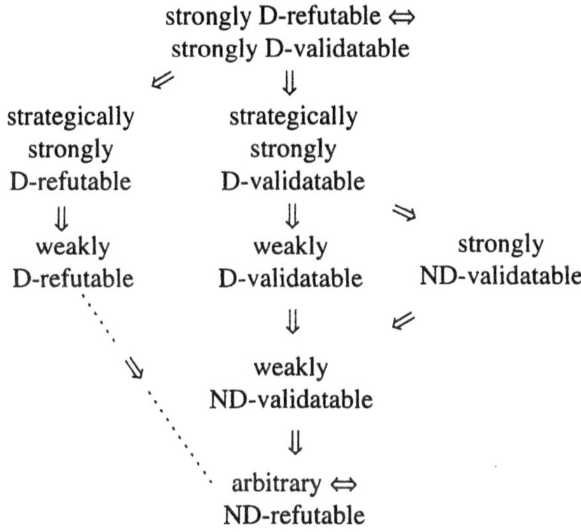

Figure 3   Implications between experimental and strategic testability properties.

As in most practical specification contexts systems may exhibit non-deterministic and infinite behaviour and validating outcomes do not exist, the practically most relevant testability property is ND-refutability. This is the weakest testability property and is held by any non-void specification.

It is worth future efforts to dedicate the testability notions for practical specification contexts and to quantify them.

Acknowledgments

We gratefully acknowledge the very helpful discussions with Helmut Wiland.

References

[1] B. Baumgarten. *Die algebraische Spezifikation von Prozessen.* Course notes, WS 1992/93, TH Darmstadt.

[2] B. Baumgarten. The Observational Significance of System Requirements. In preparation, cf. [4].

[3] B. Baumgarten and H. Wiland. Qualitative notions of testability. In: A. Petrenko and N. Yevtushenko (eds.), *Testing of Communicating Systems,* pp. 349-364, 1998, Kluwer Academic Publishers.

[4] B. Baumgarten, O. Henniger, and H. Wiland. Web pages on specification theory in the ASPEKTE project. 1999, http://www.darmstadt.gmd.de/~baumgart/aspekte.html.

[5] B. Beizer. *Software Testing Techniques,* Second edition. 1990, Van Nostrand Reinhold.

[6] E. Brinksma et al. A Formal Approach to Conformance Testing. In: J. de Meer, L. Mackert, W. Effelsberg (eds.), *Protocol Test Systems,* pp. 349-363, 1990, Elsevier/North-Holland.

[7] R. Nahm. *Conformance testing based on formal description techniques and message sequence charts.* PhD thesis, 1994, University of Bern, Switzerland.

[8] A. Petrenko, R. Dssouli, and H. König. On Evaluation of Testability of Protocol Structures. In: O. Rafiq (ed.), *Protocol Test Systems VI,* pp. 111-123, 1994, Chapman&Hall.

[9] J.M. Voas and K.W. Miller. Software Testability: The New Verification. *IEEE Software,* May 1995.

[10] S.T. Vuong, A.A.F. Loureiro, S.T. Chanson. A Framework for the Design for Testability of Communication Protocols. In: O. Rafiq (ed.), *Protocol Test Systems VI,* pp. 89-108, 1994, Chapman&Hall.

III

TESTING DISTRIBUTED SYSTEMS

5

CONFIGURATION AND EXECUTION SUPPORT FOR DISTRIBUTED TESTS

Theofanis Vassiliou-Gioles, Ina Schieferdecker,
Marc Born, Mario Winkler and Mang Li
GMD FOKUS, Kaiserin-Augusta-Allee 31, D-10589 Berlin
Tel. + 49 30 3463 7346, Fax + 49 30 3463 8346
email: { vassiliou I schieferdecker I born I winkler I m.li } @fokus.gmd.de

Abstract   This paper presents means to test the functionality, scalability and performance of distributed telecommunication applications based on CORBA ORBs. It presents generic tools for testing the functionality, performance, robustness and scalability of distributed systems. The tools cover the test suite simulator TSsim for validating concurrent test suites, the TTCN/CORBA gateway TCgate for automated execution of the test cases (including a generic coder/decoder component), and the test manager TTman to setup and parameterize the test configuration and to control the test execution. The emphasis of this paper is on TTman and its scripting facilities. Exemplarily, a TINA access session which is part of the TINA platform developed by GMD FOKUS is considered.

Keywords: Distributed Tests, Test Scripting, TINA, TSP1

1.   MOTIVATION

Due to the highly increasing complexity of new telecommunication services and the need for more scalable and manageable as well as flexible, run-time configurable execution environments for telecommunications services (telecommunication platforms) new technologies for such platforms are needed. Current telecommunication platforms are mostly based on intelligent networks (IN) technology and do not meet the new requirements any more. In the last few years a lot of research efforts have been made in the research labs all over the world to find new solutions to fit the new requirements of today's telecommunication market. The next generation of telecommunication platforms is based on distributed object technology - a key enabling factor for future telecommunication systems. In order to define a general framework for all kinds of telecommunication and information retrieval services based on distributed object technology most of the large telecommunication companies in all over the world founded the Telecommunications Information Networking Architecture Consortium (TINA-C). Technologically, TINA systems can be implemented on top of Object Request Brokers (ORB) of OMG CORBA.

This paper presents means to test the functionality, scalability and performance of distributed telecommunication applications based on CORBA ORBs. It presents generic tools for testing the functionality, performance, robustness and scalability of distributed systems. Exemplarily, a TINA access session which is part of the TINA platform developed by GMD FOKUS is considered. The test tools allow to start and configure any number of TINA test clients simultaneously on any network node and to collect the results of them after the tests have been finished.

RM-ODP describes principles for conformance assessments of ODP specifications and implementations so that implementations of different vendors can interoperate. These conformance assessments made in RM-ODP are essential and the testing of the implementations has to be performed for increasing the probability that applications can operate in the environment and interoperate with each other. Depending on the objectives of testing, distributed applications can be tested with a local or distributed test configuration. Looking at networking testing similar issues can be identified. For example, if the routing capabilities of a network has to be tested, access to the network at the remote side is needed. The distribution of the test components to the remote side can give the desired access.

In the area of Open Systems Interconnection (OSI) protocols, the Conformance Testing Methodology and Framework (CTMF) is well established and widely used. It is defined in the multipart standard IS 9646 [8]. CTMF was not aimed at describing tests for distributed systems but rather for OSI communication protocols. Although multiple upper and lower testers are supported, one assumption in CTMF is that the Implementation Under Test (IUT) is not distributed. This leads to the fact that the test coordination procedure needed between the several testers are not explicitly specified. However, concurrent TTCN (C-TTCN) which enables the use of independent and concurrent test components (TC) in a test description, gives direct raise to distributed test setups. A Main Test Component (MTC) communicates with Parallel Test Components (PTC) over Coordination Points exchanging Coordination Messages (CM). C-TTCN gives convenient means to describe abstract tests for distributed systems, however, distributed test setups and the synchronization of distributed test components are subject of current research.

The paper is structured as follows: Section 2. discusses synchronization aspects for distributed test components, gives an overview on the Test Syn-

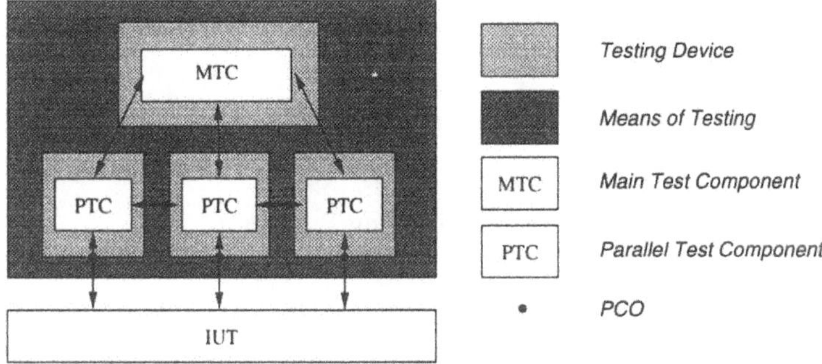

Figure 1   Generic Distributed Test Architecture

chronization Protocol TSP1 and describes an implementation of TSP1 called TTman. Section 3. identifies shortcomings of TTCN test suites in terms of dynamic test case selection and test group verdict assignments and proposes a solution to this: the Test Session Specification Language TSSL. In Section 4. an application example (a TINA Access Session) is described: its conformance tests are defined in C-TTCN and their execution make use of TTman and TSSL. Conclusions finish the paper.

2. A TEST MANAGER FOR DISTRIBUTED TESTING

Testing distributed applications results in general in distributed test setups, where test components have to be distributed to gain access at the remote ends of the tested application. The distribution of the test components is also needed because they should not influence each other - this cannot be guaranteed if the amount of test components becomes too high on a single node, what is e.g. of particular importance for performance tests. The test system in a distributed test context itself forms a distributed application which has to be managed.

Figure 1 presents the generic configuration of a distributed test system. A distributed test is realized by a set of parallel test components performing the individual test behaviour such as e.g. the emulation of client behaviour for a service under test, and by a main test component, which controls and coordinates the other parallel test components. Every test component and the test manager, i.e. every test entity, may reside on a separate tester. No resource sharing except of sharing of communication links can take place.

Therefore, two synchronization aspects have to be covered in a distributed test setup: time and functional synchronization. Between and during the execution of test cases it has to be assured that communication links are available and local and remote test components are still "on duty" at the testing devices.

For example the resource time, one of the most important resources can not be shared between two entities that do not reside on the same testing device.

Time synchronization needs to take place, where the following techniques could be used:

- synchronization via the Low-Frequency transmitter of PTB or similar institutions (DCF77) [7]

- synchronization via Global Positioning System [6]

- synchronization via Network Time Protocol [11].

In the following, we will concentrate on functional synchronization aspects. Functional synchronization[1] is needed to perform:

- test setup, maintenance and clearing

- test execution and

- test reporting.

Test setup is required to bring all involved entities, like communication channels, testing devices, test components, etc. into a well defined state, so that the test operator is able to execute the test. Possibly, a set of parameters required for proper execution of a test suite have to be distributed to the test components. The test execution is controlled via coordination messages such as 'start test' and 'report test results'. After a test suite has been finished, the testing devices and the communication channels have to release occupied resources, so that the testing devices are able to perform another test session. The process of gathering results produced by a test or a test campaign is denoted by the term test reporting. A test operator can request traces produced by the test components at the testing devices. This information has to be delivered to the test operator using the desired granularity. Either all testing devices have to report the traces, or only a specific one. The test result is considered to be transmitted to the test operator via the notification of a completed test case. At any stage during the test execution it has to be assured that all test components are in a known and stable state.

2.1   OVERVIEW OF TSP1

ETSI MTS has defined the Test Synchronization Protocol 1 (TSP1) for synchronization issues in distributed test setups [1]. The Architecture of TSP1 is presented in Figure 2.

The purpose of the TSP1 protocol is to achieve functional coordination and time synchronization between two or more Test Synchronization Architectural Elements (TSAEs). TSAEs are Front Ends (FE), Test Components (TC) and the System Supervisor (SS). For example, in a multi-party testing method (MPTM) according to [8], each lower tester can be defined by a single Parallel

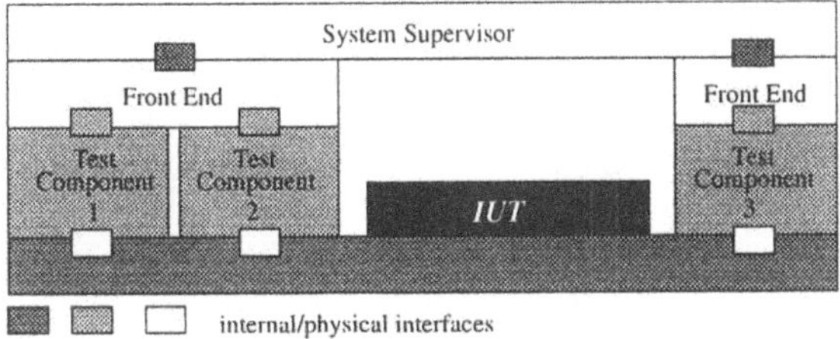

Figure 2   Architecture of TSP1

Test Component (PTC) and the Lower Testing Control Function (LTCF) can be specified by the Main Test Component (MTC). Coordination points between the MTC and the PTC enable communication between the MTC and the PTCs. With TSP1, the System Supervisor has the function of the Lower Tester Control Function (LTCF) and each test component (TC) is a Lower Tester (LT).

Physical interfaces are used where the two communicating entities do not reside on the same hardware. Mainly this will be the case between the System Supervisor and the Front Ends and the case between the executable test components and the IUT. Internal interfaces will be used were the communicating entities reside on the same hardware. In fact this is true between the Front End and the Executable Test Component (ETCO). Normally, the Front End will run as a server process on the testing device where the local Executable Test Components also reside.

The exchange of Coordination Messages via Coordination Points (CP) as defined in an abstract test suite is managed via the Front Ends in conjunction with the System Supervisor. Every executing test component needing to exchange coordination messages with an other ETCO sends them to the Front End. The Front End then forwards the CM to the System Supervisor if the message's destination is an ETCO not controlled by the Front End. The System Supervisor then takes care of distributing the message to the right Executable Test Component via the appropriate Front End. In fact, the complete test configuration and the distribution of the test components is only known to the System Supervisor.

The functionality of the System Supervisor includes the management of the test execution. It does not provide any support for implementing the necessary configuration on the testing device. The test configuration, i.e. the distribution and availability of testing devices must be known and identified in advance, and set up manually or using a Telecommunication Management Network (TMN). The System Supervisor (SS)

- manages the address table list of the test components, i.e. the mapping between each test component and its FE,

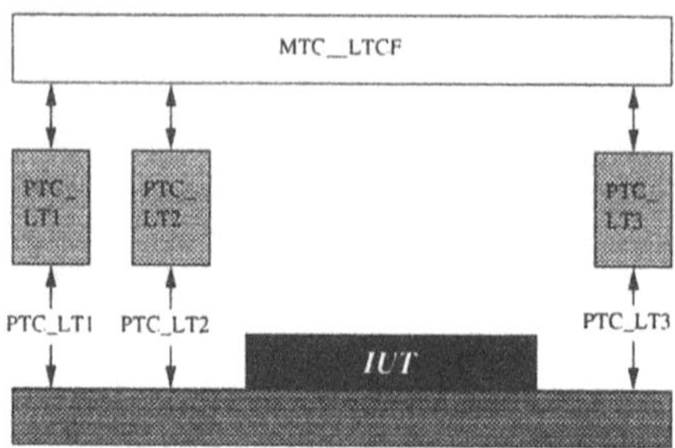

Figure 3   Multi-Party Testing Method, [8]

- has routing capabilities towards the FEs,

- communicates with the FEs, and

- manages the test session.

The Front End has two functions. Firstly, it decodes and translates received messages from the System Supervisor to the target testing device. Secondly, it distributes only those messages which are destined for test components not controlled by the Front End. If messages are received from a Test Component local to the Front End and having as destination a Test Component controlled by the same Front End, these messages stay local to the Front End, i.e. they are not sent to the System Supervisor. The Front End has to

- have routing capabilities towards its Test Components;

- communicate with the SS, and to

- communicate with the testing device.

Executable Test Components are able to handle the test interfaces and are executing the (logical) test component. They operate in the environment provided by the testing device.

2.2   THE IMPLEMENTATION OF TSP1

For implementing the Test Session Manager TTman on the basis of the TSP1 protocol, the library concept (see also Figure 4) is chosen as portability TSP1 was one of the implementation goals: A TSP1 implementation, either the System Supervisor or the Front Ends must be able to run on different operating systems (OS). Additionally, a TSP1 implementation should be able to use different communication media and/or different transport mechanisms for the communication between the System Supervisor and the Front Ends. Intentionally, [1] does not define the communication services that should be used for carrying the TSP1 PDUs as the choice of the service depends on availability but also on performance requirements. The only requirement that is formulated by [1] is that the underlying service provider shall be reliable. If performance aspects have not the highest priority but low costs of test setup are desired, TCP/IP connections might be chosen as transport layer if IP connectivity exists. If performance requirements have the highest priority, e.g. running performance tests, an ATM connection or an ISDN connection are more appropriate to provide the required service quality.

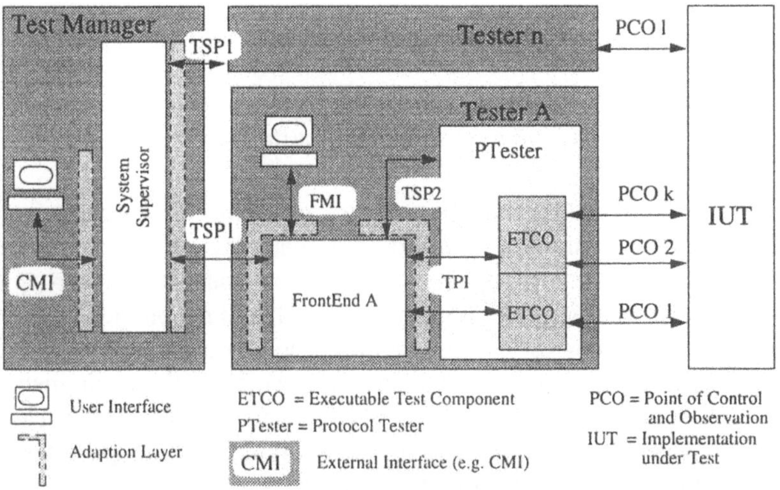

Figure 4   TSP1 Library structure

As monolithic solutions that incorporate communication services, adaption to the testing devices, etc. are very inflexible, the TSP1 protocol was, firstly, split into a System Supervisor side and a Front End side and, secondly, only the dynamic behaviour of the protocol was formulated in C. Well defined external interfaces provide access to the TSP1 library. The TSP1 library can be written without containing any OS or machine depended code. It has been designed to be compilable at any ANSI C-capable OS. The TSP1 library was already successfully compiled on a SUN ULTRA 10 running Solaris 5.6 and a PC running LINUX with the kernel version 2.0.35. On both systems the GNU C-compiler (GCC, ver 2.8.1 (SUN) and ver. 2.7.2 (LINUX)) was used. The interface at the system supervisor interface is depicted in Figure 5.

3.   TEST SCRIPTING FOR EFFICIENT TEST EXECUTION

Testing is a very time consuming process. Often it happens that test cases fail where their successful execution is a prerequisite for other, depending test cases. That means that other test cases have to fail, or at least come to an inconclusive verdict if the prerequisite test cases fail. The execution of such dependent test cases is most often a waste of resources and time.

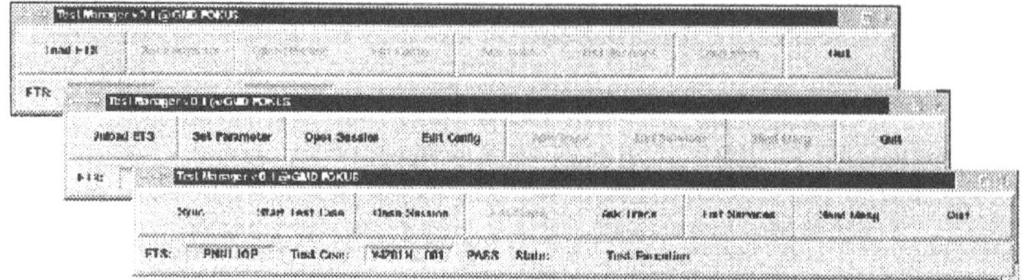

Figure 5   Control Management Interface of TTman

Another aspect paid attention in testing is reusability of test cases, i.e. that test cases might be reused in a context other than they were written primarily for. For reusing a set of test cases (i.e. a test suite) a selection and regrouping of appropriate test cases might be necessary. TTCN does not provided any means for controlling the progress of a test campaign in dependence of test results of already executed test cases. Only the dependence on static requirements, the so called selection expressions, can be used to select and deselect test cases for execution. Also, no means for different views (in terms of groups of test cases and hierarchies of test groups) of a test suite are provided by TTCN.

Finally, TTCN does not support the assignment of verdicts to test groups, which could be defined to be the accumulated verdicts of the contained test cases and/or test groups. Hence, also the assignment of the overall verdict to test suite execution in a test campaign is done in course of a subsequent evaluation of the test case verdicts, rather than via a final calculation of the verdicts of the top most test groups. In particular, there are no specified rules how to interpret the individual test case verdicts with respect to the overall test suite verdict, what opens the door to ambiguous and even contradicting overall assessments of tested systems. In order to overcome these deficiencies, a scripting language Test Session Specification Language (TSSL), will be presented that accommodates the need for dynamic test selection and execution and that enables the test suite to produce different views on the same set of test cases by simply changing the associated TSSL script. The combined use of TSSL and TTman leads to an increased grade of automated test execution and to an improved support for the evaluation of test verdicts.

The use of TSSL is not limited to TTCN test suites, but its application to TTCN is straightforward. TSSL is described subsequently in combination to TTCN in order to ease the reading. TSSL is designed to reside on top of a test suite, i.e. a TSSL script controls the execution of a TTCN test suite. TTCN is used to formulate the test cases, to group them and to define static requirements like parameters. TSSL is an add-on to an existing test suite. It does not replace the test suite or any concepts in the test suite, like selection expressions.

TSSL is based on the concept of GROUP objects which are constituted by the test suite itself, test groups and test cases (which are singleton GROUP objects). The latter two can be declared and modified. The test suite object is implicitly instantiated when the script is executed.

A test session script, that is a script written in TSSL, consists of two parts. A declaration part and a dynamic part, which define the dynamic behaviour of the test session which is based on a test suite. The first declaration of a TSSL script is the reference to the associated test suite. The import of a test suite has two effects. At first, all test cases and groups defined in the test suite can be referenced. Secondly, a predefined variable of type VERDICT, called verdict is instantiated. An instance of a GROUP object has three attributes:

- `[ref | list]`
- `verdict`

An object is either defined by a reference to an existing group in the test suite or by an explicit list of test groups/test cases. A reference to an existing group references all test cases in this group in the order in which test cases appear in the test suite (provided that they are selected due to their static selection expressions). The list attribute can be used to group test cases or groups to a new group not specified in the test suite for easier execution afterwards. In verdict the actual test group verdict is stored. It it accessible from outside the group, after the execution of the test group has finished.

Two methods are provided for GROUP variables. The first one is the exec method (shorthand for execute). The method is run if the test group is executed. Inside the execute method, individual test cases can be executed, the group verdict can be set, and decision and loop constructs enable the granular application of execution and verdict assignment. For example, a group verdict should only be set to PASS if all test cases in the group have been successfully executed a defined number of times. Also, the selection and execution of other test cases based on the result of previously executed test cases can be performed. If no execute method is defined a default execute method applies, i.e. every selected test case in the group will be executed. The group verdict will be calculated according the TTCN rules for a verdict assignment. The verdict can get only "worse", not better.

The second method that can be defined for a GROUP object is the eval method (shorthand for evaluate). This method defines a rule that will be evaluated after each and every executed test case of the group. Depending on the conditions in eval, the execution of the group can be aborted. Using this possibility, the test effort can be reduced as not every possible executable test case needs to be executed. Maybe, the execution of every test case might be not desirable if more than say for example 75% percent of the executable test cases of a group have already failed. TSSL has a loop and a decision construct. They can be used to control the test group and test case execution in either from the main level (i.e. for the test suite) or in the exec method of a GROUP object. A decision can be constructed with an if..else expression, a loop using a while style construct. Boolean expressions in a decisions and loops can consist of relations between the standard data types and verdict attributes. Two different types of functions can be used to control the execution of a TSSL script:

- execution of an object

- gathering of information about an object

An object is executed by use of its exec methods and yields its verdict attribute. If an object has no explicit exec methods, the default behaviour applies, which is to execute every test case within this group where the selection expressions hold, and where the test case was not executed before.

In a given exec method, each test case to be executed is named explicitly and will be executed despite of the verdict it has. If the test suite designer wants to avoid the repeated execution of a test case he can check the test case verdict. The exec method of a test group containing other test groups has to execute each test case of the contained groups explicitly if the groups are included by reference. Only if a test group is included by a test group variable and not by a reference it can be executed according to the test groups exec method. If every test case of a test group is deselected by its selection expressions, a NONE verdict is returned.

The attribute verdict of an object returns the basic information of an object, its verdict. Other functions exist that give a more abstract view on a GROUP object:

- `count_tc`
- `count_verdict`
- `r_count_verdict`

The function `count_tc` returns the number of test cases in an object. The `count_verdict` function return the number of test cases having the specified (i.e.different to NONE) verdict. The function `r_count_verdict` returns the percentage of test cases in an object having a specified verdict. As basis for the calculation either all test cases in an object can be used or only test cases whose selection expression hold. The default basis for calculation are the selected test cases.

4.   TINA PLATFORM UNDER TEST

This section describes the TINA platform implementation [3] which was the system under test outlined in this paper. The platform was designed according to the TINA architecture and consists in principal of access session and subscription components. They are implemented in C++ and run under Windows NT 4.0. The communication between the distributed components is done by means of CORBA mechanisms which are provided by Visibroker 3.2. The following subsections describe the structure of the implementation of both components and their environment in more detail.

4.1   ACCESS SESSION AND SUBSCRIPTION

The access session component is of major concern in this paper. It is forming one process running on Windows NT and consists of implementations for the computational objects defined in the TINA architecture like Initial Agent (IA) and User Agent (UA). That means there is a decomposition of these objects in several C++ class declarations and definitions. Furthermore these computational objects are supporting interfaces according to the Retailer Reference Point (RET-RP) defined by TINA-C like *i_RetailerInital* and *i_RetailerNamedAccess* as well as proprietary interfaces which are used internally (see figure). In order to fulfil its task the access session needs information from subscription. Therefore another process is running on the same node containing the subscription component. It contains implementation for several computational objects whereas one of them the Subscription Coordinator (SC) is of main interest for the access session. It supports an interface which provides all the necessary information to the access session. Subscription itself retrieves these information from an object-oriented database realized with Versant.

4.2   ENVIRONMENT

In order to make some interface references from subscription known to the access session and known to the test component a CORBA Naming Service (NS) has to be executed. In the relevant test configuration the NS coming with Visibroker for C++ 3.2 was used. It is running on the same node like the other components under test. The access session uses another component (UADB) to get access to the already mentioned object-oriented database, where all user information are stored. This component runs in a separate process on the same node and is also implemented in C++. Figure 6 shows the configuration of the platform.

As a precondition for the whole platform the Visibroker 3.2 Smart Agent has to run on the node as well as the Versant demon to use the database which also runs on the same node. This is not depicted in the figure.

Testing distributed applications encompasses two steps:

- In a first step the functional aspects of the system under test is verified, i.e. it is checked whether the system behaves in the target environment like expected and whether it is conform to reference points.

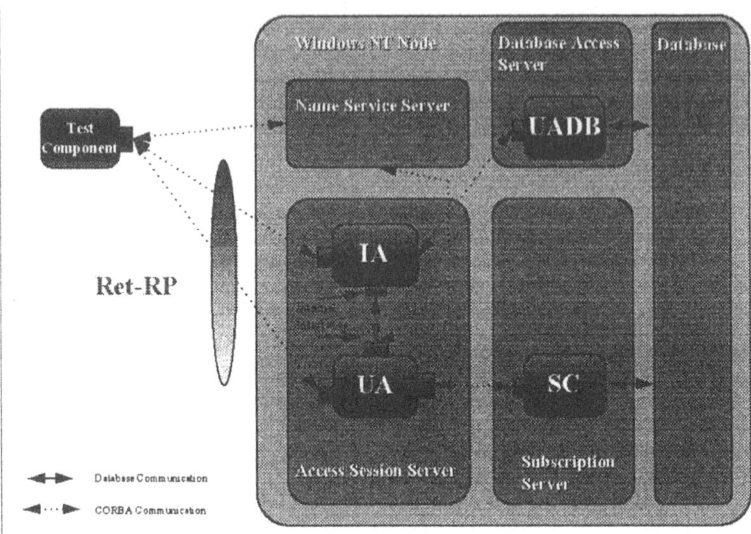

Figure 6   Configuration of the TINA Platform under Test

- Once the conformance of the system under test is checked, performance and robustness tests can be performed to determine whether the system also behaves correct under load.

The conformance tests for the TINA platform under test have been made first, the performance tests in a second step. This papers describes in more detail the use of TSSL for the efficient execution of the conformance tests, the performance tests and their results are described in [15].

Figure 7 depicts the test behaviour as a Message Sequence Chart (MSC) diagram in parallel to the following description:

1. The test component (TC) resolves a name context at the NS to retrieve the interface reference (*i_RetailerInitial* interface) to the Initial Agent (IA).

2. This interface reference is used to call the *requestNamedAccess* operation at that interface. The parameter *userId* has the value *anonymous*, the password is an empty string. This operation request causes the IA to initiate a database request to the UADB object to get some properties for that user (*userDescription*). In the case that the userId is anonymous the IA instantiates a new User Agent (UA), initializes the UA with the user description and returns the interface reference (*i_RetailerNamedAccess* interface) of the UA to the TC. In its initialization phase the User Agent resolves a name context at the NS to retrieve the interface reference to the Subscription Coordinator (SC).

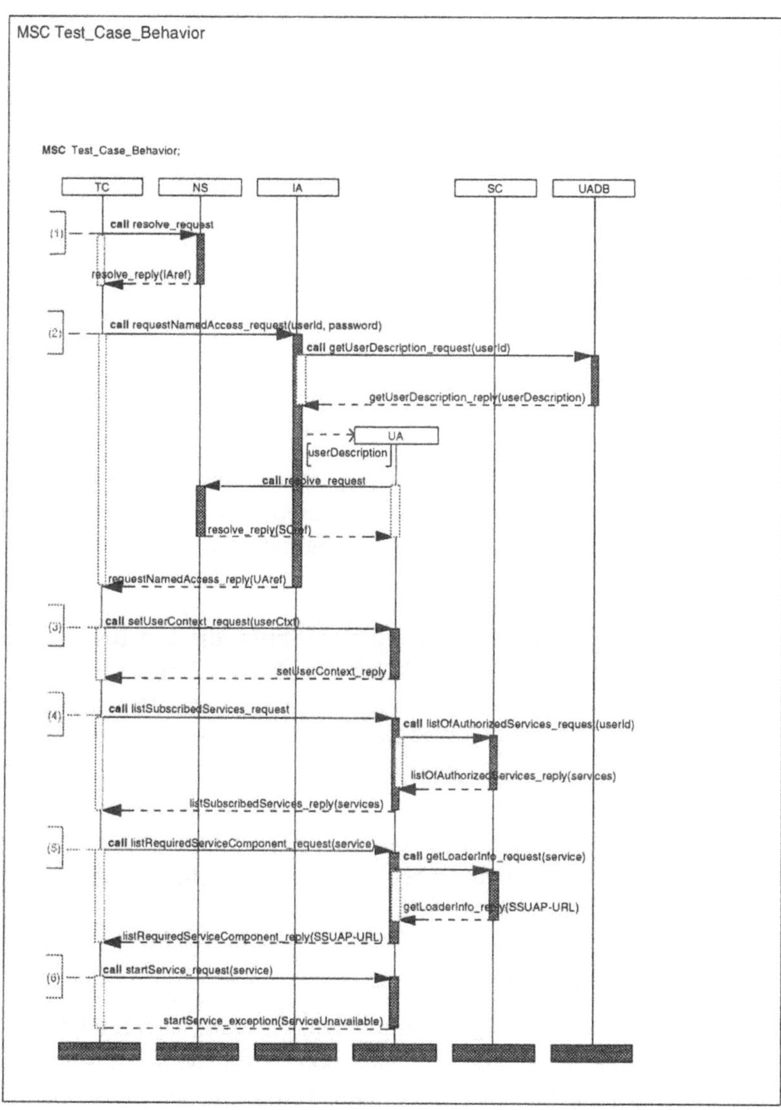

Figure 7 The behaviour of the example test case

3 The TC calls the operation *setUserContext* at the *i_RetailerNamedAccess* interface.

4 In order to retrieve the available services for that anonymous user the TC calls the operation *listSubscribedServices* at the *i_RetailerNamedAccess* interface. Then the UA sends the request *listOfAuthorizedServices* to the SC which provides the information with the help of the underlying database back to the UA. The UA replies the list back to the TC.

5 The TC which acts like a Provider Agent in the TINA architecture needs some information about the service specific user application of the selected service in order to start the service. Therefore it calls the operation *listRequiredServiceComponent*. This causes the UA to send the request *getLoaderInfo* to the SC which retrieves this information from the database. In the current implementation this information consists of an URL to a JAVA applet which implements the service specific user application.

6 After the TC has got the information about the service specific user application it calls the *startService* operation for the selected service. Since no service is running on the platform the UA responds with a *ServiceUnavailable* exception.

The test cases make use of PTCs which describe the test behaviour with respect to the individual interfaces of NS, IA, and UA. TTman is used to setup and parametcrize the executable test components in the distributed test setup.

The test cases for Figure 7 are defined step-wise so that every test case makes use of test cases which reflects the preceding operation requests. For example, the test case for *RequestNamedAccess* makes use of the test case for *Resolve_Request*. This makes the usefulness of test case execution depending on the success of the respective preceding test cases, what is reflected in the TSSL script given in Figure 8.

5.   CONCLUSION

The presented paper investigated the synchronization of parallel and distributed test components in order to obtain means for the synchronization of distributed test components. There is no standardized framework for distributed testing, however different testing techniques such as conformance testing, performance and interoperability testing require in general distributed test setups for testing distributed systems. The time and functional synchronization of distributed test components are still only partially covered issue in research. The main focus is on functional synchronization of distributed testing components. The test synchronization protocol TSP1 was described, its relevance

```
Basic_capabilities_Access_Session = {
  IMPORT "AS_Tests";
  GROUP InitTest, AccessTest, ServiceStartTest, ...;
  TESTCASE Resolve_Request, RequestNamedAccess, SetUserContext, ...;
  INTEGER tries;
  Resolve_Request.ref = "AS_Tests/GENERAL/Init_Service/Valid/U0";
  RequestNamedAccess.ref = "AS_Tests/GENERAL/Init_Service/Valid/U1";
  SetUserContext.ref = "AS_Tests/GENERAL/Init_Service/Valid/U2";
  ...;
  InitTest.list = { Resolve_Request };
  AccessTest.list = { RequestNamedAccess, SetUserContext, ...};
  ServiceStartTest.list= { InitTest, AccessTest };
  ServiceStartTest.exec()
  {
    if( InitTest.exec ())
     if (AccessTest.exec()) verdict = PASS;
     else { verdict = FAIL; break; }
    else { verdict = FAIL; break; }
  };
 main () {
   tries = 1;
   /* tries 10 times service setup */
   while((ServiceStartTest.verdict) && tries =< 10)
   {
     if (NOT ServiceStartTest.exec())
     { verdict = FAIL; break; }
     tries = tries + 1;
   }
   verdict = PASS;
  }
}
```

Figure 8   The Example TSSL Script

to distributed testing and the testing notation TTCN was analysed. The Test Session Manager TTman, which is based on TSP1, has been described.

Testing might be a time consuming process. In distributed testing, the optimization of testing time is especially desired as connections have to be maintained for the synchronization of the distributed test components and test personal have to operate at the remote site. Therefore, a careful selection of test cases that have to be executed has to be done. As no standardized notation for specifying the dynamic test case selection exists, a test session specification language TSSL was presented.

The usage and application of TTman and TSSL are demonstrated for an exemplarily test for a TINA Access Session implementation. The use of test setup and synchronization features of TTman to the conformance tests of the TINA Access Session eased the test execution. TTman's features are of particular importance for repeated test executions as in e.g. regression tests. Due to the small number of test cases for the Access Session, the performance gain in test execution from the TSSL script was not that high, but the script helped to control the test execution. The test suite as well as the scripts will be further extended in order to investigate the applicability, usefulness and efficiency of scripting techniques in distributed test executions. The translation of a TSSL script into a target language has still to be performed by hand. Automated support for the use of TSSL is proposed. Incorporating a TSSL interpreter into the presented test session manager TTman will increase the usability of TSSL.

Notes

1. The fact that the coordination message exchange may cross the boundaries of a single tester requires internetworking between the single testers. Reliability of the internetwork is assumed.

References

[1]   ETSI TC-MTS: Methods for Testing and Specification (MTS); *Test Synchronization; Architectural reference; Test Synchronization Protocol 1 (TSP1) specification*, ETSI Technical Report ETR 303, Sophia Antipolis, January 1997.

[2]   Farley, P.; Hogg, S.; Kristiansen, L. et al.: *Ret Reference Point Specifications*, Snapshot 1, Version 0.4, TINA-Consortium, May 14, 1997.

[3]   Project "TINA Platform" by GMD FOKUS/Deutsche Telekom Berkom, http://www.fokus.gmd.de/research/cc/platin/projects/, 1998.

[4]   TINA-C: Service Architecture Version 4.0, Oct. 1996.

[5]   Eurescom Project 412: *Methodology and Tools For ISDN Network Integration and Traffic Route Testing*, Deliverable 3, Test Specifications for ISDN Network Integration Testing, EURESCOM, Heidelberg, August 1996.

[6]   *Global Positioning System Standard Positioning Service Signal Specification*, Second Edition, U.S. Department of Defense at the U.S. Coast Guard Navigation Center, Alexandria, VA, June 1995.

[7]   Hetzel, P.: *Zeitinformation und Normalfrequenz von der PTB › Über den Telekom-Langwellensender DCF77*, telekom praxis, Heft 1, 1993, pp. 25-36.

[8]   ISO/IEC 9646: *Information technology - Open systems interconnection - Conformance testing methodology and framework*, International Standard, Geneva,1991.

[9]   ISO/IEC 9646-3: *Information technology - Open Systems Interconnection - Conformance testing methodology and framework - Part3: The Tree and Tabular Combined Notation*, International Standard, Second Edition, Geneva, 1997.

[11]   RFC1305 (Mills, D. L.): *Network Time Protocol (Version 3) Specification, Implementation and Analysis*, DARPA Network, University of Delaware, March 1992.

[12]   Ousterhout, J.K.: *Tcl and the Tk Toolkit*, Addison-Wesley, April 1994.

[13]   ITU-T Z.100: *CCITT specification and description language (SDL)*, ITU-T Recommendation, Geneva, March 1993.

[14]   ITU-T Z.120: *Message Sequence Chart (MSC)*, ITU-T Recommendation, Geneva, 1996.

[15]   Born, M.; Hoffmann, A.; Winkler, M.; Schieferdecker, I.; Vassiliou-Gioles, Th.: *Performance Testing of a TINA Platform.* - Accepted to Appear in TINA'99, Hawai, May 1999.

6

PRINCIPLES AND TOOLS FOR TESTING OPEN DISTRIBUTED SYSTEMS *

Mohammed Benattou, Leo Cacciari, Régis Pasini and Omar Rafiq
Laboratoire TASC, University of Pau, France

Abstract   With emergence of new models, architectures and midlleware such as ODP, TINA and CORBA for developing open distributed systems, testing technology requires adaptation for use within conformance assessment in such systems. All these frameworks are object-based and aim at creating open distributed environments supporting interworking, interoperability, and portability, in spite of heterogeneity and autonomy of the related systems. In this context an open distributed system may be viewed as a system providing standardized distributed interfaces for interacting with other systems. Conformance of such a system can be assessed by attaching a related tester at each provided interface. However, many problems of controllability and observability influencing fault detection during the testing process arise if there is no coordination between the testers. In this paper, we show how to cope with these problems by using a distributed test method derived from OSI conformance testing. Afterwards, a CORBA prototype is designed, realized and experimented with a forum application.

1.   INTRODUCTION

ODP (Open Distributed Processing) [3], is a joint ISO/ITU standard which provides a generic framework for building open distributed systems. CORBA (Common Object Request Broker Architecture) [7], which is a more implementation-oriented architecture with similar objectives to those of ODP, has been defined by the international consortium OMG (Object Management Group). Several implementations of CORBA are already commercially available. TINA (Telecommunications Information Networking Architecture) [10], designed by the international Consortium TINA-C, is a kind of future telecommunications architecture based on ODP and CORBA. ODP, CORBA and TINA are then

---

*This work has been supported by CNET-France Télécom under Grant 98 1B 221 as part of the CTI programme.

complementary frameworks and they are currently recognized as constituting a real trilogy of open distributed processing.

In this context an open distributed system may be viewed as a system providing standardized distributed interfaces for interacting with other systems. To deal with this interfacing problem in conformance testing, ODP has already defined the concept of reference point as part of the reference model. It is a specific location of an open distributed system at which one or more interfaces can be localized, so that interactions at these interfaces can be observed. Hence, some subset or all possible reference points may be declared as conformance points, stating the requirements to be met at each of them. Some other work has also been done on the subject [1, 5, 6]; however, much effort and considerable work are required before reaching a workable and accepted conformance testing methodology for open distributed processing.

Based on testing of OSI communicating systems, conformance of an open distributed system can be assessed by attaching a related tester at each provided interface. However many problems of controllability and observability, influencing fault detection during the testing process arise if there is no coordination between the testers. In this paper, we show how to cope with these problems by using a distributed test method derived from OSI conformance testing [2, 9]. Afterwards, a CORBA prototype is designed, realized and experimented with a forum application.

2.   TESTING DISTRIBUTED APPLICATIONS

Concurrent with academic research activities, efforts have been made within industry and standardization bodies (ISO and ITU) since the 80s to develop approaches for conformance testing of OSI protocols. After a decade of hard work, a standardized methodology, tools, experiences, and theories exist today with a certain degree of maturity [4]. This section is dedicated to extending results from protocol testing to deal with testing of an implementation under test (IUT) of an open distributed system.

2.1   TESTING ARCHITECTURE

Testing may be seen as a means to execute an IUT by carrying out test cases, in order to observe whether it has a certain behavior or not. An IUT is a kind of black-box with points of control and observation (PCOs) at which control of input events and observation of output events may take place during the testing process.

Testing architecture is a description of the environment within which an IUT is tested. Figure 1 gives an abstract view of what could be an ODP testing architecture. It is a synthesis of the architectures that have been used or proposed until now in the context of testing ODP systems [1, 5, 6]. It consists of the IUT, its PCOs that could be reference points or not, and a test system which may be seen as an implementation of test cases. The test system has particularly to execute given test cases by communicating with the IUT via its PCOs and to observe the test results. It may consist of one or several components called testers. The example of Figure 1 is made up of three testers.

Dealing with this architecture similarly to OSI conformance testing methodology [2], one can distinguish four kinds of test method that could be defined in terms of coordination between the testers in the testing process. If the test system consists of one tester, the related test method is called a centralized test method. If it consists of several testers located at different sites, we have three cases. If there is no coordination between the testers and each of them behaves with its corresponding PCO, independently of the others, the method is called a remote test method. If the testing process has to deal with the coordination of the testers, the related method is called a distributed test method. Finally, if the coordination has to follow given specific procedures, the method is called a coordinated test method and therefore is a particular case of the distributed one.

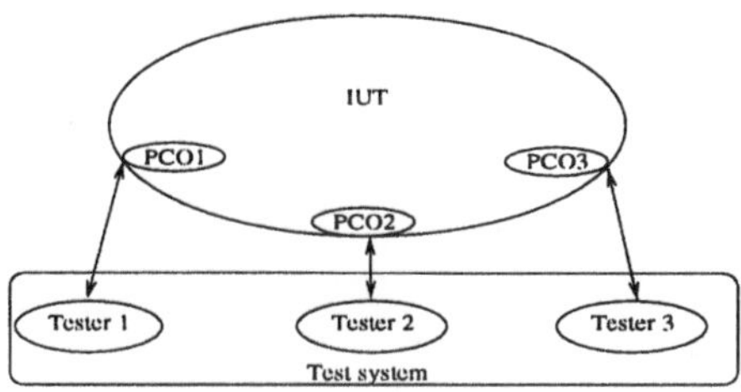

Figure 1. ODP testing architecture

2.2   MODELING CONCEPTS

Generally speaking, specifications and related implementations are described in different formalisms. In order to be able to reason about the testing process in a formal setting, the specification and IUT must be modeled by comparable concepts. Therefore, conformance of an IUT to its specification may be defined by means of relations between the IUT model and specification model [4].

I/O FSM (Input/Output Finite State Machines) are widely used in the communication protocol area [4], and may be easily adapted with some extensions for modeling distributed systems. In a communication protocol, a protocol entity communicating with a peer entity, is described by an I/O FSM with one input queue and one output queue. Distributed applications are supposed however to communicate with multiple partners. This leads to the notion of a multi-port FSM [6] which may use several input/output queues called ports.

Definition 1. *A multi-port FSM with n ports (np-FSM) $\mathcal{A}$ is a 6-tuple $(Q, \Sigma, \Gamma, \delta, \lambda, q_0)$, where: Q is the finite set of states of $\mathcal{A}$; $q_0 \in Q$ is a distinguished state, the initial state of $\mathcal{A}$; Σ is a n-tuple $(\Sigma_1, \Sigma_2, \ldots, \Sigma_n)$ where Σ_k is the input alphabet of port k, and $\Sigma_i \cap \Sigma_j = \emptyset$ for $i \neq j$. We write $\bar{\Sigma}$ for the input alphabet $\Sigma_1 \cup \Sigma_2 \cup \cdots \cup \Sigma_n$ of $\mathcal{A}$; Γ is a n-tuple $(\Gamma_1, \Gamma_2, \ldots, \Gamma_n)$ where Γ_k is the output alphabet of port k, and $\Gamma_i \cap \Gamma_j = \emptyset$ if $i \neq j$. ε being the empty symbol, we write $\vec{\Gamma}$ for the output alphabet $(\Gamma_1 \cup \{\varepsilon\}) \times (\Gamma_2 \cup \{\varepsilon\}) \times \cdots \times (\Gamma_n \cup \{\varepsilon\})$ of $\mathcal{A}$; δ is the transition function, it is a partial function $Q \times \bar{\Sigma} \to Q$; λ is the output function, it is a partial function $Q \times \bar{\Sigma} \to \vec{\Gamma}$. Moreover, $\lambda(q, \alpha)$ is defined if and only if $\delta(q, \alpha)$ is.*

A transition of np-FSM $\mathcal{A}$ is a 4-tuple $\mathbf{t} = (q, \alpha, \gamma, q')$ where $q, q' \in Q$, $\alpha \in \bar{\Sigma}$ and $\gamma \in \vec{\Gamma}$ are such that $\delta(q, \alpha) = q'$ and $\lambda(q, \alpha) = \gamma$.

Example 1. Figure 2 gives an example of 3p-FSM with set state $Q = \{q_0, q_1, q_2, q_3, q_4, q_5\}$, q_0 being the initial state, $\Sigma_1 = \{a\}$, $\Sigma_2 = \{b\}$, $\Sigma_3 = \{c\}$, and $\Gamma_1 = \{w, x\}$, $\Gamma_2 = \{y\}$, $\Gamma_3 = \{z\}$. Transitions are represented by arrows.

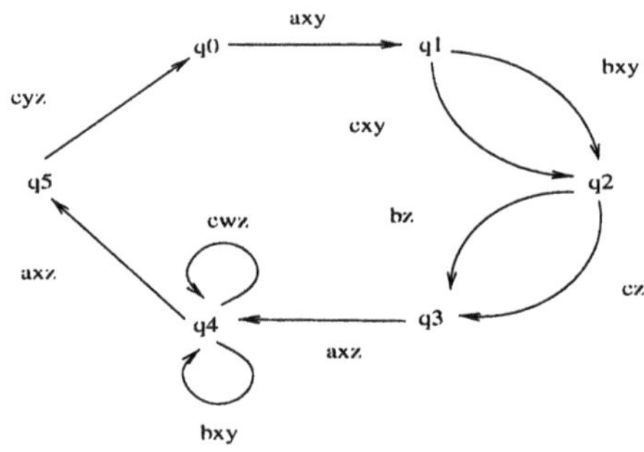

Figure 2. An example of 3p-FSM

Definition 2. *A test sequence of np-FSM $\mathcal{A}$ is a sequence ω in the form: $!x_1?y_1!x_2?y_2 \cdots !x_t?y_t$ where, for $i = 1, 2, \ldots, t$, x_i belongs to $\bar{\Sigma}$ and y_i is a subset of $\cup_{k=1}^n \Gamma_k$ such that, for each port k, $|y_i \cap \Gamma_k| \leq 1$, i.e. y_i contains at most one symbol from the output alphabet of each port of $\mathcal{A}$. $!x_i$ means sending message x_i to the IUT and $?y_i$ means receiving the messages belonging to y_i from the IUT.*

Example 2. A test sequence for Example 1:

$$\omega = !a?\{x,y\}!b?\{x,y\}!c?\{z\}!a?\{x,z\}!b?\{x,y\}!c?\{w,z\}!a?\{x,z\}!c?\{y,z\} \tag{1}$$

Test sequences are generally generated from the IUT specification and characterized by their fault coverage. Several methods exist for generating test sequences from I/O FSM specifications. They are mainly for detecting two basic types of fault: output faults and transfer faults [8]. An edge with an incorrect output has an output fault which is generally observable. Any generation method providing a test suite traversing each transition of an FSM at least once is capable of detecting such faults. An edge with an incorrect starting state or ending state has a transfer fault. Since states are not directly observable, these faults are relatively more difficult to detect. No specific test generation method is considered in the following. Test sequences may be any.

2.3 CONTROLLABILITY AND OBSERVABILITY ISSUES

If the centralized method does not pose any particular problem, the others raise several issues in terms of controllability and observability which are fundamental features of conformance testing. In this context, controllability may be defined as the testing power or the capability of the test system to realize input events at corresponding PCOs in a given order, and observability may be defined as the testing power or the capability of the test system to determine the output events and the order in which they take place at the corresponding PCOs.

It is easy to see that controllability and observability related to the chosen test method have a great influence on several aspects of the testing activity such as the execution of test sequences, the observations that can be made during the test execution, and the interpretation of the testing results.

Example 3. Figure 3 gives an example of faulty IUT related to Example 1. It is easy to see that test sequence (1) in the centralized method leads to a fail verdict. In the remote method however projections of (1) on port alphabets are required to get the test sequence related to each tester. Projections ω_1, ω_2 and ω_3 of ω on port 1, port 2 and port 3 are: $\omega_1 = !a?x?x!a?x?x?w!a?x$, $\omega_2 = ?y!b?y!b?y?y$ and $\omega_3 = !c?z?z!c?z?z!c?z$.

Let us examine three situations in applying these sequences to the IUT of Figure 3 by using the remote method. The test starts in q_0.

Situation 1. When the IUT is in q_4, it gets both b on port 2 and c on port 3. Then, either it follows the intended path reading b before c, or it reads c before b. In the first case, tester 1 receives w before x and it detects an output fault. However, if the IUT decides to read c before b then none of the testers is able to detect this fault.

Situation 2. When the IUT is in q_4, it receives a and then it sends y and z instead of x and z and goes to q_5. In q_5, it receives c and sends x and z instead of y and z and goes to q_0. One can notice that there are two successive output faults and none of the testers is able to detect them.

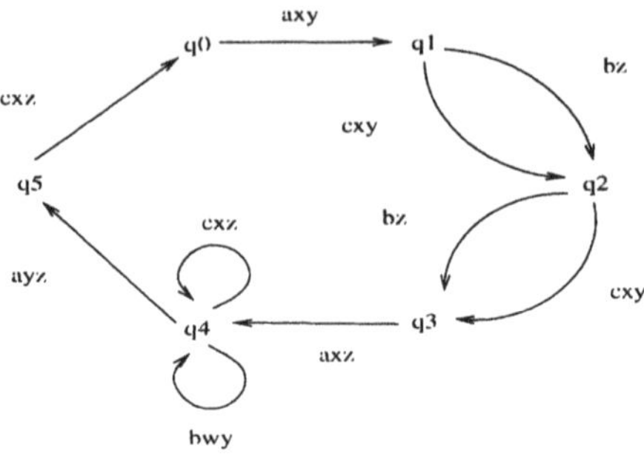

Figure 3. Example of faulty IUT for Example 1

Situation 3. When the IUT is in q_1, it gets both b on port 2 and c on port 3. Then, either it follows the intended path reading b before c, or it reads c before b. In both cases, tester 1 receives x, tester 2 receives y and tester 3 receives z as specified by the test sequences of Example 3. However, one can notice that the IUT has in fact two successive output faults: on receiving b in q_1, IUT sends z instead of x and y and on receiving c in q_2, IUT sends x and y instead of z.

In situation 1, there is a control problem because the input events b and c may be taken into account by the IUT in any order. In situation 2, there is an observation problem as the testers are unable to determine the order in which output events take place. Finally, in situation 3, there is both a control problem and an observation problem. All these situations have already been analyzed in [6]. They are due to the fact that, in the remote method, there is no relationship between the testers except the IUT itself. This implies, among others, that the fault coverage of a test sequence may be lower in the remote method than in the centralized one.

3. DISTRIBUTED TEST METHOD

In this section, we are designing a configuration for the distributed method that is able to avoid control and observation problems by using auxiliary messages called coordination messages.

3.1 ARCHITECTURE

The basic idea is to coordinate the testers by using a communication service parallel to the IUT through a multicast channel. Each tester interacts with the IUT only through the port to which it is attached and communicates with the other testers through the multicast channel. It is an extension of the Astride approach that we have already designed for protocol testing [9].

The multicast channel is mainly to allow each tester to send a coordination message to the other testers during testing. We are aware of the importance of temporal considerations and their coherence in testing distributed systems. To cope with that in this contribution, we make a simplifying and realistic hypothesis. We assume that the time required for a coordination message to travel from a tester to another tester is of the same order of magnitude as the reaction time of the IUT, i.e. the time elapsed between the reception of a message by the IUT and the sending by the IUT of the corresponding response. Figure 4 depicts an architecture of our distributed test method.

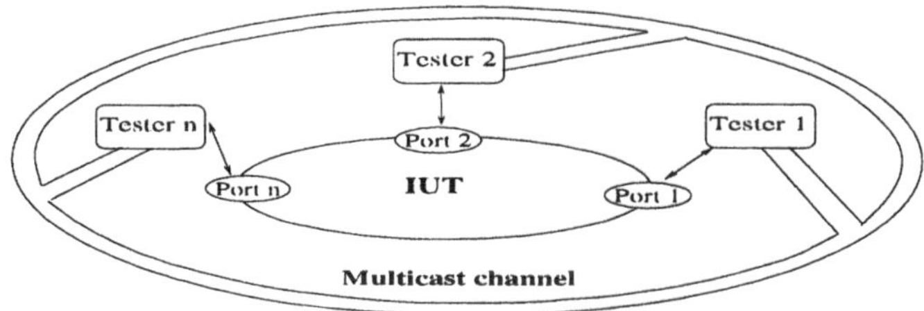

Figure 4. Distributed test architecture

3.2   LOCAL TEST SEQUENCES

In the distributed test method, each tester executes a local test sequence constructed from the complete test sequence of the IUT. A *local test sequence* is in the form $\alpha_1 \alpha_2 \cdots \alpha_t$, where each α_i is either:

— $!x$, sending of message $x \in \Sigma_k$ to the IUT,
— $?y$, receiving of message $y \in \Gamma_k$ from the IUT,
— $-c_{h_1,\ldots,h_r}$, sending of coordination message c to testers $h_1, \ldots, h_r$,
— $+c_h$, receiving of coordination message c from tester h.

For each α_i, if it is a sending, either of a message to the IUT or of a coordination message, the tester sends it. If α_i is a receiving, either of an output from IUT or of a coordination message, then the tester waits for a message. If no message is received, or if the received message is not the expected one, the tester gives a **fail** verdict. If the tester reaches the end of its sequence, then it gives a **pass** verdict. If all testers give a **pass** verdict, then the test system ends the test by giving a **pass** verdict.

If the IUT has n ports, Algorithm 1 is dedicated to compute the n related local test sequences from a complete test sequence of IUT. $\backslash$ is to denote the set difference and Δ is to denote the symmetrical difference: $A \Delta B = (A \backslash B) \cup (B \backslash A)$. Function Port gives the port corresponding to a given message. For a set y of messages, function Ports is defined by: $\text{Ports}(y) = \{k | \exists a \in y \text{ s. t. } k = \text{Port}(a)\}$.

The local test sequences are basically projections of the complete test sequence over the port alphabets. In fact, Line 4 adds $!x_i$ to the sequence $\omega_{\text{Port}(x_i)}$

and the loop of Line 5 adds the reception of messages belonging to y_i to the appropriate sequences. Coordination messages are added to the projections to get the same controllability and observability when using the complete test sequence in centralized method.

Algorithm 1 uses two kinds of coordination messages: C coordination messages for guaranteeing controllability, and O coordination messages for guaranteeing observability.

As pointed out in [6], control problem may arise when the tester sending x_{i+1} is neither the one sending x_i, nor one of those receiving a message belonging to y_i. In this case (Line 6), we add $+C$ and $-C$ to the appropriate local test sequences. $+C$ is added to ω_h, where h is the tester sending x_{i+1}. $-C$ is added to the sequence of a tester receiving a message belonging to y_i, if $y_i \neq \emptyset$ (Lines 8 and 9), if not $-C$ is added to the sequence of the tester sending x_i (Line 7). The former case does not introduce any control problem because the reception of C by tester h means that x_i has been received by IUT. However, in the latter case, one may still have a control problem if this exists in the centralized method because x_{i+1} may be sent to the IUT before that x_i has been received by it. Therefore, we have the same controllability as in centralized method.

Remark 1. Let us already notice that Lines 8 and 9 may also reduce the number of O coordination messages. This is due to the fact that for sending C, Algorithm 1 considers if there exists a tester that receives a message belonging to y_i and does not receive any message belonging to y_{i+1}.

Algorithm 1. Generating local test sequences.

> **input:** $\omega = !x_1?y_1!x_2?y_2\cdots!x_t?y_t$: a complete test sequence
> **output:** $(\omega_1,\ldots,\omega_n)$: local test sequences
> **for** $k = 1,\ldots,n$ **do** $\omega_k \leftarrow \varepsilon$ **end for**
> **for** $i = 1,\ldots,n$ **do**
> > $k \leftarrow \mathrm{Port}(x_i)$
> > **if** $i > 1$ **then**

1:      $\mathrm{Send\_To} \leftarrow (\mathrm{Ports}(y_i)\Delta\mathrm{Ports}(y_{i-1}))\backslash\{k\}$
2:      **if** sender $\neq 0$ **then** $\mathrm{Send\_To} \leftarrow \mathrm{Send\_To}\backslash\{\text{sender}\}$ **end if**
3:      **if** $\mathrm{Send\_To} \neq \emptyset$ **then**

> > > > $\omega_k \leftarrow \omega_k \cdot -O_{\mathrm{Send\_To}}$
> > > > **for all** $h \in \mathrm{Send\_To}$ **do** $\omega_h \leftarrow \omega_h \cdot +O_k$ **end for**
> > > > **end if**
> > > **end if**

4:      $\omega_k \leftarrow \omega_k \cdot !x_i$
5:      **for all** $a \in y_i$ **do** $\omega_{\mathrm{Port}(a)} \leftarrow \omega_{\mathrm{Port}(a)} \cdot ?a$ **end for**
> > **if** $i < t$ **then**
> > > $h \leftarrow \mathrm{Port}(x_{i+1})$
> > > sender $\leftarrow 0$

6:      **if** $h \notin \mathrm{Ports}(y_i) \cup \{k\}$ **then**
> > > > **select**

7:         **case** $y_i = \emptyset$:

$$\omega_k \leftarrow \omega_k \cdot -C_h$$
$$\omega_h \leftarrow \omega_h \cdot +C_k$$
$$\text{sender} \leftarrow h$$

8:     **case** $\text{Ports}(y_i)\backslash\text{Ports}(y_{i+1}) \neq \emptyset$:
       choose $l \in \text{Ports}(y_i)\backslash\text{Ports}(y_{i+1})$
       $$\omega_l \leftarrow \omega_l \cdot -C_h$$
       $$\omega_h \leftarrow \omega_h \cdot +C_l$$
       $$\text{sender} \leftarrow l$$

9:     **otherwise:**
       choose $l \in \text{Ports}(y_i)$
       $$\omega_l \leftarrow \omega_l \cdot -C_h$$
       $$\omega_h \leftarrow \omega_h \cdot +C_l$$
       $$\text{sender} \leftarrow l$$

    end select
   end if
  end if
 end for
End Algorithm 1

To avoid observation problems, each tester receiving a message a belonging to y_{i-1} should be able to determine that a has been sent by the IUT after the IUT has received x_{i-1} and before the IUT receives x_i. This can be achieved by using an O coordination message. $+O$ is added to the local test sequences of those testers receiving either a message belonging to y_i or a message belonging to y_{i-1}. $-O$ is added to ω_k where k is the tester sending x_i. It can be shown by induction [1] that this avoids observation problems. Finally, let us notice that Line 2 avoids sending O to the tester having sent C to tester k, reducing in this way the number of O coordination messages as was said in Remark 1.

Example 4. Applying Algorithm 1 to test sequence (1), we get the following local test sequences:

$$\left\{ \begin{array}{rcl} \omega_1 & = & !a?x?x+O_3+C_3!a?x?x?w!a?x+O_3 \\ \omega_2 & = & ?y!b?y-C_3+C_3!b?y-C_3+O_3?y \\ \omega_3 & = & +C_2-O_1!c?z-C_1?z-C_2+C_2!c?z?z-O_{\{1,2\}}!c?z \end{array} \right. \qquad (2)$$

Using these sequences in the distributed method, one can detect all the faults of Figure 3.

4.   CORBA TESTING TOOLS

The objectives of CORBA and ODP are quite similar. They aim to describe systems implemented in terms of configurations of distributed objects. To do that, they identify common infrastructure functions that support the operation of such systems and simplify their implementation. However if ODP is a generic model, CORBA is more implementation-oriented. Therefore, several implementations of CORBA are already commercially available. After introducing basic CORBA elements, this section is experiments with implementation of the distributed test method through a prototype within the CORBA infrastructure.

4.1   CORBA

To provide a standard for interoperability of distributed objects in heterogeneous environments, OMG has in fact developed two architectures: OMA (Object Management Architecture) and CORBA [7]. OMA defines its object terminology and various facilities required for object-oriented distributed processing. CORBA provides the mechanisms necessary to identify, locate, access and manipulate the OMA objects.

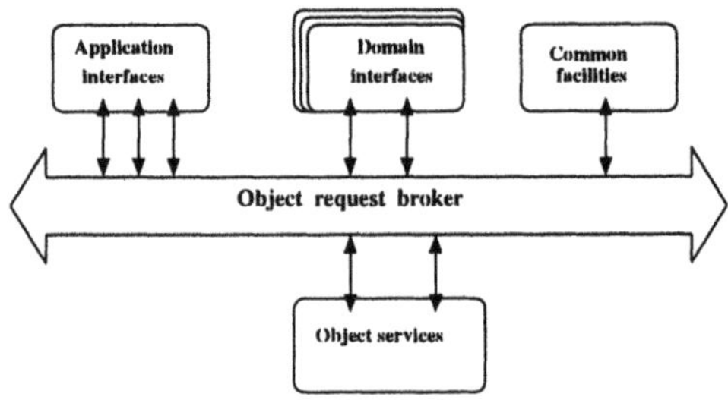

Figure 5. OMA reference architecture

OMA architecture.   OMA is made up of an object model and a reference architecture. The object model defines how objects distributed across an heterogeneous environment can be described, while the reference architecture deals with interactions between those objects. Figure 5 shows the components of the OMA reference architecture, which are detailed below.

Object Request Broker (ORB): this is a kind of a message bus for communicating between objects of a distributed application regardless of their location and implementation.

Object services: These are domain-independent interfaces that provide fundamental assistance to application developers such as the naming and trading services. The naming service allows clients to find objects based on names while the trading service allows them to find objects based on their properties. They also include many other services such as event notification, transaction management and security.

Common facilities: These are a set of interfaces providing commonly required services to applications such as printing and compound documents.

Domain interfaces: A set of interfaces that are oriented towards specific application domains such as telecommunications manufacturing, and so on.

Application interfaces: These are interfaces developed specifically for a given application.

CORBA specification. CORBA specification details the interfaces and characteristics of an ORB that conveys requests for invocation of objects operations from CORBA client to CORBA object implementation. Figure 6 shows the different an ORB interfaces. In CORBA, the terms client and server are used to specify the role played by a component in a distributed application. A client makes a request of some other component. A server provides an implementation of a component that a client uses. Servers may act as clients to other component. A pure client does not play the server role. Each component communicates with others on a peer-to-peer basis.

The key feature of an ORB is transparency. Usually, an ORB hides the following characteristics of objects: object location, object implementation, object execution state, and object communication mechanisms, by using object references. When a CORBA object is created, a corresponding object reference is also created. In other words, a creation request returns an object reference for the newly-created object to the client. To make a request, a client specifies the target object by using its object reference which is immutable and opaque.

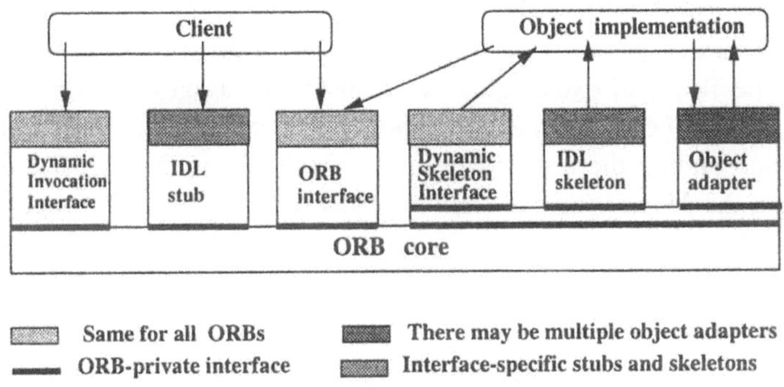

Figure 6. ORB interfaces

In CORBA, objects are defined only by their interfaces by using the OMG Interface Definition Language (IDL). An object interface specifies the operations and types that the object supports. Therefore, IDL is a purely declarative language. It is based on C++ with additions for distribution. It does not provide features like control constructs, leaving one free to construct objects using different programming languages. OMG has defined several language mappings like C, C++, Smalltalk, Ada 95, Java and so on. These mappings are automatically implemented by IDL compilers, which generate code in the given programming language for the types specified in IDL and facilitate communication between clients and servers wherever they may be located.

An ORB is drawn as a single entity in the OMA. Figure 6 shows that in fact it is implemented in several parts. In addition to generating programming language types, IDL compilers generate stubs and skeletons supporting what is called static invocation. A stub is a mechanism that creates and issues requests on behalf of client, while a skeleton delivers requests to the object implementation. CORBA also provides dynamic invocation through the dynamic invocation interface and dynamic skeleton interface, without generating stubs and skeletons at compile time. The ORB interface is to provide a variety of services from the ORB to clients and servers, and because the ORB core may be implemented in different ways an object adapter is required to provide standard interfaces to servers.

4.2   PROTOTYPE STRUCTURE

In developing a prototype for the distributed test method, three main testing aspects may be distinguished: preparation, execution and conclusion. Preparation consists of designing a test sequence for the IUT from its specification. Execution consists of connecting the testers to the IUT ports and providing them the related local test sequences. The testing can then be executed according to the behavior described in Section 3. Finally, the conclusion consists of collecting the testers verdicts to determine the global verdict. Figure 7 describes the prototype structure. It is made up of a set of testers, attached to the ports of the IUT and exchanging coordination messages through a multicast channel, called the *Coordination Channel*. A *Management System* is used for starting and ending the test as well as for determining the global test verdict. To do that, the *Management System* interacts with the testers through a specific channel called the *Management Channel*.

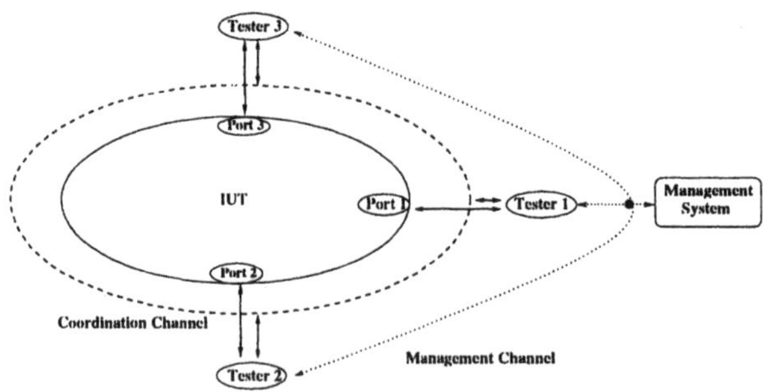

Figure 7. Prototype of distributed method

4.3   PROTOTYPE REALIZATION

This section is dedicated to describe the main objects of the prototype, which are grouped into three categories: information objects, computing objects and communication objects.

Information objects.   There are two types of information object: test sequences and verdicts.

A *Test sequence* object encapsulates the sequence of inputs to be applied to the IUT by a given tester as well as the expected results. It also includes the coordination messages to be sent to, or to be received from, other testers.

A *Verdict* object encapsulates information allowing the *Management System* to determine the global verdict. Each *Tester* object provides a local verdict to the *Management System* object.

Computing objects.   There   are   two   kinds   of   computing   object: *Tester* object and *Management System* object.

A *Tester* object applies a test sequence to the IUT interface related to it and provides a local verdict. A *Tester* object behaves as described in Section 3.2.

A *Management System* object configurates the test application, dispatches the local test sequences, starts and ends the test by providing the user with a global verdict deduced from the local test verdicts.

Communication objects.   Three kinds of communication are considered: communication between the *Management system* and the testers, communication between the testers, and communication between the testers and the IUT.

In communication between the *Management system* and the testers, three kinds of information are exchanged: commands to start and to end the test, local test sequences and local verdicts. It should be noticed that the local test sequences and the local verdicts are exchanged on a 1 to 1 basis: each tester receives its local test sequence and sends back its verdict. These exchanges can then be realized by method invocation through the ORB. Commands starting and ending the test are sent on 1 to n basis: the management system must send them to each and every tester at the same time. This is realized through a broadcast channel (*Management Channel*) using the CORBA event service, which allows CORBA objects to produce events to be consumed by other objects. Commands are events produced by the *Management System* and consumed by the testers. Herein, the event service is used in *push* mode in which, when the producer puts an event into the channel, the consumers are automatically notified. Communication between testers is 1 to n: a tester may send a coordination message to some or all other testers. This is realized through the *Coordination Channel* using the CORBA event service. Figure 8

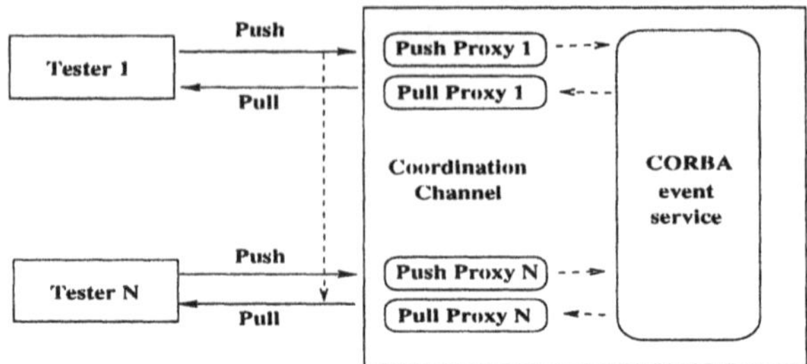

Figure 8. Coordination channel

gives the layout of the *Coordination Channel*. To send a coordination message to testers $h_1, \ldots, h_r$, tester k invokes the Push method on object *Push Proxy k*. To receive it, tester h_i invokes the Pull method on object *Pull Proxy h_i*. The coordination message is transfered from *Push Proxy k* to *Pull Proxy h_i* using the CORBA event service.

Communication between testers and the IUT consists mainly of me-thod invocation. To get a generic test application, adapting itself to the interfaces of the IUT, we introduce an *Adapter* object which provides a *Tester* object with methods to send and to receive actions. An *Adapter* object communicates with the associated IUT interface through the CORBA dynamic invocation mechanism.

5.   TESTING CARRYING OUT

This section describes a testing example of a CORBA application using the test prototype. The application and its interfaces are briefly presented. Related local test sequences are constructed from a complete test sequence, and then test results are given and discussed.

5.1   FORUM APPLICATION

The application to be tested is a *forum application*, similar to a newsgroup where only *subscribers* can post and retrieve messages. The application has three interfaces:

— a User interface allowing subscribers to post and to retrieve messages;
— a MessageManager interface which is an extension of the User inter-face, allowing suppression of messages;
— a Manager interface allowing one to add and to remove subscribers.

The IDL specification of these interfaces is given in Figure 9.

The np-FSM describing the behavior of the application is not given here, due to the space limitation in the paper.

```
interface User {                              interface MessageManager : User {
  /* Exceptions */                              /* Methods */
  exception not_allowed{};                       void Suppres_message(in id_mess ref)
  exception cant_send{string reason;};             raises(no_such_message);
  exception no_such_message{};                 };
  /* Methods */                                interface Manager {
  void Deposit(in Person p,in Message m)         /* Exceptions */
    raises(not_allowed,cant_send);               exception cant_create {string reason;};
  short How_many_Messages(in Person p)           exception no_such_subscriber{};
    raises(not_allowed);                         /* Methods */
  HeadersList Get_Headers_List(in Person p,      void Subscribe(in Person p)
  in short min,                                    raises(cant_create);
                      in short max)              void UnSubscribe(in Person p)
    raises(not_allowed);                           raises(no_such_subscriber);
  Message Get_Message(in Person p,               PeopleList Subscribers();
                  in id_mess ref)              };
    raises(not_allowed,no_such_message);
};
```

Figure 9. IDL interfaces of the forum application

5.2   TEST EXECUTION

To illustrate how local test sequences are constructed by the prototype, we use the test sequence of Figure 10. This sequence checks that a user who is not subscriber is not allowed to submit message. The notation used in Figure 10 is self explaining.

```
!Manager:Unsubscribe(PersonA)
!Manager:Subscribers
?Manager:PeopleList, PersonA ∉ PeopleList
!User:Deposit(PersonA,Message)
?User:not_allowed
```

Figure 10. Test sequence of forum application

Local sequences corresponding to the sequence of Figure 10 are given in Figure 11. Notice that a coordination message is used to avoid `PersonA` sending a message before the `Unsubscribe` message has been received by the IUT.

```
         Manager                              User

!IUT:Unsubscribe(PersonA)          ?CC:coordinate_message(Sender),
!IUT:Subscribers                      Sender == Manager
?IUT:PeopleList,                   !IUT:Deposit(PersonA,Message)
   PersonA  does not belongs to    ?IUT:not_allowed
PeopleList
!CC:coordinate_message(User)
```

Figure 11. Local test sequences of Figure 10

Each tester uses two interfaces: `IUT`, related to the IUT interface associated with the tester, and `CC` related to the interface with the *Coordination Channel*. These sequences come from testing experiments with the prototype on students realizations of the forum application.

6.   CONCLUSION AND FUTURE WORK

With emergence of modern models and architectures such as ODP, CORBA and TINA, for developing new distributed applications, a related conformance testing methodology is required for getting working distributed systems. Our goal is to go from our experience with protocol testing to contributing to the design of such a framework. In this paper, we showed how OSI conformance test methods could be extended to take into account distributed systems providing more than two access points. After that, a distributed test method was designed and analyzed in terms of controllability and observability that play an important role in fault detection. A CORBA prototype of this method has been realized and experimented on students applications. Our work is now oriented to develop testing environments that are more adapted for tackling real-world applications.

References

[1] L. Cacciari and O. Rafiq, Controllability and observability in distributed testing, To appear in: *Information and Software Technology*.

[2] ISO/IEC, OSI conformance testing methodology and framework, 9646, Parts 1–7, 1991.

[3] ISO/IEC, ODP, Reference model, 10746, Parts 1-4, 1995.

[4] ISO/IEC, Information retrieval, transfer and management for OSI, Framework: formal methods in conformance testing, 1996.

[5] P. F. Linington, J. Derrick, and H. Bowman, The specification and testing of conformance in ODP systems, in: B. Baumgarten, H. J. Burkhardt and A. Giessler, eds, Testing of Communicating Systems, Chapman&Hall, London, 1996, pp. 93–114.

[6] G. Luo, R. Dssouli, G. v. Bochmann, P. Venkataram, and A. Ghedamsi, Test generation with respect to distributed interfaces, in: Computer Standards and Interfaces 16, 1994, pp. 119–132.

[7] OMG, The common object request broker: architecture and specification, Version 2.2, 1998.

[8] A. Petrenko, G. v. Bochmann, and M. Yao, On fault coverage of tests for finite state specifications, in: Computer Networks and ISDN Systems 29, 1996, pp. 81–106.

[9] O. Rafiq, The Astride testing approach : principles, tools and carrying out, in: Computer Standards and Interfaces 11, 1990, 85–94.

[10] TINA-C, Overall concepts and principles of TINA, version 1.0, Document Label: TB_MDC.018_1.0_94, 1995.

7

ARCHITECTURES FOR TESTING DISTRIBUTED SYSTEMS

Andreas Ulrich[a], Hartmut König[b]

a   *Siemens AG, Corporate Technology, ZT SE 1, 81739 München, Germany*
    E-mail: andreas.ulrich@mchp.siemens.de

b   *BTU Cottbus, Department of Computer Science, PF 101344, 03013 Cottbus, Germany*
    E-mail: koenig@informatik.tu-cottbus.de

Abstract   Stabilizing network infrastructures has moved the focus of software system engineering to the development of distributed applications running on top of the network. The complexity of distributed systems and their inherent concurrency pose high requirements on their design and implementation. This is also true for the validation of the systems, in particular the test. Compared to protocol testing the test of distributed systems and applications requires different methods for deriving test cases and for running the test. In this paper, we discuss architectures for testing distributed, concurrent systems. We suggest three different models: a global tester that has total control over the distributed system under test (SUT) and, more interestingly, two types of distributed testers comprising several concurrent tester components. The test architectures rely on a grey-box testing approach that allows to observe internal interactions of the SUT by the tester. In order to assure a correct assessment of the test data collected by the distributed tester components, the tester has to maintain a correct global view on the SUT. This can be achieved either by the use of redundant points of control and observation or by test coordination procedures. We outline the features of each approach and discuss their benefits and shortages. Finally, we describe possible simplifications for the architectures.

Keywords: Distributed systems, concurrency, test architectures, specification-based testing, concurrent transition tour.

1   INTRODUCTION

Designing, implementing, and validating complex distributed systems requires a deep understanding of the communication taking place and the order of communication messages exchanged between the individual components to avoid unwanted behavior during runtime. This applies in particular to the testing of these systems. Compared to protocol testing the test of distributed systems and applications requires different methods for deriving test cases and for running the tests. In [12] we introduced the concept of a concurrent transition tour an approach for providing test cases for distributed systems. The derivation of concurrent transition tours was represented in [13]. In this paper we continue the approach with the discussion of possible test architectures to run these tests.

We assume that a distributed system consists of a collection of components, each of them realizing a sequential flow of execution. Each component has an interface that defines its incoming and outgoing Messages. The test of distributed systems usually follows the steps of single component tests, integration tests of components, and finally the system test. Especially, the integration test is mainly aggravated by the following properties of distributed applications:

- true concurrency between components,
- absence of a global clock, and
- message exchanges between components that are unobservable by a tester.

We assume that a test architecture for distributed systems consists of the following basic elements:

- the system under test (SUT), i.e. the executable implementation of the distributed software system to be tested,
- the tester that implements a test case and also assesses the results of a test run,
- points of control and observation (PCOs) between tester and SUT, and
- possibly test coordination procedures which coordinate the actions between different tester components.

The paper defines requirements to the test architecture that need to be matched to support the execution of test runs and their correct assessment afterwards. These requirements address the SUT, which must be prepared in a suitable way to perform a test run. Requirements in this category are referred commonly to rules of "design for testability". Further requirements are suited to design testers that correctly assess the results of a test run. Note that our discussion is confined for the testing of functional aspects. We do not consider quality-of-service and performance aspects of distributed systems that are discussed in [15].

Two principal test architectures are presented: a global tester that observes and controls all distributed components of a SUT in a central manner, and a distributed tester that consists of a number of distributed, concurrent tester components, each of them collecting only partial information about the execution progress in components of the SUT. Especially, a distributed tester is of interest since it makes better use of system resources and supports concurrency within the SUT directly. Whereas the global tester can be a performance bottleneck during a test run.

Assuming a grey-box testing approach and the right choice of points of control and observation (PCOs) between tester and SUT, we show that also a distributed tester works correctly and brings up the expected result. The PCOs must be provided in the implementation of the SUT to allow the tester to observe necessary information for assessing a test run. A common means for this purpose is the insertion of software probes into the SUT.

The paper is organized as follows. In Section 2 we introduce some basic notions as well as an example which are used in the following discussion. Section 3 describes the proposed test architectures. Section 4 discusses possible simplifications for the application of these architectures. The final remarks summarize the results and give an outlook on needed further research.

2   PRELIMINARIES

2.1   BASIC NOTATIONS

We suppose a specification of a distributed system as a parallel composition $\mathfrak{S} = M_1 \parallel \ldots \parallel M_k$ of interacting components. Each component realizes a certain function of the distributed system, e.g. in the form of a client and/or server. It is described by a sequential automaton (*labeled transition system*, LTS). Components communicate with each other solely via interaction points. The communication pattern used is synchronous communication and non-blocking send based on interleaving semantics. Transmitting messages and their receipt through interaction points are referred to *actions*.

Definition 1. A *labeled transition system* (LTS or *machine* for short) M is defined by a quadruple $(S, A, \rightarrow, s_0)$, where S is a finite set of states; A is a finite set of actions (the alphabet); $\rightarrow \subseteq S \times A \times S$ is a transition relation; and $s_0 \in S$ is the initial state.

To distinguish the different kinds of communication, we denote all inputs and outputs of the distributed system implementation from/to the environment as *external* (reactive system), analogously all inputs and outputs belonging to the inter-component communication as *internal*. Events appearing only inside a module are not considered.

Consider the simple system $\mathfrak{S} = A \parallel B$ whose LTSs are given in Figure 1. Under the assumption that actions a and c in each LTS synchronize, removal of parallel operator ‖ by applying interleaved-based semantics rules yields the composite machine $C_{\mathfrak{S}}$.

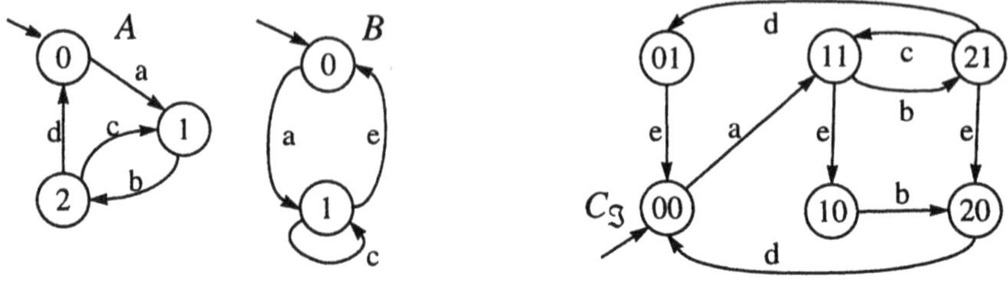

Figure 1. Example system $\mathfrak{S} = A \parallel B$ and its composite machine $C_{\mathfrak{S}}$.

2.2   TEST SUITES FOR CONCURRENT SYSTEMS

A conventional approach to test suite generation starts from the composite machine. The generation of a transition tour from the interleaving model of the composite machine has its limitations since concurrent events are serialized. Due to a lack of controllability during testing, this approach is not feasible. The resulting order of concurrent events in a test run could not be predicted. The order of events is, however, essential to assess whether an implementation is correct or not.

Possible test cases of our example derived from the composite machine in Figure 1 are the following sequences:

$$\sigma_1 = a.b.c.b.d.e.a, \quad \sigma_2 = a.b.c.b.e.d.a, \quad \sigma_3 = a.b.c.e.b.d.a.$$

All three sequences can be equally employed in a test run. They possess the same fault detection capability [7]. They differ only in different orders of the interleaving actions b, d and e. However, this order cannot be predicted in a sequential test run.

In [12], we extended the notion of a *transition tour* [8] to a *concurrent transition tour (CTT)* and applied it as a test suite for distributed systems. In the context of distributed systems, a transition tour is extended to a CTT such that all transitions in all modules of the system are visited at least once on the shortest possible path through the system. A CTT takes concurrency among actions of different modules into account. It is depicted graphically as a time event sequence diagram where nodes are events and the directed arcs define the causality relation between events. A feasible construction algorithm of a CTT is presented in [13].

Definition 2. A *lposet (labeled partially ordered set)* is defined by the quadruple $(E, A, \leq, l)$, where E is a set of event names; A is a set of action names;

≤ is a partial order expressing the causality information between events, i.e. e ≤ f if event e precedes event f in time; $l: E \rightarrow A$ is a labeling function assigning action names to events. Each labeled event represents an occurrence of the action labelling it, with the same action possibly having multiple occurrences. A *pomset* (*partially ordered multiset*) is an isomorphism class over event renaming of an lposet, denoted $[E, A, \leq, l]$.

Definition 3. (CTT) A *concurrent transition tour* of a concurrent system $\mathfrak{S}$ is the least pomset $CTT = [E_{\mathfrak{S}}, A_{\mathfrak{S}}, \leq, l]$ such that all actions of $\mathfrak{S}$ are covered at least once in the pomset, i.e. $a \in A_{\mathfrak{S}}$ for all actions in $\mathfrak{S}$ and $E_{\mathfrak{S}}$ is minimal.

Figure 2 depicts the CTT of the example system $\mathfrak{S} = A \parallel B$.

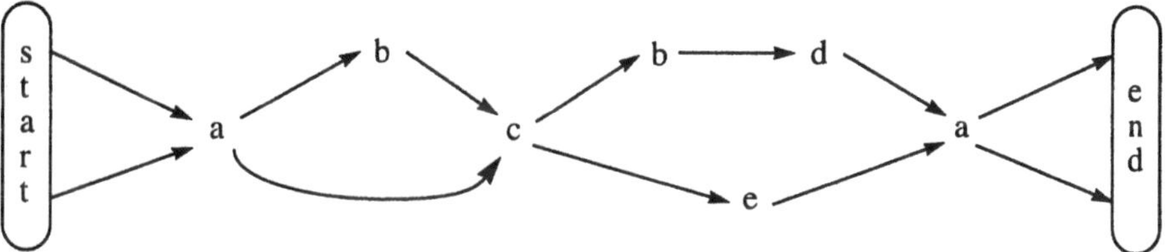

Figure 2. A CTT of system $\mathfrak{S} = A \parallel B$.

The CTT is used as a test case to test a distributed system. The CTT assumes that also the internal actions within the SUT are visible to the tester. Therefore, it supports the grey-box testing approach. To guarantee full coverage of all executable transitions of a distributed system, usually a set of CTTs is required that makes up the test suite for the distributed system.

2.3 ASSUMPTIONS ON A TEST ARCHITECTURE

A *test architecture* is defined as a parallel, synchronous composition of the *system under test (SUT)* of the distributed system $\mathfrak{S}$ and its tester. If a test suite comprises several test cases, a separate tester is built for each test case. It is required that the number of LTSs in the SUT is constant when the test run takes place (static system). Furthermore, the structure of the system $\mathfrak{S}$, i.e. its composition of n LTSs, is preserved in the implementation. From the testing perspective, it should be stated what actions can be observed and controlled. To avoid nondeterministic execution of the SUT due to nonobservability, we assume that all internal and external actions are observable during testing (grey-box testing approach).

In testing distributed systems the information about action names observed during a test run is not sufficient to assess conformance between specification and SUT. Due to the existence of multi-rendezvous among components of the SUT, the tester must also know what components participate in a specific multi-rendezvous. Furthermore, the issue of nondeterminism within the SUT requires that the tester has the power not only to observe an action of the SUT, but must also control the occurrence of a multi-rendezvous. Thus, the crucial point in testing concurrent systems is to perform a *deterministic test run*.

The problem is addressed by applying *instant replay techniques* used for debugging concurrent systems [10]. The proposal in [10] assumes a global controller that is asked for permission by the components of the SUT before they are allowed to interact with each other. To achieve this, control code, called *probes*, is added into the source code of the components before each interaction invocation. Only after the tester has received all requests to execute a certain interaction from participating components, it grants this interaction. If not or if a wrong component asks for permission, an error in the SUT has occurred.

3   TEST ARCHITECTURES FOR TESTING DISTRIBUTED SYSTEMS

3.1   GLOBAL TESTERS

In this section, we describe possible test architectures that can be employed for the test of distributed systems. We first start with the simplest model, *the global tester*, that entirely simulates the environment of the STU as a sequential process during a test run. The global tester centrally collects information of the distributed SUT and derives the test verdict.

A global tester is implemented as a sequential machine T_G. The tester runs in parallel with the distributed system observing and controlling if necessary all external and internal actions of the SUT (grey-box test):

$T_G \parallel SUT$

After starting the test run the tester executes the events of the test sequence σ step by step. This leads to a rendezvous or multi-rendezvous communication between the tester and the SUT. The tester participates in the execution of a test event and records it together with the components of the SUT participating in this test event.

The global tester performs an action sequence derived from the composite machine of the distributed system as a test case. This action sequence contains a certain, assumed interleaving order of concurrent events. This assumed order must be validated to exist in the SUT during the test run. However since the interleaving sequence consists of concurrent test events with no causal dependency between them, the action sequence assumed in the global tester is only observed by chance during a test run.

The global tester itself is modeled as an acyclic machine $T_G = (S_T, A_{CCT}, \rightarrow_T, S_{T0})$ with $|\sigma|+1$ states. The transitions in T_G are determined by the sequence of actions $\sigma = a_1.a_2. \ldots .a_n$ used as the test case:

$$S_{T0} -a_1 \rightarrow S_{T1}, S_{T1} -a_2 \rightarrow S_{T2}, \ldots, S_{Tn\text{-}1} -a_n \rightarrow S_{Tn}$$

For assigning the test verdict, a verdict label is attached to each state: *pass* to the final state and *fail* to all other states. If the tester reaches the final state after correctly executing σ, the test verdict *pass* is assigned to the test run. In all other cases the test run results in a *fail* verdict. The tester may not reach the final state if a desired component of the SUT is not able to participate in a certain test event; the test architecture including the tester and the SUT dead-locks in this case.

Figure 3 shows the test architecture with a global tester to test the system $SUT_3 = A \parallel B$. Note again that the tester has access to the internal actions of the SUT. The model of a global tester implementing $\sigma_1 = a.b.c.b.d.e.a$ as a test case is given in Figure 4.

The advantage of the global tester approach is its simplicity. Due to its global view on the distributed system under test, it can register the processed test events immediately as a unique sequence which preserves the correct causal dependencies between the actions of the SUT. A severe drawback of the global tester is however that it requires strict control over the execution of concurrent test events. This control might heavily intervene in the original behavior of the SUT. The question is whether there are other options to real-ize a test architecture that takes attention to the concurrency of actions in a SUT.

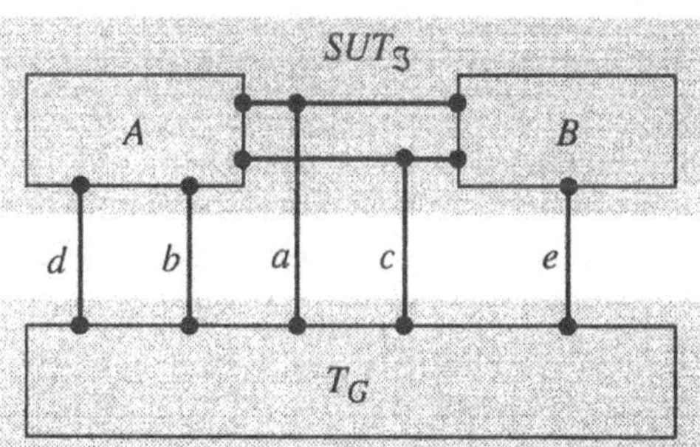

Figure 3. The test architecture with a global tester T_G for system $SUT_3 = A \parallel B$.

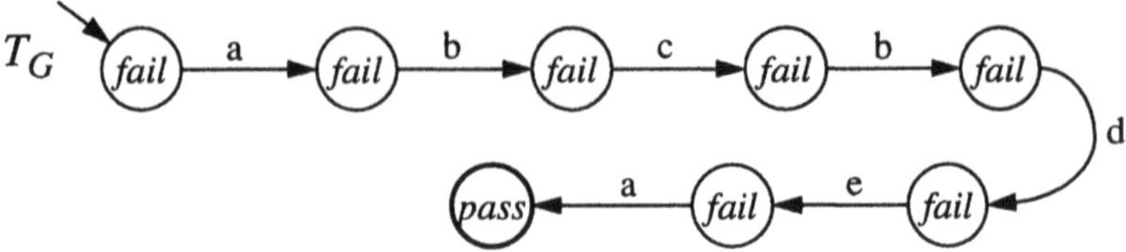

Figure 4. A global tester to test the system $SUT_3 = A \parallel B$.

3.2   DISTRIBUTED TESTERS

3.2.1   General Characteristics

A distributed tester is characterized by the following properties:
- It consists of several concurrently operating tester components which process together, but independently a *global test case* (TC). The tester can be described in the same way as the SUT, e. g. as a set of communicating machines.
- Each tester component executes a *partial test case* (PTC). A PTC is projection of the global test case TC which comprises only those events which can observed at the particular tester component. The causal dependencies between the test events of a PTC are determined by the global TC.
- Each tester component observes a subset of the set of all PCOs. Their selection and assignment to a tester component is a decision taken by the designer of the test architecture.
- The behavior of the tester components is controlled by a *Test Coordination Procedure* (TCP).

The general issue in distributed tester design is that the tester may assign a successful verdict to a test run although the SUT contains faults. This is possible because there is no global view on the SUT if a distributed tester is used. This problem can be tackled by a TCP between the distributed tester components, either by introducing synchronization events into the partial test cases of tester components or by the use of redundant PCOs to observe internal events of the SUT simultaneously by several tester components. For example, if a component of the SUT sends a message to another component, one tester component observes the send event of this communication, whereas another tester component observes the resulting receive event.

We can describe a distributed tester T_D by means of a set of concurrent tester components

$$T_D = T_{D1} \parallel T_{D2} \parallel \ldots \parallel T_{Dn}$$

which are controlled by a TCP. Each tester component processes a sequential partial test case in an acyclic machine that might be augmented with synchronization events. We further assume the same markings of *fail* and *pass* to assign a test verdict as discussed for the global tester.

In a test run, the distributed system $(T_{D1} \parallel T_{D2} \parallel \ldots \parallel T_{Dn}) \parallel SUT$ is executed. The tester components and the SUT participate in the respective test events and transfer from one local state to the next. If a test event or synchronization event of a TCP cannot be executed, the whole test architecture deadlocks. Only if all tester components reach their final state, the verdict *pass* is assigned to the test run.

3.2.2   Distributed Testers with Synchronization Events

A common distributed test architecture that is often used in testing distributed telecommunication systems uses synchronization events to implement a TCP. We discuss this type of distributed testers first.

We develop a distributed tester ^1T_D for the example system in Figure 1 and assume two tester components: $^1T_{D1}$ observes the actions *b* and *d* of the SUT, while $^1T_{D2}$ tests the actions *a*, *c* and *e*.

$$^1T_D = {}^1T_{D1} \parallel {}^1T_{D2}$$

Additionally, the tester components exchange the synchronization event *sync*. Figure 5 shows this test architecture.

The derivation of test cases for the tester components is done as a projection of test events observable by the particular tester component from the global test case.

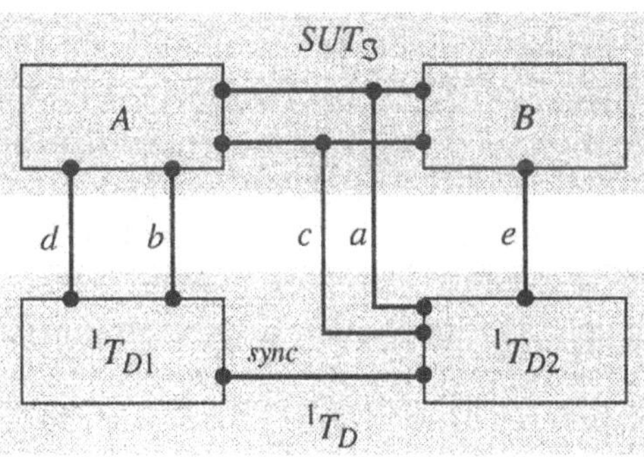

Figure 5. Test architecture with distributed tester ^1T_D and synchronization event *sync*.

Let us assume the CTT in Figure 2 as the global test case for system 3. The projected and linearized partial test cases for the two tester components are the following:

$^1\sigma_1 = b.b.d$ for tester component $^1T_{D1}$ and
$^1\sigma_2 = a.c.e.a.$ for tester component $^1T_{D2}$.

These sequences must be supplemented with synchronization events. Synchronization events are included when the control goes over from one tester component to the other.

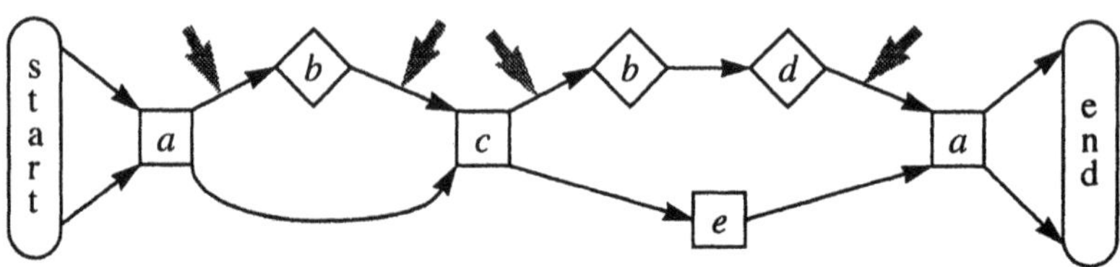

Figure 6. The CTT of the example system with arrows denoting necessary synchronization events within the partial test sequences.

We consider Figure 6. It depicts the global CTT of our example system. The actions of $^1T_{D1}$ are represented by squares   , those of $^1T_{D2}$ by diamonds   . The grey arrows mark the change from one tester component to the other one. At these points a synchronization event must be included to guarantee the correct global view on the SUT. Consequently, the tester components run the following partial test cases:

$^1\sigma_1' = sync.b.sync.sync.b.d.sync$
$^1\sigma_2' = a.sync.sync.c.sync.e.sync.a$

Note that the synchronization events are indispensable for assuring the correct global view on the SUT. If, for instance, the first appearance of action b in component A of the SUT follows action c, a distributed tester without a test coordination procedure would not detect this fault.

Assuming that a tester component is assigned to each component of the SUT, the distributed tester can fully exploit the concurrency among test events. For example, the total order of the interleaving actions b, d and e is not important for assessing a test run. Any test run is correct if only the causal dependency "b before d" is assured. Action e can interleave at any time instant within the constraints of the *sync* events.

Thus, the distributed tester does not need to control the execution of the test case entirely. Each tester component can run independently its partial test case. Only coordination between the tester components is necessary to guarantee the correct global view on the SUT.

3.2.3   Distributed Testers with Redundant Observation of Internal Events

The insertion of synchronization events is not the only possibility to realize coordination among tester components. Another and a more elegant option is the redundant observation of internal actions of the SUT by several tester components. This type of distributed test architecture is discussed again using the example from Section 2. It consists of two tester components

$$^2T_D = {}^2T_{D1} \parallel {}^2T_{D2}.$$

Each tester component controls and observes all interactions of a single component of the SUT. Thus, the tester represents an inverted image of the SUT. Figure 7 shows the test architecture of tester ^2T_D.

Partial test cases for the tester components are derived in the same manner as already described for the first type of distributed testers in Section 3.2.2 by projecting and linearizing the global test case of a CTT. We also assume the same marking of states in the tester components with *fail* and *pass* as discussed for the global tester. Using the CTT in Figure 2 as the basis for the derivation of partial test cases, the tester components $^2T_{D1}$ and $^2T_{D2}$ need to implement the following partial test cases:

$^2\sigma_1 = a.b.c.b.d.a$ for tester component $^2T_{D1}$ and
$^2\sigma_2 = a.c.e.a$ for tester component $^2T_{D2}$.

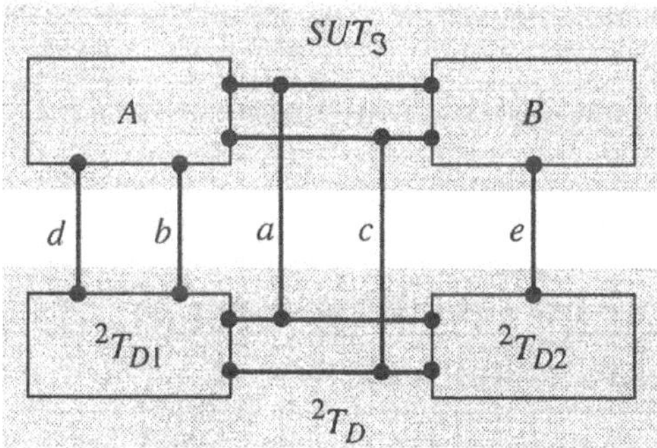

Figure 7. Test architecture with the distributed tester ^2T_D and redundant observation of test events.

Since both tester components observe the internal actions of the SUT, they are able to coordinate each other to maintain the correct global view on the system. In case of message passing, a practical solution to implement the redundant observation of internal communication is the observation of the send event of a message by one tester component and the observation of the related receive event by another one. The causal dependency between the send and the receive event for the same communication action (message passing) can then be reconstructed.

In a test run the distributed system $(T_{D1} \parallel T_{D2} \parallel ... \parallel T_{Dn}) \parallel SUT$ is executed. The tester components and the SUT participate in the respective test events. A multi-rendezvous between the SUT components and the tester components takes place when internal actions of the SUT are executed as test events. In this case, the participating tester components synchronize to preserve the global view on the SUT. If a test event cannot be executed, the tester components associated to this test event deadlock. Only if all tester components reach their final state, the verdict *pass* is assigned to the test run.

Note also here that concurrency within the SUT is supported to a large extend if a tester component is provided for each component of the SUT. The advantage of this second distributed test architecture is that no additional synchronization events are needed to coordinate the tester components.

4   SIMPLIFICATIONS OF THE PROPOSED TEST ARCHITECTURES

The test architectures presented above contain several restrictive elements. For certain applications, the following test assumptions may be too restrictive:

- the control and observation of all actions of the SUT including the internal actions, and
- inclusion of additional synchronization events.

In the following discussion we shortly sketch some conditions which permit simplifications of the test architecture. More details are given in [14].

4.1   OBSERVATION OF INTERNAL COMMUNICATION

The complete observation of the internal communication (grey-box test) was introduced to force a deterministic test run. By observing the internal communication of the SUT by means of a *controller* according to [10] the SUT is forced to follow a certain execution order defined by the CTT. To (partially) omit the internal observation, it must be proved whether the behavior of the SUT does not lead to a nondeterministic behavior which cannot be detected by the tester. This requires that we need to take into consideration the following types of nondeterminism[*]:

- nondeterminism within a component, i.e., it exists the possibility of two or more outputs in the same local state, and
- race conditions, i.e., it exists a global state with at least two inputs to the same component enabled.

If the analysis of the specification of the SUT shows that these types of non-determinism do not appear, additional means to force a deterministic test run can be omitted. This approach requires however that the components of the SUT are correctly implemented. To assure this a step-wise test method as proposed in [4] can be applied.

4.2   RENUNCIATION OF TEST COORDINATION PROCEDURES

In cases in which the number of additional synchronization events that are needed to implement the distributed tester of Section 3.2.2 is very high or in cases in which the redundant observation of internal actions of the SUT as required for the distributed tester of Section 3.2.3 is not possible, it might be desired to renounce the test coordination procedures. To synchronize the then asynchronously acting tester components, the following possibilities exist:

Simulation of a global clock by means of a synchronization protocol
There exist several such protocols in literature. A well-known one is the *Network Time Protocol* (NTP) [5]. It supports a synchronization of a few milliseconds divergence for local area networks and some 10 milliseconds for wide area networks. This precision is scarcely acceptable for many practical applications as the following measurements of the transmission time of RPC messages show. RPC (*Remote Procedure Call*) is a network service for the transparent execution of procedures on remote machines. It is the basic service used by the middleware platform CORBA. The transmission time of an about 8 kilobyte RPC message over an Ethernet network is 7,2 msec. For a 100 Byte message, the time is 0,29 msec only. In an ATM network with a 155 Mbps transfer rate, these times shorten to 0,5 and 0,02 msec, respectively. These measurements indicate that within the precision limits of NTP (1 msec in the best case), several messages can be sent over an Ethernet such that the causal dependencies between messages cannot be correctly established in a test run. For ATM and other high speed networks, the situation is even worse.

---

* Another kind of nondeterminism is interleaving, since the selection among two concurrent actions is arbitrary and unpredictable. We do not consider interleaving as nondeterminism here, because we allow concurrent actions in a test run to be carried out if a distributed tester is used.

Use of logical clocks in the test architecture
Logical clocks [2] are a very reliable and robust means to synchronize actions in a distributed system. It requires, however, that logical clocks are available in all components of the distributed system, i.e. also in the SUT. That means, the software of the SUT must be augmented such that logical clock values are piggybacked for each message exchanged between SUT and tester and between internal SUT components.

Emulation of a synchronous execution by means of a synchronizer
In our current discussion we assumed an asynchronous execution between the components of the SUT. A synchronized execution can be forced by introducing a *synchronizer* that determines the time for the execution of a communication event [1]. However, a synchronous execution of components in a distributed system cannot be assumed for most implementations. Moreover, the introduction of a synchronizer for the purpose of testing only would change the implementation enourmously. An approach that works similar to a synchronizer is discussed in [3].

Generation of synchronizable test suites
Synchronizable test suites which can be separately executed on different tester components without synchronization events are an attractive solution for this problem. Existing approaches [6], [11] run short because they can only be applied if a global composite machine for the distributed SUT is available. Approaches that support communicating machines as the description model for the SUT do not exist up to now.

5   FINAL REMARKS

In this paper we have suggested different architectures for the test of distributed systems. We have discussed rules for the positioning of PCOs as well as means to ensure a deterministic test run. Two principle test architectures have been considered: a global and a distributed tester. The global tester is a sequential component that observes all actions of the SUT at its PCOs, whereas the distributed tester consists of several concurrent tester components that observe partial behavior of the SUT only. These tester components must synchronize each other by means of a test coordination procedure to assure a correct and consistent global view on the SUT.

The presented approaches base on a synchronous communication paradigm between the components of the SUT and the test architecture. The assumption of synchronous communication is required for two reasons. First it restricts the state space of the distributed systems to a finite set, and secondly it allows to preserve the correct causal dependencies between the test events. Using an asynchronous communication paradigm, these properties would not hold anymore. The state space would be infinite and the causal dependencies between test events could not be correctly assessed. Nevertheless, it is necessary also to pursue research within this direction, because asynchronous communication is common in practice.

References

[1]   C. T. Chou, I. Cidon, I. Gopal, S. Zaks: *Synchronizing asynchronous bounded-delay networks*; IEEE Transactions on Communications, vol. 38, no. 2 (Feb. 1990); pp. 144–147.

[2]   C. J. Fidge: Logical Time in Distributed Computing Systems; IEEE Computer, vol. 24, no. 8 (August 1991); pp. 28-33.

[3]   C. Jard, T. Jeron, H. Kahlouche, C. Viho: *Towards Automatic Distribution of Testers for Distributed Conformance Testing*; IFIP Joint Int'l Conference on Formal Description Techniques, and Protocol Specification, Testing, and Verification (FORTE/PSTV'98); Paris, France; 1998.

[4]   H. König, A. Ulrich, M. Heiner: *Design for Testablility: A Step-wise Approach to Protocol Testing*; Proceedings of the IFIP 10th Int'l Workshop on Testing of Communicating Systems (IWTCS'97), Seoul, Korea; 1997; pp. 125-140.

[5]   D. L. Mills: *Precision Synchronization of Computer Network Clocks*; ACM Computer Communication Review, vol. 24, no.2 (April 1994; pp. 28-42.

[6]   G. Luo, R. Dssouli, G. v. Bochmann, P. Venkataram, A. Ghedamsi: *Test generation with respect to Distributed Interfaces*; Computer Standards & Interfaces, vol. 16, no. 2 (June 1994); pp. 119–132.

[7]   A. Petrenko, A. Ulrich, V. Chapenko: *Using partial-orders for detecting faults in concurrent systems*; Proceedings of the IFIP 11th Int'l Workshop on Testing of Communicating Systems (IWTCS'98), Russia, 1998.

[8]   Sidhu, D. P.; Leung, T. K.: *Formal methods for protocol testing: a detailed study;* IEEE Trans. on Software Eng. 15 (1989) 4, 413–426.

[9]   K. C. Tai, R. H. Carver: *Testing of distributed programs*; in A. Zomaya (ed.): *Handbook of Parallel and Distributed Computing*; McGraw Hill; 1995; pp. 956-979.

[10]   K. C. Tai, R. H. Carver, E. E. Obaid: *Debugging concurrent Ada programs by deterministic execution*; IEEE Trans. on Software Eng., vol. 17, no. 1 (Jan. 1991); pp. 45-63.

[11]   K. C. Tai, Y. C. Young: *Port-synchronizable test sequences for communication protocols*; 8th Int'l Workshop on Protocol Test Systems

(IWPTS'95); Paris, France; 1995; pp. 379–394.

[12] A. Ulrich, S. T. Chanson: *An approach to testing distributed software systems*; 15th PSTV 1995; Warsaw, Poland; pp. 107-122; 1995.

[13] A. Ulrich, H. König: *Specification-based testing of concurrent systems*; IFIP Joint Int'l Conference on Formal Description Techniques, and Protocol Specification, Testing, and Verification (FORTE/PSTV'97); Osaka, Japan; Nov. 18-21, 1997.

[14] A. Ulrich: *Testfallableitung und Testrealisierung in verteilten Systmen*; Dissertation (in German), Magdeburg University; Shaker Verlag, 1998.

[15] T. Walter, I. Schieferdecker, J. Grabowski: *Test architectures for distributed systems: state of the art and beyond*; 11th Int'l Workshop on Testing of Communicating Systems (IWTCS'98), Russia, 1998.

8

DECISION ON TESTER CONFIGURATION FOR MULTIPARTY TESTING

Maria Törö

Computer Science Department
The University of British Columbia
2366 Main Mall, Vancouver, B.C., V6T 1Z4 Canada
toeroe@cs.ubc.ca

Abstract   Communication systems are composite systems which can conduct simultaneous communications with their environment. Such systems may require multiparty testing to determine their conformity.

To develop an appropriate test suite, one should establish the required test method first. To decide whether multiparty testing is necessary, that is, whether the system truly accepts simultaneous signals from the environment, we describe the interconnection of system interfaces with internal buffers by a matrix, assuming the system structure and the way of communication of SDL. From this matrix we try to select independent signals, which can be accepted simultaneously by the system. The criterion of signal independence is that signals should be offered to different buffers. From sets of independent signals we generate composite inputs. We suggest to use these composite signals as selection criteria to develop the set of test purposes and apply them to generate a test suite. Accordingly we suggest a modification to the strong reasonable environment technique, most often used to generate test suites. We demonstrate the method on the foreign agent part of the IP Mobility Support.

Keywords: SDL, communicating finite-state machines, interface-buffer interconnection matrix, required test method, reduced reachability tree

1.   INTRODUCTION

Real protocol implementations are typically composite systems and can communicate simultaneously with several entities. However, most of the theoretical works done on conformance testing, were restricted to a single or, if multiple entities were involved, to a sequential communication with those entities. Only such sequential test cases could be specified in the first version of TTCN [CTMF], where all the communication between the tester and the system under test (SUT) was described as a single sequence, even if it involved multiple entities connected through different PCOs (point of control and observation). This lack of concurrency on the tester side resulted in the introduction of the methodology of multiparty testing. Accordingly TTCN was extended to concurrent TTCN [TTCNe] that would allow to specify multiple test components that makes the tester capable to send and receive simultaneous events. multiparty testing is still an open research area.

In all cases, when a specification defines several interfaces between a system and its environment one must consider multiparty testing, because an implementation of such a system might be able to accept and to process concurrent signals. Only multiparty testing can provide simultaneous inputs toward the SUT, although it can still not guarantee that these events will occur indeed concurrently.

On the other hand, having several interfaces does not mean in all cases that signals provided simultaneously through them are indeed processed concurrently by the system. To decide whether multiparty testing is really necessary, one needs to take a closer look at the internal structure of the system specified, and the way it communicates to its environment. This greatly depends on the underlaying model of the formal description technique used for specification, provided that implementations follow both the specified structure and features of the model.

A system internally can be specified as a single-entity system; or it can be distributed, and composed of multiple entities working in parallel and communicating with each other and possibly with the environment.

A component's behavior can be sequential, i.e. it accepts only one stimulus at a time; or it can synchronize on its inputs, i.e. waits until all required stimuli have been received. An example of the first case is the (extended) finite-state machine model, which defines the behavior as a reaction on a single input received in a given state. An example of the second case is the Petri-net model. In Petri-nets a transition may have several input places, and the transition may fire only when each input place holds at least one token.

The communication between the entities of a system can be synchronous, such as in LOTOS, where participating processes are blocked until the required synchronization occurs; or asynchronous as it is in SDL [SDL], where signals are buffered in queues until consumption.

All these and the specified system itself determine whether multiparty testing is a must. These factors define the required test method, which in turn, guides the test suite generation.

The goal of our paper is to take a closer look at SDL and systems specified in SDL to decide whether multiparty testing is inevitable. We present a matrix which helps us to decide what test method and tester configuration should be applied to a system. Then the result is used for selection of test purposes and in the generation of test cases. The novelty of our approach is that it takes a structural approach. It investigates first the structure of the system specified, then uses the results in the subsequent steps of test suite generation. Until now the question of determining of the required test method was hardly ever discussed further than enlisting the abstract test methods of appropriate standards [CTMF, FMCTa].

According to SDL, we assume a compound system of asynchronously communicating finite state automata. To determine the required test method, we describe the interconnection of the system interfaces and the internal buffers of the system by a matrix. From this matrix we derive the independent signal sets, from which we generate composite inputs used to control multiparty testing. The main criterion of signal independence is that signals are offered to different buffers. According to the methodology of multiparty testing, for each signal in the set we need to provide a separate test component in the tester. Also, the test generation method should take into account these simultaneous events. Hence we suggest to build a reduced reachability tree by applying a composite input (i.e. a set of simultaneous signals at the selected interfaces) at a time, whenever the system is in a stable global state.

Our paper presents this approach as follows: First, we give an overview of test generation methods used for systems specified in SDL. We conclude this overview with our approach to the test suite generation. In the third section, we introduce a system of communicating finite-state machines (CFSM) as the basis of our further discussions. In section four, we analyze the features of SDL to determine the required test method for a system specified in SDL. We introduce the interface-buffer interconnection matrix, from which we derive sets of composite inputs are for multiparty testing. In section five a modification of the algorithm of the generation of the reduced reachability tree is suggested. Finally, section six presents an example: the model and the reduced reachability tree for the foreign agent part of the IP Mobility Support protocol.

2. TEST GENERATION METHODS FOR COMPOSITE SYSTEMS SPECIFIED IN SDL

For standardized FDTs draft standards [FMCTa, FMCTb] suggest several test generation methods. Here we discuss only methods suggested for SDL.

Most of the methods suggested for SDL, are based on a predefined set of test purposes given in MSC (Message Sequence Chart) [FMCTb, Ek97]. These methods generate the reachability tree of a system specified, then they compare it with the set of test purposes. Whenever a part of the reachability tree covers a given test purpose scenario a test case is generated. The pre- and postambles, and the alternative behaviors are derived from the reachability tree. The *pass* verdict is assigned to the branch covering the test purpose, *inconclusive* verdicts are assigned to other behaviors derivable from the reachability tree, and a *fail* verdict is assigned to everything else, i.e. to behaviors which cannot be derived from the specification. According to the literature and tool descriptions, implementations of this method take into consideration mostly single-process systems, but never more than a single input queue between the system and its environment (although this input queue may be connected to several interfaces) [FMCTb, Fern96b, Grab97]. In this methodology the set of test purposes is derived from the requirements specified for the protocol in the standard.

Given a (standard) protocol specification in SDL, it is already a requirement specification that an implementation should satisfy. Indeed, it is the most detailed set of requirements that ever would be imposed by a standard. Any set of test purposes will only be a subset of this specification if it does not cover the full specification (in which case there is no reduction of complexity). The question, whether a subset of requirements is sufficient to determine conformity, rises all the time. In addition, there is no established methodology to derive the test purposes from a system specification and it is mostly done by experts:

A protocol/system lifecycle starts with the specification of the requirements, for example, in MSC. These message sequence charts are then used (1) to produce a skeleton of the system specification in SDL which will be refined then until the necessary level of details is achieved. The requirements also (2) compose the basis for test purpose definitions. Unfortunately this branching in the lifecycle means that the original set of requirements and the specification might not be consistent any more. The refinement of the specification introduces new parts to the system behavior which might not be covered by the original requirements. That is, the original set of requirements cannot be used directly as set of test purposes. They have to be re-generated from the SDL specification, or at least validated and verified against the final system specification. Hence the question is: how to select the appropriate (sub)set of requirements from the specification?

[Luo94a, Tan96] describe a different approach of test suite generation for SDL: A transformation of the SDL system into a system of communicating finite-state machines. Unfortunately the transformation of the SDL system cannot be generalized. Not all SDL systems can be transformed into an equivalent system of communicating finite-state machines, even when only the control part of protocols is considered.

[Fern96a, Fern96b, Grab96] apply verification techniques to test suite generation.

The method presented in [Fern96a, Fern96b] again uses test purpose specifications to generate the test suite. The algorithm interleaves the events at different PCOs, which is not necessary when one is allowed to define parallel test components. The goal of the introduction of concurrent TTCN and parallel test components was to avoid this complexity. The method, as presented, deals primarily with the control part of a specification though it does not exploit this assumption.

[Grab96] discusses different methods used for verification of SDL specifications to reduce the complexity of the state space exploration, and their application to test generation methods. In SDL the buffer size is unlimited and may be infinite. This feature makes the state space infinite, therefore a reasonable reduction technique is required. [Grab96] concludes that the most effective reduction technique for SDL is the strong reasonable environment, i.e. the application of a single external stimulus and only at a stable global state of a system. Similarly as it is used in the random walk methods (guided [Lee96], weighted [Kang97]). Obviously, this technique excludes simultaneous inputs what is necessary for multiparty testing. Therefore we suggest to modify this technique to adjust it to multiparty testing:

That is, we allow at most one external signal at a time *for each point of control*. This means that if there are several PCOs, instead of a single external signal, a set of external signals is applied to the system at a time, but no more than one signal at each PCO. We apply a single set of input signals only to a stable global state of the system, i.e. all input queues are empty, so does the original method. Note that this requirement puts a limitation on the delay introduced by delaying channels of SDL, since a stable global state is reached only when all these buffers are emptied. The question is what sets of external inputs should be used to build the reachability tree. Is there any possibility to reduce this set compared to the set of all possible combinations of inputs?

In addition random walk methods avoid the state space explosion by not creating the composite automaton for a CFSM system. They make the assumption, that stable global states of the implementation are observable. This may not be the case at the test campaign, however, one can safely assume the same at the test suite generation without introducing any limitation regarding the implementation. At test suite generation, one can observe internal queues and states of the specification and determine the stable global states. Since these events are invisible and uncontrollable for a tester at the test campaign, one may only observe and should not control the internal communication even at the test suite generation. Thus we have to evaluate all possible ordering of simultaneous signals in internal input queues.

Note: It is not our goal to introduce any new test suite generation technique. We aim merely at answering the question whether multiparty testing is necessary and if it is, what abstract test method should be used. We suggest to use the result to select test purposes and to produce a test suite by existing methods. We only demonstrate our approach on the method producing a reduced reachability tree and specifying a test case for each branch of the tree. For our demonstration we use a model of communicating finite-state machines as follows.

3. COMMUNICATING FINITE-STATE MACHINES WITH EXTERNAL INTERFACES

We simplify the SDL model to a system of n communicating finite-state machines, which has m external interfaces: $CFSM = (\{M_1, M_2, ..., M_n\}, \{I_1, I_2, ..., I_m\})$. An automaton can have a dedicated interface or multiple interfaces or share an interface with other automata.

Automata communicate asynchronously, thus each automaton has a single input queue, which collects all inputs signals for that given automaton. An automaton accepts input signals from automata of the system and from the system environment.

An automaton of the system is described as $M = \langle S, E, O, tr, s_0, q \rangle$, where

S - the finite set of states of the automaton;

E - the finite set of input signals of the automaton. It is composed of two sets

$\quad E = L \cup E^i$ where L is the set of external input signals and E^i is the set of internal input signals;

O - the finite set of output signals of the automaton; similarly to input signals, it is composed of two sets $O = O^e \cup O^i$ where O^e is the set of external output signals and O^i is the set of internal output signals;

tr - the transition function of the automaton;

s_0 - the initial state of the automaton, $s_0 \in S$;

q - the input queue of the automaton.

The transition function is given as $tr = (s_s, \iota, \{q_i{:}\omega_l ..., q_x{:}\omega_x\}, s_e)$, and it determines for the (s_s, ι) state-input pair $s_s \in S$, $\iota \in E$ the resulting outputs $\omega_i, ..., \omega_x \in O$ appended to appropriate queues $(q{:}\omega)$, and the next state $s_e \in S$ of the automaton. An input from the environment is put directly to the input queue of the appropriate automaton. To model delaying channels of SDL we allow spontaneous transitions, i.e. no input signal is required to trigger the transition, which leads to a new state $(tr_\varepsilon = (s_s, \varepsilon, \{q_j;\omega_j ..., q_x;\omega_x\}, s_e)$, where $s_s \neq s_e)$. An automaton modeling a delaying channel is shown in Figure 1a. It has two states: s_0 - delivering state (initial) - and s_1 - blocking state. The transitions between them are spontaneous transitions. In the blocking state the automaton introduces a delay; in the delivering state it forwards signals. A timer can be modelled in a similar way (see Figure 1b). It is an automaton with two states: s_0 - timer does not run - and s_1 - timer runs. From s_0 state the timer is set by the set signal and moves to the s_1 state, from which a spontaneous transition or the reset signal leads back to the s_0 state. In s_1 state the set signal triggers a null transition.

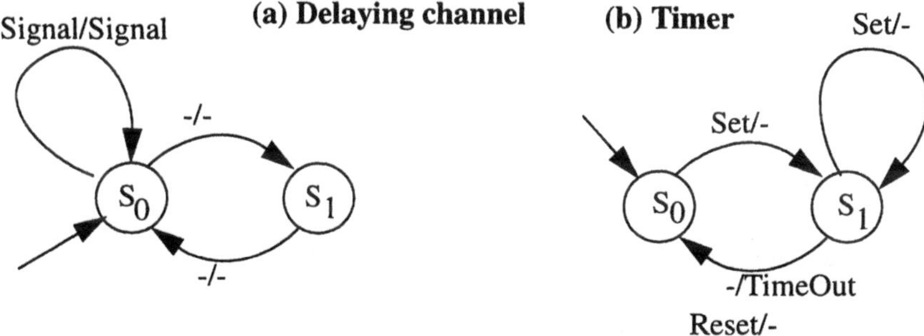

Figure 1   The timer and the channel automata

An interface of the system is composed of two parts, one for each direction $I = (Inp, Out)$. The input part $Inp = \{(q_i, L_i)|1 \leq i \leq n, L_i \subseteq E_i\}$ is the list of queues to which the interface delivers signals and the set of carried signals for each queue. The output part $Out = (q, O^e)$ is the output queue and the set of output signals at this interface $O^e \subseteq \bigcup_{i=1}^{n} O_i^e$.

Such a system may accept simultaneous stimuli. In the next section we discuss whether the application of concurrent inputs, i.e. multiparty testing is necessary.

4.   DETERMINING THE REQUIRED TEST METHOD

The required test method depends on the structure and the method of communication of the system to be tested. The abstract test method determines the required test suite generation method. Since we focus on SDL we consider of the structure and the way of communication of such systems. We also take into account assumptions made by TTCN about the tester configuration:

According to concurrent TTCN for each simultaneous thread of communication with the system under test, one needs a dedicated parallel test component, because each parallel test component is able to control only a single thread of events (though these events may occur at different PCOs). Thus we have to determine the possible concurrent events in the communication with the system.

We assume that a test campaign can and should use only the channels specified between the SDL system and its environment as PCOs. We refer to these channels as *interfaces*. In general a tester offers signals only at interfaces and it observes signals received only at interfaces. It cannot influence or even observe directly any internal communication. Therefore we focus on signals offered at interfaces. They also play an essential role in the test suite generation: They determine the tester configuration required to *control* the test campaign as each simultaneous signal requires a separate test component. The configuration required to *observe* test events is derived from the expected output events, after a test generation method has been applied (e.g. reachability tree) to the specification.

All these suggest that to cover all the possible behavior, one should offer all the combinations of valid input signals to the system at its interfaces. However after a closer look at the system structure it is possible to reduce this estimation:

An SDL process has a single input queue and consumes a single signal at a time from its input queue. All signal routes leading to the process deliver signals to this single input queue, where simultaneous signals are ordered arbitrarily and buffered until consumption. The behavior of the process is sequential. Hence a single test component is sufficient to control a single process if its input queue is exposed directly to the tester. The number of interfaces leading to this process, that is to its input queue, has no effect on this. The question is whether an input queue is exposed directly to the tester.

A process, if applicable, is connected to the system environment by at least a channel and a signal route. Signal routes and non-delaying channels, deliver signals immediately. Therefore one may assume that they deliver simultaneous signals concurrently, i.e. they preserve the concurrency of interfaces. This also means that if a process connected to the environment only by non-delaying channels, it exposes its input queue directly to the environment. Thus the process itself serializes the incoming signals via buffering them in its input queue.

In contrast, a delaying channel acts like an input queue itself. A delaying channel arbitrarily orders simultaneous signals at reception, then it delivers them one by one, after a non-deterministic delay. As we suggested in the previous section, delaying channels can be represented as predefined automata with a single input queue, thus a single test component is sufficient to control a delaying channel and all the processes connected via it.

That is, we need to investigate the interconnection of interfaces and buffers (input queues of processes and delaying channels) to determine what combination of events may occur concurrently at the interfaces.

Figure 2 shows the three basic cases of interface-buffer interconnections and a combination of them:

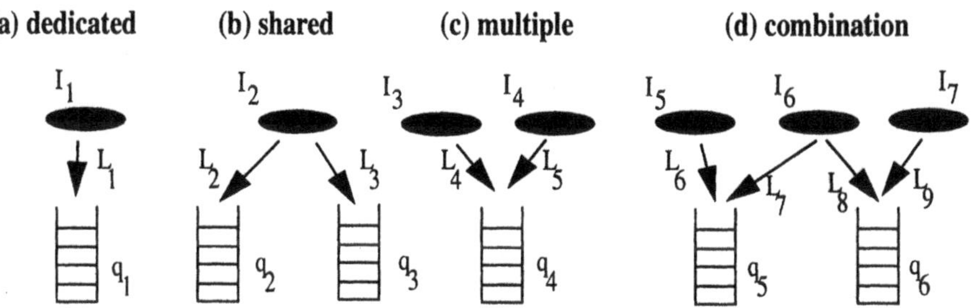

Figure 2   Interface-buffer interconnections

(a) *Dedicated interface*: buffer q_1 have a dedicated interface I_1.

(b) *Shared interface*: buffers q_2 and q_3 share a common interface I_2. We may or may not assume that a single interface can deliver simultaneous signals.

(c) *Multiple interfaces*: buffer q_4 has two interfaces I_3 and I_4. Although the two interfaces can deliver signals simultaneously, the signals will be ordered by the buffer, that is the interfaces exclude each other, i.e. they act sequentially.

(d) *Combination*: To buffers q_5 and q_6 three interfaces deliver signals: I_5, I_6 and I_7. I_5 delivers exclusively to q_5, but I_6 is shared between q_5 and q_6. I_7 is dedicated to q_6. This means that if I_6 delivers a signal either to q_5 or to q_6, only one of the other interfaces can act in parallel: I_7 in the first case, I_5 otherwise. If I_6 is not used I_5 and I_7 can deliver independently from each other. Whether I_6 can deliver to both q_5 and q_6, and exclude the other interfaces is a further issue

We describe the interface-buffer interconnection (R) of a system by a matrix columns of which mean interfaces and rows of which represent buffers. If an interface delivers signals to a given buffer, we put down the appropriate signal list (L) in the cell. Otherwise we put down the $\emptyset$ symbol for the empty set. For the example of Figure 2(d) the matrix is :

$$
\begin{array}{c}
\textbf{interfaces:} \\
I_5 \quad I_6 \quad I_7
\end{array}
$$

$$
R = \begin{bmatrix} L_6 & L_7 & \varnothing \\ \varnothing & L_8 & L_9 \end{bmatrix} \begin{array}{c} \textbf{queues:} \\ q_5 \\ q_6 \end{array}
$$

To select independent signals, one needs to select independent elements of the matrix.

(1) Since input queues serialize signals, elements of the same row are dependent. To select independent sets one should chose only one signal at a time for each buffer. This means that at most one element per row can be selected.

(2) If we also make the assumption that each interface delivers only one signal at a time, that will further restrict our selection of matrix elements. That is, in addition to the previous restriction of one element per row, one is allowed to select at most one element per column.

These selections result in matrices, rows and columns of which linearly independent from each other.

From matrix R we derived the following matrices:

$$
R_1 = \begin{bmatrix} L_6 & \varnothing & \varnothing \\ \varnothing & L_8 & \varnothing \end{bmatrix}, \; R_2 = \begin{bmatrix} L_6 & \varnothing & \varnothing \\ \varnothing & \varnothing & L_9 \end{bmatrix}, \; R_3 = \begin{bmatrix} \varnothing & L_7 & \varnothing \\ \varnothing & \varnothing & L_9 \end{bmatrix}, \; R_4 = \begin{bmatrix} L_6 & \varnothing & \varnothing \\ \varnothing & \varnothing & \varnothing \end{bmatrix},
$$

$$
R_5 = \begin{bmatrix} \varnothing & L_7 & \varnothing \\ \varnothing & \varnothing & \varnothing \end{bmatrix}, \; R_6 = \begin{bmatrix} \varnothing & \varnothing & \varnothing \\ \varnothing & L_8 & \varnothing \end{bmatrix}, \; R_7 = \begin{bmatrix} \varnothing & \varnothing & \varnothing \\ \varnothing & \varnothing & L_9 \end{bmatrix}, \; R_8 = \begin{bmatrix} \varnothing & \varnothing & \varnothing \\ \varnothing & \varnothing & \varnothing \end{bmatrix}
$$

Each of the derived matrices represent a tester configuration for control, since the simultaneous signals can be offered only by parallel test components.

In our example R_1, R_2 and R_3 represent configurations of multiparty testing. Each of them requires simultaneous use of two interfaces: I_5 and I_6, I_5 and I_7, and I_6 and I_7 respectively. Configurations of R_4, R_5, R_6, and R_7 offer a single external signal to the system. Finally in R_8 no interface is selected for control, that allows to observe a behavior triggered by internal events, such as keep-a-live signals, etc.

If we allow multiple signals at an interface, that is we do not make our second assumption about the interfaces, an additional matrix is derived:

$$
R_9 = \begin{bmatrix} \varnothing & L_7 & \varnothing \\ \varnothing & L_8 & \varnothing \end{bmatrix}.
$$

We construct composite inputs from the possible independent inputs by selecting one signal from each selected independent signal list. Thus we create the Cartesian product of the selected signal lists of the matrices and create the union of these sets. The set of composite inputs for our example is:

$$\Lambda = (L_6 \times L_8) \cup (L_6 \times L_9) \cup (L_7 \times L_9) \cup L_6 \cup L_7 \cup L_8 \cup L_9 \cup \varnothing .$$

For R_9 this set is extended with the set of $(L_7 \times L_8)$.

5. GENERATION OF THE REDUCED REACHABILITY TREE

As we suggested, we relaxed the strong reasonable environment technique to generate a reduced reachability tree. That is, we build the reduced reachability tree by offering a composite input at a time to the system in a stable global state. We consider a system state as a tuple of the component automata's states coupled with their input queues and output queues of the system:

$$\sigma = \langle \langle s_1, q_1 \rangle, \langle s_2, q_2 \rangle, ..., \langle s_n, q_n \rangle \rangle ,$$ where S_i is the state, q_i is the input queue of the ith automaton.

A state is a stable global state when all of the input queues of automata are empty, hence such a system state is described as an tuple of the automata's states, i.e. $\sigma^{st} = \langle s_1, s_2, ..., s_n \rangle$.

States having spontaneous transitions are stable if the automaton's input queue is empty. However at the generation of the reachability tree one has to treat spontaneous inputs as well. We calculate the successor states for the spontaneous transition i.e. we add to the set of reachable states, states reached by firing the spontaneous transition.

We generate the reduced reachability tree by the following algorithm:

begin
$\qquad \Sigma = \varnothing ;$ *// reached stable global states*

$\qquad \sigma_0^{st} = \langle s0_1, s0_2, ..., s0_n \rangle ;$ *// the initial stable global state*

$\qquad d = 0 ;$ *// the current depth of the reachability tree*

$\qquad \Sigma_0 = \{ \sigma_0^{st} \} ;$ *// stable global states at depth 0*

$\qquad do\{$

$\qquad\qquad\qquad \Sigma = \Sigma \cup \Sigma_d ;$ *// add newly reached stable global states*

$\qquad\qquad\qquad \Sigma_{d+1} = \varnothing ;$

$\qquad\qquad\qquad for\, (i = 1, i \le |\Sigma_d|, i = i + 1)$ *// for all stable global states of depth d*

$\qquad\qquad\qquad\qquad for\, (j = 1, j \le |\Lambda|, j = j + 1)\, \{$ *// for all composite inputs*

$\qquad\qquad\qquad\qquad\qquad$ *// apply the composite input to the stable global state*

```
                                // and generate the reachable stable global states:
```

$$\sigma_i^{st} \xrightarrow{l_j} \{\sigma^{st} | \sigma^{st} = \langle \langle s_1, \varnothing \rangle, \langle s_2, \varnothing \rangle, ..., \langle s_n, \varnothing \rangle \rangle \}$$

where $l_j \in \Lambda$, $\sigma_i^{st} \in \Sigma_d$

```
                // store a stable global states iff it has not been reached before:
```

$$if\,(\sigma^{st} \notin \Sigma)\ then\ \Sigma_{d+1} = \Sigma_{d+1} \cup \{\sigma^{st}\}\,;$$

```
          }
      d = d + 1 ;
  } while (Σ_d ≠ ∅);                    // until new stable global states have been reached
```

end

To generate all reachable stable global states from a stable global state $\sigma_i^{st} \in \Sigma_d$ by applying a composite input $l_j \in \Lambda$, we do a breadth first exploration (similarly as it is presented in [ObjG]). For each automaton we interleave simultaneous signals from different senders and take into account possible spontaneous transitions. We also interleave signals at the same output queue. According to the algorithm we check each newly reached stable global state with the set of stable global states reached before and if it is already in the set, we truncate the reachability tree.

Note that until now we discussed the tester configuration only with respect to composite inputs, i.e. from the point of view of control. We determine the configuration for observation from the generated output events each time we reach a stable global state: We define an observer at each output queue which contains output signals. Any time a control point and a point of observation belong to the same interface, they are merged into a single test component. This way we assign a tester configuration to each arc of the reduced reachability tree.

Undoubtedly the presented change increases the state space exploration. However it will determine test cases which would not be generated otherwise. As small example of two automata is shown in Figure 3. By application of the two external inputs a and d, one at a time one can generate only the right part of the reachability (solid lines), since the first automaton will go always through state S2 and generate the output o1. Only when the two signals are applied simultaneously the automaton will move to S3 and produce output o2.

In the following section we demonstrate our approach on the example of the foreign agent of the IP Mobility Support.

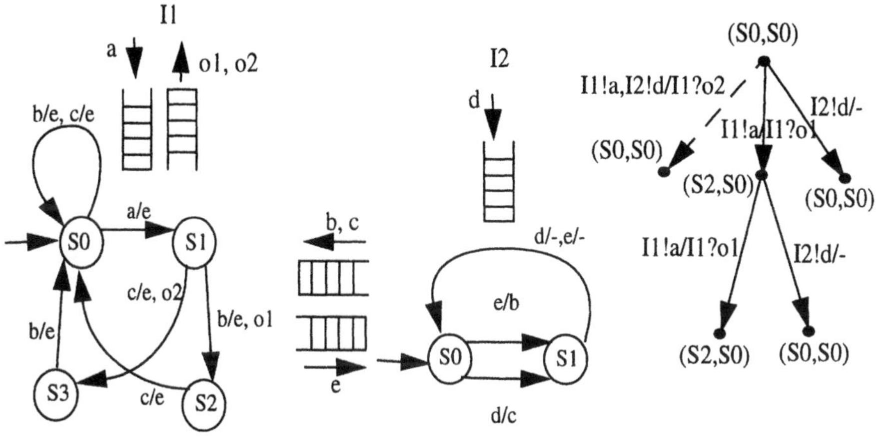

Figure 3   Example CFSM system

6.   IP MOBILITY SUPPORT

The IP Mobility Support provides host mobility on the Internet. Accordingly, a mobile host obtains a temporary IP address for the time it visits a foreign subnetwork. This temporary care-of address is used then to re-route packets sent to the mobile host. The mobile host registers its care-of address with a foreign agent of the visited subnetwork and with its home agent. After registration the home agent intercepts packets addressed to the mobile host, and tunnels them to the registered care-of address of the mobile host.

Since the three interoperating components: the mobile host, the foreign agent and the home agent likely to have different ownerships and vendorships we develop a separate test suite for each of them. Here we present the test suite generation only for the foreign agent.

6.1   Foreign Agent

We modeled the case when the foreign agent serves at most one mobile host at a time, i.e. the first registration request is accepted, any other new registration request is rejected until this registration is valid.

The foreign agent is composed of four communicating finite-state machines of Figure 4, two of which are timers and have no external interface (Figure 4b, 4c). The third automaton (Figure 4a) advertises the foreign agent. It generates the appropriate agent advertisement indicating the agents availability. Accordingly the transition between these states are initiated by the

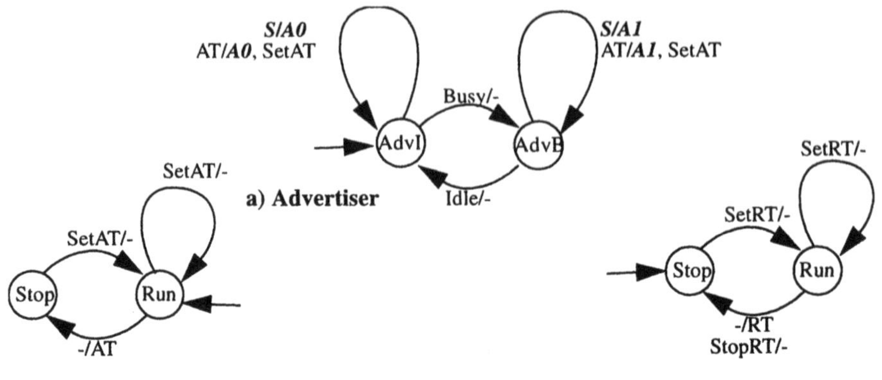

b) Advertisement timer

c) Registration timer

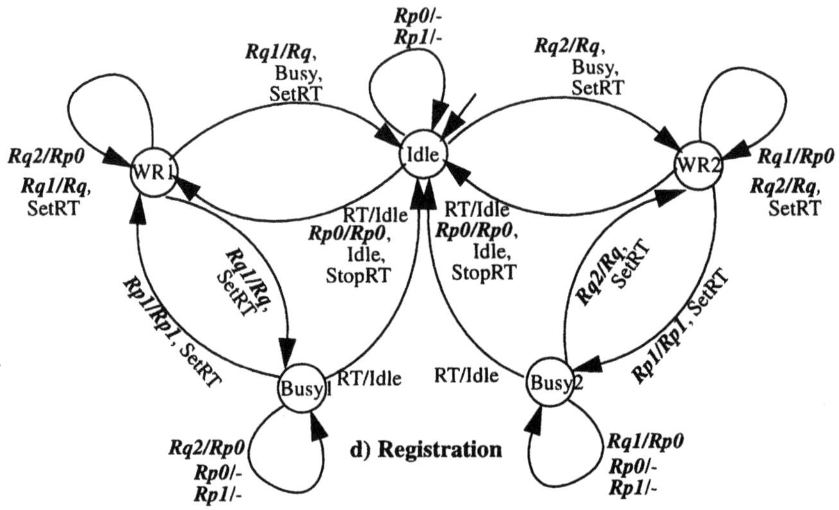

Figure 4   The four automata of the foreign agent

signal of the fourth automaton, which handles the registration (Figure 4d). This automaton becomes busy after forwarding the registration request to the home agent and becomes the end point for tunneling of data packets to the registered mobile host (we neglected the loop transition of forwarding data packets). The automaton distinguishes requests coming from the registered and from a non-registered mobile hosts. We do not model delaying channels for the foreign agent.

Signals occurring at interfaces - both inputs and outputs - are set in bold-italic in Figure 4.

The foreign agent has three interfaces: two toward mobile hosts and one toward a home agent as it is shown in Figure 5. Only the Advertiser and the Registration automata have external inputs, i.e. communicate through these interfaces, so only their input queues are represented.The matrix of the interface-buffer interconnection is: $R = \begin{bmatrix} \{Rp0, Rp1\} & \{Rq1\} & \{Rq2\} \\ \varnothing & \{S1\} & \{S2\} \end{bmatrix}$

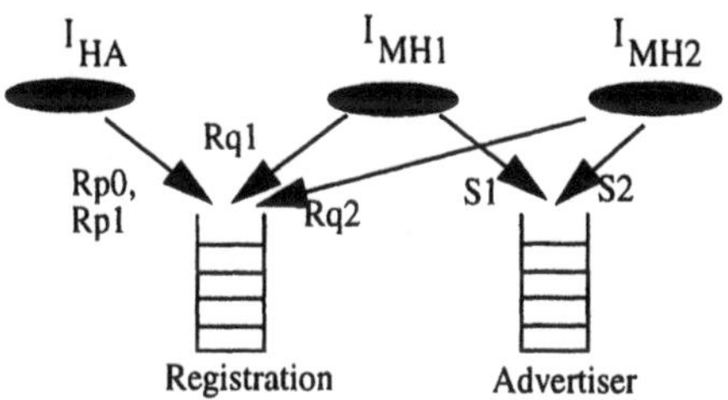

Figure 5   The interface-buffer interconnection in the foreign agent

We do not allow multiple signals at any interface. The matrices of independent signal lists are:

$$R_1 = \begin{bmatrix} \{Rp0, Rp1\} & \varnothing & \varnothing \\ \varnothing & \{S1\} & \varnothing \end{bmatrix}, \quad R_2 = \begin{bmatrix} \{Rp0, Rp1\} & \varnothing & \varnothing \\ \varnothing & \varnothing & \{S2\} \end{bmatrix}, \quad R_3 = \begin{bmatrix} \varnothing & \{Rq1\} & \varnothing \\ \varnothing & \varnothing & \{S2\} \end{bmatrix},$$

$$R_4 = \begin{bmatrix} \varnothing & \varnothing & \{Rq2\} \\ \varnothing & \{S1\} & \varnothing \end{bmatrix}, \quad R_5 = \begin{bmatrix} \{Rp0, Rp1\} & \varnothing & \varnothing \\ \varnothing & \varnothing & \varnothing \end{bmatrix}, \quad R_6 = \begin{bmatrix} \varnothing & \{Rq1\} & \varnothing \\ \varnothing & \varnothing & \varnothing \end{bmatrix},$$

$$R_7 = \begin{bmatrix} \varnothing & \varnothing & \{Rq2\} \\ \varnothing & \varnothing & \varnothing \end{bmatrix}, \quad R_8 = \begin{bmatrix} \varnothing & \varnothing & \varnothing \\ \varnothing & \{S1\} & \varnothing \end{bmatrix}, \quad R_9 = \begin{bmatrix} \varnothing & \varnothing & \varnothing \\ \varnothing & \varnothing & \{S2\} \end{bmatrix}, \quad R_{10} = \begin{bmatrix} \varnothing & \varnothing & \varnothing \\ \varnothing & \varnothing & \varnothing \end{bmatrix}.$$

We use at most two interfaces in any of these case. The set of composite inputs is: $\Lambda = \{<S1, Rp0>, <S1, Rp1>, <S2, Rp0>, <S2, Rp1>, <S2, Rq1>, <S1, Rq2>, <Rp0>, <Rp1>, <Rq1>, <Rq2>, <S1>, <S2>, <>\}$.

The generated reduced reachability tree of the CFSM model with use of the Λ set of composite inputs is shown in Figure 6. The notation used in the reachability tree is the following:

- Nodes represent stable global states of the system. Each of them denote a 4-tuple composed of the states of automata in the form <Advertiser, Registration, Advertisement timer, Registration timer>.
- Arcs have labels of three parts: <inputs> / timer events / <outputs>
    - the set of inputs which initiated the transition, in angle brackets;
    - between slashes internal timer related events if applicable;
    - the generated outputs again in angle brackets.

Listing the timer events in the graph let us derive timer events for test cases if necessary.

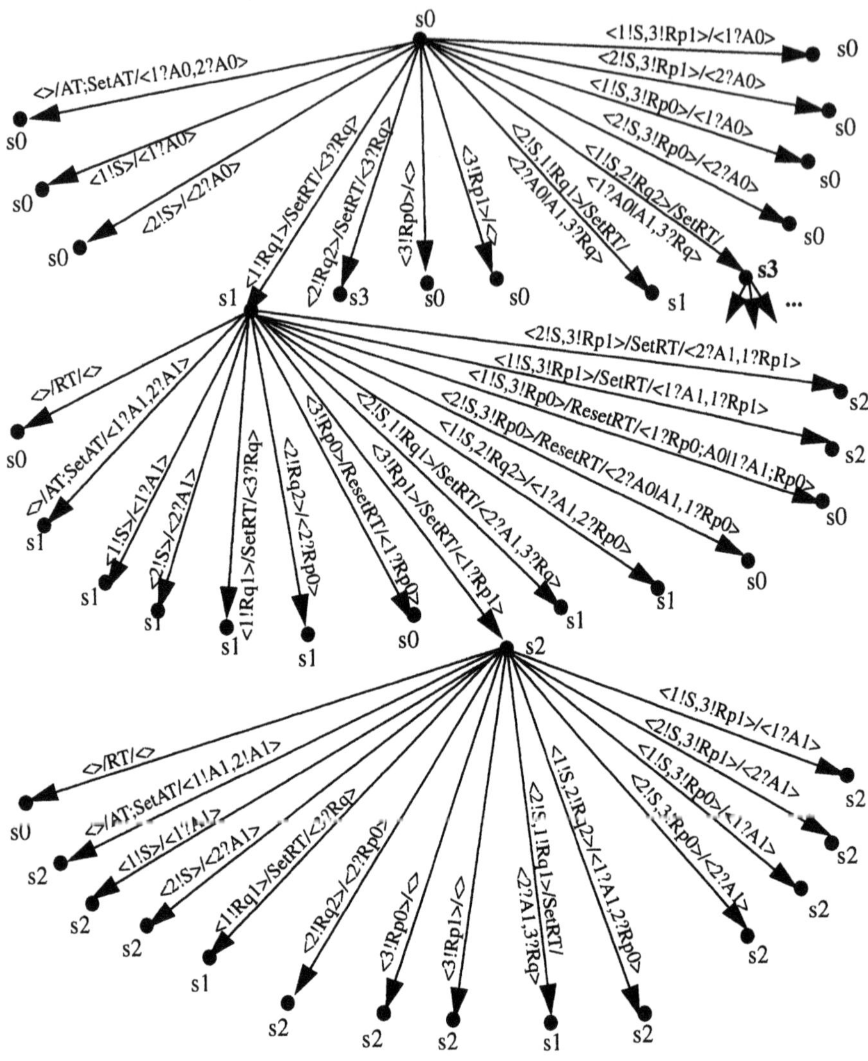

Timers: AT - Advertiser Timer, RT - Registration Timer

Interfaces: 1 - I_{MH1}, 2 - I_{MH2}, 3 - I_{HA}

States: s0 -<AI, I, R, S>, s1 - <AB, WR1, R, R>, s2 - <AB, B1, R, R>,
s3 - <AB, WR2, R, R>, s4 - <AB, B2, R, R>

Advertiser: AI - AdvIdle	*Registration:* I - Idle	*Timers:* S - Stop
AB - AdvBusy	WR1,2 - WaitForReply1,2	R - Run
	B1,2 - Busy1,2	

Figure 6   The reachability tree of the foreign agent

- In front of each input/output event we indicate the interface at which it takes place. Simultaneous input/output events of different interfaces are separated by comma.
- Alternate outputs at the same interface are separated by a vertical bar "|".
- A sequence of output events is concatenated by semicolon ";".

In the reachability tree we indicated but did not expand a subtree rooted in state "s3", which is symmetrical to the subtree rooted in the stable global state "s1". We derived test cases for each arc of the reachability tree, except the subtree rooted in "s3". A certain care had to be taken regarding the timers and the coordination of the test components. One can notice in the reachability tree, that once the foreign agent accepted a registration the only way back to the initial state of the system is the expiration of the registration timer. This means no actions from the tester side, i.e. it has to wait at least for the amount of time the registration was accepted. Therefore these - no input from the tester - branches of the reachability tree were selected "to reset" the IUT to the initial state. 41 test cases and 30 test steps were derived manually from the reachability tree for the MIP foreign agent. 18 of the test cases take into account simultaneous inputs, and they require 16 of the test steps. This means that the application of composite inputs doubled the size of the generated test suite compared to the approach of the strong reasonable environment with a single input signal at a time.

Table 1: Sample test step: Preamble

Name	Preamble (A: PCO, B: PCO)		
Label	Dynamic Behavior	Verdict	prev. notation
	A ! RegistrationRequest, Start (RegistTimer)		Rq
	B ? RegistrationRequest	PASS	Rq
	B ? otherwise	FAIL	

Table 2: Sample test step: Parallel Test Component_1

Name	ParallelTestComponent_1 (A: PCO)		
Label	Dynamic Behavior	Verdict	prev. notation
	A ! RegistrationReply(reject)	(pass)	Rp0

Table 3: Sample test step: Parallel Test Component_2

Name	ParallelTestComponen_2 (A: PCO)		
Label	Dynamic Behavior	Verdict	prev. notation
	A ! Solicitation		S
	A ? RegistrationReply(reject), Cancel (RegistTimer)		Rp0
	A ? AgentAdvertisement(idle)	PASS	A0
	A ? AgentAdvertisement(busy)		A1
	A ? RegistartionReply(reject)Cancel (RegistTimer)	PASS	Rp0
	A ? Otherwise	FAIL	

Table 4: Sample test case

Label	Dynamic Behavior	Verdict	prev. notation
	+ Preamble (MH1, HA)		
	create (PTC_1: ParallelTestComponent_1 (HA) PTC_2: ParallelTestComponent_2 (MH1))		
	[R=pass]	PASS	
	[R=fail]	FAIL	

An example test case presented in Tables 1-4 which was derived from the consecutive arcs: s0 <1!Rq> /SetRT/ <3?Rq> s1 and s1 <1!S, 3!Rp0> / ResetRT/ <1?Rp0;A0 I 1?A1;Rp0> s0. The first arc is used as the preamble for the second one.

The test case specifies two parallel test components: PTC_1 which has a PCO toward the HA(3) interface (Table 2) and PTC_2 which has a PCO connected to the MH1 (1) interface (Table 3). The main test component starts with the preamble then creates the two parallel test components and finally evaluates the verdict (Table 4).We found that testing of this mobile protocol does not differ from stationary protocols except that it requires multiparty testing and the timers play essential role in the protocol, so testing of them is inevitable. We also found that the timer handling is ambiguous in the extension of TTCN in the case of parallel test components: Each newly created test component receives a copy of all running timers, however it is not clear how a timer is stopped if only one of the components detects the appropriate event.

7. CONCLUSION

We approached the problem of test suite generation for composite systems from a structural point of view. We investigate the question of signal dependency to determine the required test method. I.e. we establish, for a system specified in SDL, the sets of signals that can happen simultaneously, therefore need to be offered concurrently to verify the system's conformity. To offer simultaneous signals one has to apply multiparty testing.

In SDL the communication happens via queues buffering, and serializing concurrent signals. Therefore our main criterion of signal independence is being submitted to different buffers. We use the interface-buffer interconnection matrix to determine signal independence. In the matrix signal lists of the same row belong to the same buffer, so only one of them is selected at a time. Additionally one may assume that an interface delivers only a single signal at a time, which restricts further the selection by allowing at most one element in each column. After the selection of independent signal lists we create the set of composite inputs taking at most one signal from each list. Subsequently we use these composite inputs to generate the reduced reachability tree of the system. This approach can be interpreted as an adaptation of the strong reasonable environment reduction technique to multiparty testing.

We have demonstrated this method on the example of the foreign agent part of the IP Mobility Support protocol. The work was done manually with the use of an SDL simulator. The next step will be to implement the algorithm. The critical point is the derivation of the synchronization messages of test components. We found that timer handling of parallel test components is ambiguous in the extension of TTCN. In the MIP protocol due to mobility, timer events could be detected by a different test component than the one, which sent the message setting the timer. At the moment, we had to specify explicitly the synchronization between relevant test components. However this work was very cumbersome and errorprone, at the same time testing of these features for mobile protocols is essential. Therefore an implicit synchronization of timer events between test components provided by TTCN standard would be highly beneficial.

REFERENCES

[CTMF] ISO/IEC 9646 IT-OSI, OSI Conformance Testing Methodology and Framework

[Ek97] A.Ek, J.Grabowski, D.Hogrefe, R.Jerome, B.Koch, M.Schmitt: Towards the Industrial Use of Validation Techniques and Automatic Test Generation Methods for SDL Specifications, Institute for Telematics, University of Lübeck, Technical Report A-97-03, 1997

[Fern96a] J.C.Fernandez, C.Jard, T.Jéron, L.Nedelka, C.Viho: Using on-the-fly verification techniques for test generation of test suites, INRIA Technical Report No 2987, France, 1996

[Fern96b] J.C.Fernandez, C.Jard, T.Jéron, L.Nedelka, C.Viho: An Experiment in Automatic Generation of Test Suites for Protocols with Verification Technology, INRIA Technical Report No 2923, France, 1996

[FMCTa] Revised Working Draft on "Framework: Formal Methods in Conformance Testing", JTC1/SC21/WG1/Project 54.1, January 1995

[FMCTb] Revised Working Draft on "FMCT guidelines on Test Generation Methods from Formal Descriptions", JTC1/SC21/WG1/Project 54.2, February 1995

[Grab96] J.Grabowski, R.Scheurer, D.Toggweiler, D.Hogrefe: Dealing with the complexity of state space exploration algorithms for SDL, Proceedings of the 6th GI/ITG techn. meeting on FDTs for Distributed Systems, University of Erlagen, Germany, 1996

[Grab97] J.Grabowski, R.Scheurer, Z.R.Dai, D.Hogrefe: Applying SAMSTAG to the B-ISDN Protocol SSCOP, Institute for Telematics, University of Lübeck, Technical Report A-97-01, 1997

[Kang97] D.Kang, M.Kim, S.Kang: A Weighted Random Walk Approach for Conformance Testing of a System Specified as Communicating Finite State Machines, FORTE/PSTV'97

[Lee96] D.Lee, K.K.Sabnani, D.M.Kristol, S.Paul: Conformance Testing of Protocols Specified as Communicating Finite State Machines - A Guided Random Walk Based Approach, IEEE Trans on Communications, Vol.44, No. 5, May 1996

[Luo94a] G.Luo, A.Das, G.v.Bochmann: Software Testing Based on SDL Specifications with Save, IEEE Trans. on Software Engineering, Vol. 20, No. 1, 1994

[ObjG] ObjectGEODE on-line documentation, Verilog, 1998

[SDL] ITU-T Z.100, Specification and Description Language (SDL)

[Tan96] Q.M.Tan, A.Petrenko, G.v.Bochmann: A Test Generation Tool for Specifications in the Form of State Machines, ISBN-0-7803-3250-4/96, IEEE 1996

[TTCNe] Amendment 1 to ISO/IEC 9646-3:1992 TTCN extensions

IV

NEW FIELDS OF PROTOCOL TESTING

9

PROTOCOL-INSPIRED HARDWARE TESTING

Ji He and Kenneth J. Turner

Computing Science and Mathematics, University of Stirling, Scotland FK9 4LA
jih@cs.stir.ac.uk, kjt@cs.stir.ac.uk

Abstract

The relevance of protocol conformance testing techniques to hardware testing is discussed. It is shown that the *ioconf* (input-output conformance) approach used in protocol testing can be applied to generate tests for a synchronous hardware design using its formal specification. The generated tests are automatically applied to a circuit by a VHDL testbench, thus giving confidence that the hardware design meets its high-level formal specification. Case studies illustrate how the ideas can be applied to standard hardware verification benchmarks such as the Single Pulser and Black-Jack Dealer.

Keywords: Conformance Testing, Design Validation, Hardware Description

1.   INTRODUCTION

1.1   BACKGROUND

Modern digital circuits are becoming extremely complex, requiring substantial effort to ensure design correctness prior to manufacture. DILL (Digital Logic in LOTOS [14, 15, 16, 25]) is an approach and a language for specifying digital circuits using LOTOS (Language of Temporal Ordering Specification [12]). The formal basis of LOTOS supports rigorous specification and analysis, both crucial for correct circuit design.

The inspiration for the work presented in this paper comes from conformance testing of communications protocols. The paper thus concerns testing of communicating systems, but in this case hardware devices rather than protocol entities. In formally based protocol testing, tests are derived from a formal specification and then used to check a concrete implementation. This is regarded as a black box whose operation has to be checked against its specification. Essentially this is the way that the functionality of a digital circuit is tested. Moreover, as in a communications system each part of a circuit has to communicate with the other parts. As a novel application of conformance testing, it is therefore worthwhile investigating how protocol testing ideas can be applied to validation of digital hardware.

In electronics, the equivalent term is **design verification, design validation** or **functional testing**. The term testing in this area applies to checking *manufacturing* defects in products rather than *design* problems. In the field of formal methods, the term testing has a more general meaning. The system under test might be a physical product, a formal specification or an informal specification. This paper interprets testing to mean evaluation of a requirements specification against a purported implementation specification.

1.2   APPROACH

Conformance testing uses experimentation to check an implementation against its formal specification. Tests are derived from the specification, then applied to the **IUT** (Implementation Under Test). Based on observations made during the execution of the tests, a verdict is given about the correct functioning of the implementation.

Hardware circuits are specified in this paper using LOTOS, whose semantics is given by an **LTS** (Labelled Transition System). The implementation of the same circuit is described by VHDL (VHSIC Hardware Description Language [10]). The behaviour of a VHDL program is presumed to be modelled by an **IOLTS** (Input-Output Labelled Transition System). This model is merely assumed to exist – it need not be known explicitly. Making the assumption that an implementation has a formal model is referred to as the **test hypothesis**. This makes it possible to express conformance of an implementation with respect its specification using a formal relation. One such relation, *ioconf* (input-output conformance [24]), is used as the criterion for correct hardware design.

The test suite for a circuit is generated from a LOTOS specification following an algorithm based on that proposed by Tretmans [24]. The authors have extended CADP (Cæsar Aldébaran Development Package [6]) to generate hardware test suites automatically. Each test case in the generated test suite is a sequence of input and output signals. Designing test cases as input-output sequences is close to engineering practice in hardware testing. Moreover, it allows test execution and obtaining test verdicts to be completely automated. This is achieved by a VHDL testbench that executes and evaluates the test cases. If there is an inconsistency between the formal specification and its VHDL implementation, the implementation is regarded as incorrect.

Section 2. introduces the theory for testing an IOLTS. This is followed by its application to testing digital circuits in Section 3. Section 4. examines the techniques in two case studies.

1.3   RELATED WORK

Test theories for LTSs were first studied more than a decade ago. These theories aim to define implementation relations by explicitly using external tests and observations (e.g. [4, 19]). Apart from defining an implementation relation, conformance testing involves finding a set of tests for a specification to distinguish between correct and incorrect implementations. [2] elaborates a theory for testing systems specified in LOTOS. Several test generation algorithms for an LTS and for Basic LOTOS have been proposed, e.g. [17, 20]. In [23, 24] the testing theory for an LTS is refined for communicating systems that distinguish inputs and outputs. This is a more realistic view of systems.

For validating hardware designs, simulation has been and is still the predominant method in industry. In the main, test cases for simulation are manually defined or are randomly generated. Recent developments for improving this *ad hoc* approach lie in combining formal methods with traditional simulation techniques. In [26] design verification tests are generated from behavioural VHDL programs using traditional software testing methods. In [8, 18] test generation is based on an FSM (Finite State Machine) or ECFM (Extracted Control Flow Machine) that represents the control logic of a circuit. The generated test cases are then applied to both higher level and lower level specifications in Verilog [11] or VHDL. Verdicts are obtained by comparing outputs from the two levels. These approaches *extract* a formal model from a circuit design and use techniques based on FSM testing theory. But in this paper, tests are *derived* from higher level specifications using conformance testing theory for LTSs.

In [21] test generation is based on a higher level FSM specification using a commercial tool. Tests are then applied using a VHDL simulator. Unfortunately this method cannot handle non-determinism in specifications. Its aim is to fill the gap between abstract tests and concrete test signals. Of direct relevance to the current work, the CADP toolset has a test generation tool TGV [7] under development. The implementation relation exploited by TGV is very similar to *ioconf* used in this paper. TGV was still to be released at the time of writing, and so was not available to the authors for evaluation.

2.   TESTING IO TRANSITION SYSTEMS

2.1   IOLTS AND IOCONF

Conformance testing involves checking the correctness of an implementation against its specification by external tests and observations. To formally define an implementation relation, a test hypothesis is needed that implementations can be expressed by a formal model. In traditional conformance testing theories for LTSs, both the specification and the IUT are modelled as LTSs. An IUT communicates with its environment through symmetric interactions, hence its environment as expressed by tests is also modelled as an LTS.

An **LTS** is a quadruple $\langle S, L, T, s0 \rangle$ where S is a set of states, L is a set of observable actions, $T \subseteq S \times (L \cup \{\tau\}) \times S$ is the transition relation, and $s0 \in S$ is the initial state. The class of transition systems with actions in L is denoted by $\mathcal{LTS}(L)$. A transition in T is also denoted as $s \xrightarrow{\mu} s'$ if $(s, \mu, s') \in T$. The special action $\tau \notin L$ represents an unobservable (or internal) action. The following notations are commonly used for LTSs.

Let $p = \langle S, L, T, s0 \rangle$ be an LTS with $s, s' \in S$, and let $\mu_i \in L \cup \{\tau\}, a_i \in L$, $L^\star$ denotes the set of all finite action sequences of L and $\sigma \in L^\star$. The following definitions then apply:

$$s \xrightarrow{\mu_1 \cdots \mu_n} s' \quad =_{def} \quad \exists s_0, \ldots, s_n : s = s_0 \xrightarrow{\mu_1} s_1 \xrightarrow{\mu_2} \ldots \xrightarrow{\mu_n} s_n = s'$$

$$s \xrightarrow{\mu_1 \cdots \mu_n} \quad =_{def} \quad \exists s' : s \xrightarrow{\mu_1 \cdots \mu_n} s'$$

$$s \not\xrightarrow{\mu_1 \cdots \mu_n} \quad =_{def} \quad \text{not } \exists s' : s \xrightarrow{\mu_1 \cdots \mu_n} s'$$

$$s \overset{\epsilon}{\Rightarrow} s' \quad =_{def} \quad s = s' \text{ or } s \xrightarrow{\tau \cdots \tau} s'$$

$$s \overset{a}{\Rightarrow} s' \quad =_{def} \quad \exists s_1, s_2 : s \overset{\epsilon}{\Rightarrow} s_1 \overset{a}{\rightarrow} s_2 \overset{\epsilon}{\Rightarrow} s'$$

$$s \overset{a_1 \cdots a_n}{\Longrightarrow} s' \quad =_{def} \quad \exists s_0 \ldots s_n : s = s_0 \overset{a_1}{\Rightarrow} s_1 \overset{a_2}{\Rightarrow} \ldots \overset{a_n}{\Rightarrow} s_n = s'$$

$$s \overset{\sigma}{\Rightarrow} \quad =_{def} \quad \exists s' : s \overset{\sigma}{\Rightarrow} s'$$

$$s \not\overset{\sigma}{\Rightarrow} \quad =_{def} \quad \text{not } \exists s' : s \overset{\sigma}{\Rightarrow}$$

$$\mathbf{init}(p) \quad =_{def} \quad \{\mu \in L \cup \{\tau\} \mid p \xrightarrow{\mu}\}$$

$$\mathbf{traces}(p) \quad =_{def} \quad \{\sigma \in L^\star \mid p \overset{\sigma}{\Rightarrow}\}$$

$$\mathbf{p\ after}\ \sigma \quad =_{def} \quad \{p' \mid p \overset{\sigma}{\Rightarrow} p'\}$$

Many real world systems communicate with their environment in a different way from an LTS. There is a clear distinction between the inputs and outputs of a system. The inputs of a system are always enabled and cannot refuse the actions offered by the environment. After the system consumes an input and produces its outputs, the environment has to accept the outputs. In other words, such a system will never reject inputs and its environment will never block outputs. Communication is thus no longer symmetric. In [24] this kind of behaviour is modelled as an IOLTS, which is a special kind of LTS.

An **IOLTS** (Input-Output Labelled Transition System) p is an LTS in which the set of actions L is partitioned into input actions L_I and output actions L_U such that $L_I \cup L_U = L$ and $L_I \cap L_U = \emptyset$. (The suffix $_U$ derives from the Dutch/German word for out.)

$$\text{whenever} \quad p \overset{\sigma}{\Rightarrow} p' \quad \text{then } \forall a \in L_I : p' \overset{a}{\Rightarrow}$$

Intuitively this means that input actions are always enabled in any state. The class of input-output transition systems with input actions in L_I and output actions in L_U is denoted by $\mathcal{IOLTS}(L_I, L_U) \subseteq \mathcal{LTS}(L_I \cup L_U)$.

Several implementation relations have been defined to express conformance of an implementation to its specification. In these relations, specifications are modelled as LTSs and implementations are modelled as IOLTSs. This is because an LTS can give a more abstract view of a system, while an IOLTS is closer to reality. The specification LTS can be regarded as a partially specified IOLTS in the sense that there are some states in the specification that can refuse input actions. There are two reasons to write such kinds of specifications. One is that it does not matter how implementations respond to unspecified inputs. The other is that the environment is assumed not to offer such inputs, so there is no need to specify them.

To define the implementation relation *ioconf*, several other definitions have to be introduced. Let $p \in \mathcal{LTS}(\mathcal{L_I} \cup \mathcal{L_U})$, s be a state in the LTS and S be a state set. Then:

- A state s of p is **quiescent,** denoted by $\delta(s)$, if $\forall \mu \in L_U \cup \{\tau\}$: $s \not\xrightarrow{\mu}$
- A **quiescent trace** of p is a trace σ which may lead to a quiescent state: $\exists p' \in (p \text{ after } \sigma) : \delta(p')$
- **out**(s) $=_{def}$ $\{x \in L_U \mid s \xrightarrow{x}\} \cup \{\delta \mid \delta(s)\}$
- **out**(S) $=_{def}$ $\bigcup\{out(s) \mid s \in S\}$

From the definition, a quiescent state is one that cannot perform any output transitions or an internal transition. *out(s)* defines all the output actions that a state can perform. This includes the quiescent 'action' δ that means the state cannot perform any output actions. Let $i \in \mathcal{IOLTS}(L_I, L_U), s \in \mathcal{LTS}(L_I \cup L_U)$. Then:

$$i \textbf{ ioconf } s \;\; =_{def} \;\; \forall \sigma \in \text{traces }(s) : \; out(i \text{ after } \sigma) \subseteq out(s \text{ after } \sigma)$$

This means that an implementation is correct if, after all traces σ of the specification, the implementation outputs can also be produced by the specification. An implementation cannot produce outputs which are not expected by the specification. Since this also holds for δ, the implementation may not output if the specification cannot do so.

2.2   TEST GENERATION FOR IOCONF

To support the generation of test cases for *ioconf*, an intermediate LTS termed the **suspension automaton** is built from the specification LTS. The suspension automaton Γ_p of an LTS p is obtained by adding self-loops $s \xrightarrow{\delta} s$ for all quiescent states and then determinising the resulting automaton. The important properties of a suspension automaton are that it is deterministic and for $\sigma \in L^*$, $out(\Gamma_p \text{ after } \sigma) = out(p \text{ after } \sigma)$. As will be seen later, checking *ioconf* can be easily reduced to checking trace inclusion on the suspension automaton.

A **test case** t is an LTS $< S, L_I \cup L_U \cup \{\delta\}, T, s0 >$ such that:

- t is deterministic and has finite behaviour
- S contains the terminal states *Pass* and *Fail*, with $init(Pass) = init(Fail) = \emptyset$
- for any state $t' \in S$ of the test case, $t' \neq Pass, Fail$, either $init(t') = \{a\}$ for some $a \in L_I$, or $init(t') = L_U \cup \{\delta\}$.

The class of test cases over L_U and L_I is denoted as $\mathcal{TEST}(L_U, L_I)$. A **test suite** T is a set of test cases: $T \subseteq \mathcal{TEST}(L_U, L_I)$. L_I and L_U refer to inputs and outputs from the point of view of the IUT, so L_I represents the outputs and L_U the inputs of test cases.

The following test generation algorithm is based on the suspension automaton obtained from an LTS. It is a slightly modified version of the one in [24] which generates tests according to various implementation relations. The following one is tailored for the *ioconf* relation.

Test Generation Algorithm: Let Γ be the suspension automaton of an LTS s, and let $\mathcal{F} = trace(s)$. Then a test case $t \in \mathcal{TEST}(L_U, L_I)$ is obtained by finite, recursive application of one of the following three non-deterministic choices:

Choice 1: Terminate the test case: $t := Pass$.

Choice 2: Give a further input to the implementation: $t := a; t'$. Here, $a \in L_I$ such that $\mathcal{F}' = \{\sigma \in L_I^\star \mid a \cdot \sigma \in \mathcal{F}\} \neq \emptyset$. To obtain t' the algorithm is applied recursively to $\mathcal{F}'$ and Γ', which is derived from $\Gamma \xrightarrow{a} \Gamma'$.

Choice 3: Check the next outputs of the implementation:

$$t \quad := \quad \sum \{x; Fail \mid x \in L_U \cup \{\delta\}, x \notin out(\Gamma)\}$$
$$\square \quad \sum \{x; t_x \mid x \in L_U, x \in out(\Gamma)\}$$
$$\square \quad \delta; Pass \text{ if } \delta \in out(\Gamma)$$

where t_x is obtained by recursively applying the algorithm for $\{\sigma \in L_\delta^\star \mid x \cdot \sigma \in \mathcal{F}\}$, and Γ' arises from $\Gamma \xrightarrow{x} \Gamma'$.

The first choice terminates the test generation procedure. Since specifications usually have infinite behaviour, test generation has to be stopped at some point. The second choice gives the next input to the implementation. Since inputs are always enabled, this step will never result in deadlock when an input is applied to the IUT. It is therefore not possible to reach a terminal *Pass* or *Fail* state. To avoid unnecessary non-determinism during testing, only one input is applied each time. The third step checks the next output of the implementation, i.e. for each $x \in L_U \cup \{\delta\}$ it is checked if $out(\Gamma_i \text{ after } \sigma) \subseteq out(\Gamma_s \text{ after } \sigma)$. Here, σ is the trace which has been produced so far. Any implementation producing an output x that does not belong to $out(\Gamma_s)$ will result in a *Fail* terminal state, indicating a non-conforming implementation. For all other outputs x, the generation procedure may continue. However δ does not belong to *trace(s)* so a *Pass* terminal state results and no further sequences need be checked. This test generation algorithm guarantees sound test cases with respect to *ioconf*, and the set of all possible test cases that can be obtained is exhaustive.

3.   TESTING SYNCHRONOUS CIRCUITS

3.1   SYNCHRONOUS CIRCUIT MODEL

The idea of applying the above theory to validating hardware circuits comes naturally. The DILL approach uses LOTOS to specify digital circuits, so the behaviour of circuits is expressed by an LTS. On the other hand, real hardware communicates with its environment via input and output ports. An IOLTS is a realistic model since inputs are always accepted by circuits.

In this paper, only **synchronous circuits** are considered. Synchronous circuits are also referred to as **clocked** since their operation is controlled by one or more clocks. The classical synchronous circuit model is shown in Figure 1. In this model, the combinational logic provides the primary outputs and internal outputs according to the primary inputs and internal inputs. Internal outputs are then fed into state hold components to produce the internal inputs. Changes of the internal inputs are synchronised with the clock, in other words they are changed only at a particular moment of the clock cycle (usually its transition). The internal inputs determine the state of the whole circuit.

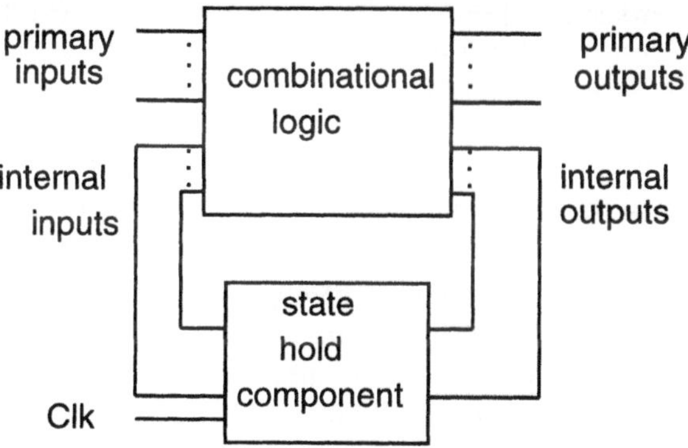

Figure 1   Synchronous Circuit Model

Because only circuit behaviour (not design) is specified in LOTOS for testing purposes, this paper addresses only behavioural modelling of synchronous circuit. Other issues such as structural modelling are discussed elsewhere [16]. A clock signal can be specified explicitly or implicitly according to convenience. LOTOS events represent signal levels during a clock cycle. Circuit behaviour is specified with reference to a clock signal. After an active clock transition, the primary outputs and internal outputs are updated according to the primary inputs and internal inputs. The rest of this section gives two illustrative examples.

A JK flip-flop is a single-bit memory element with control inputs J and K. If they are both set to 0, the flip-flop stays in the same state. If they are both set to 1, the flip-flop inverts its current value. If J and K are set to different values, the value of J is stored. The output is conventionally called Q, while its complement is NQ (not Q). The JK flip-flop specification below fixes the order in which inputs and outputs occur. This might not be a restriction of real hardware. However the order does not influence the functionality of the flip-flop, so there is no need to distinguish which input or output happens first. By restricting the event order, the state space can be substantially reduced when a component has multiple inputs and/or outputs.

```
behaviour JK [J, K, Q, NQ] (0)                          (* initial state is 0 *)
    where
        process JK [J, K, Q, NQ] (dtQ : Bit) : noexit :=
        J ?newJ : Bit; K ?newK : Bit;                   (* get new J and K *)
        ( [(newJ eq 0) and (newK eq 0)] ->              (* both 0 - same state *)
            Q !dtQ; NQ !not(dtQ);
            JK [J, K, Q, NQ] (dtQ)                       (* output current values*)
        []

            [(newJ eq 1) and (newK eq 1)] ->            (* both 1 - flip state *)
            Q !not (dtQ); NQ !dtQ;
            JK [J, K, Q, NQ] (not (dtQ))                 (* invert outputs *)
        []

            [newJ ne newK] ->                           (* both differ - take J *)
            Q !newJ; NQ !not (newJ);
            JK [J, K, Q, NQ] (newJ) )                    (* use J as input *)
        endproc (* JK *)
```

The Single Pulser [22] is a standard hardware verification benchmark. It is a clocked sequential device with a one-bit input and a one-bit output. It outputs a one-cycle pulse when there is a pulse on its input. The Single Pulser can thus be used to debounce a switch, for example. Two kinds of implementations are allowed. The output pulse may be asserted on the positive-going ⌈ or negative-going ⌊ input transition, so the specification is non-deterministic. Test generation for the Single Pulser will be covered in Section 4.1. This example is introduced now to illustrate the issues in modelling synchronous circuits.

```
process SP [Ip, Op] : noexit :=                         (* Single Pulser *)
    i; SP_P [Ip, Op] (0)                                (* +ve triggered implementation *)
[]
    i; SP_N [Ip, Op] (0)                                (*-ve triggered implementation *)
    where
        process SP_P [Ip, Op] (dtI: Bit) : noexit :=
        Ip ?newI : Bit;                                 (* get new input *)
        ( Op !1 [(dtI eq 0) and (newI eq 1)];           (* output 1 on 0->1 input *)
            SP_P [Ip, Op] (newI)
```

[]
    Op !0 [not ((dtI eq 0) and (newI eq 1))];   (* else output 0 *)
    SP_P [Ip, Op] (newI))
endproc (* SP_P *)
process SP_N [Ip, Op] (dtI: Bit) : **noexit** :=
  Ip ?newI : Bit;   (* get new input *)
  (Op !1 [(dtI eq 1) and (newI eq 0)];   (* output 1 on 1$\rightarrow$0 input *)
    SP_N [Ip, Op] (newI)

[]
    Op !0 [not ((dtI eq 1) and (newI eq 0))];   (* else output 0 *)
    SP_N [Ip, Op] (newI))
  endproc (* SP_N *)
  endproc (* SP *)

The LTSs that are observationally equivalent to the above LOTOS specifications appear in figure 2. Observational equivalence is used here since conformance testing relates only to external behaviour of circuits. The equivalence is preserves all external behaviour and has much smaller state space compared to the original specifications. Figure 3 shows suspension automata built from the LTSs. Self-loops in this figure denote δ (quiescent state) actions. Figure 4 presents several tests generated from the automata using the algorithm explained in the preceding section.

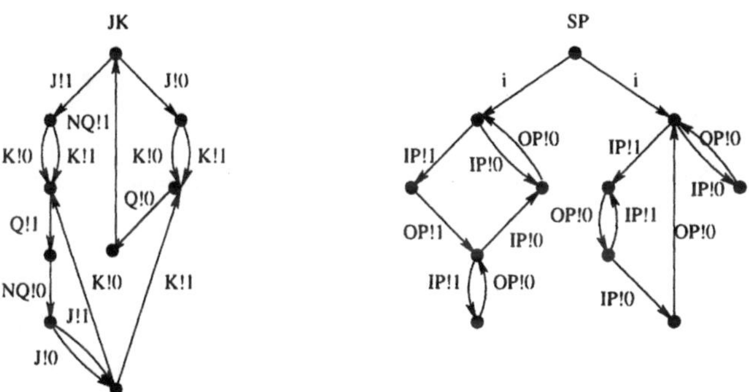

Figure 2   LTSs for the JK Flip-Flop and Single Pulser

The modelling approach discussed above has some implications for testing. Firstly, LOTOS events represents stable signal values in a specific clock cycle. When testing a circuit, applying inputs and observing outputs should also be conducted when the circuit is stable. This is not a problem for clocked circuits since the clock cycle is always chosen such that the circuit has enough time to settle down. Secondly, as stable values of inputs and outputs should appear once in every clock cycle, there is no need to worry about the δ action which indicates the absence of outputs. It is therefore less interesting to generate tests cases like *JK_t2* that check absence of outputs. They are therefore excluded from the outputs of the test generator. Finally, as discussed earlier the order

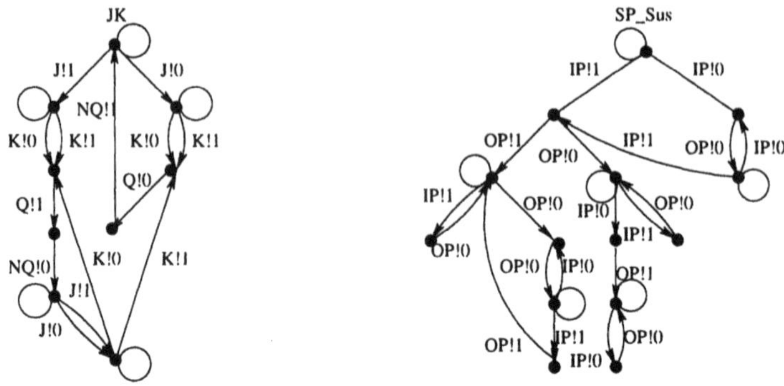

Figure 3 Suspension Automata for the JK Flip-Flop and Single Pulser

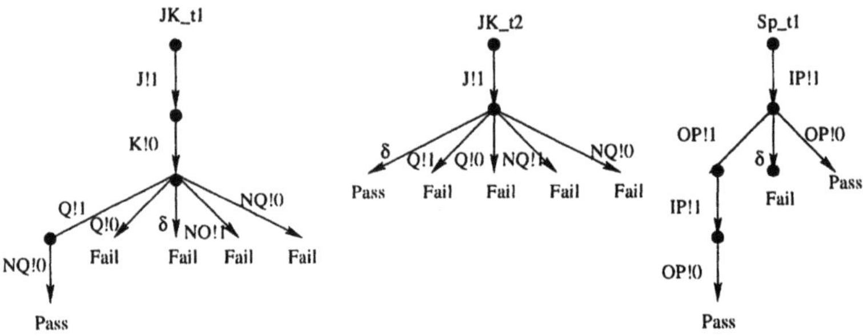

Figure 4 Several Tests of the JK Flip-Flop and Single Pulser

of inputs and outputs is fixed to restrict the state space. In test case *JK_t1*,
the test gives a *Fail* verdict when the first *NQ !0* is observed. This would not
have happened if the full state space had been generated. The way to solve this
problem is discussed in Section 3.2.

These two examples also indicate why the *ioconf* relation is a suitable implementation relation for validating synchronous circuits. If a specification is
deterministic then *ioconf* requires that, for all possible input sequences, all the
outputs of an implementation should agree with those given by the specification. This is strong enough to distinguish erroneous implementations from
correct ones. On the other hand, it also permits non-deterministic specifications
to be tested. For the example of the Single Pulser, a correct implementation
may produce the output pulse after a ⌠ or ⌡ input. This can be properly captured
by the *ioconf* relation. For example if input is initially presumed to be 0 and
then it changes to 1, the output of a ⌠ implementation should be 1 (or 0 for a ⌡
implementation). As seen in test case *Sp_t1* of Figure 4, both design decisions
can pass the test so implementation freedom is respected.

3.2   TEST GENERATION AND EXECUTION

The test cases generated from the algorithm in Section 2.2 have the form of a tree. This might have a straightforward mapping to TTCN (Tree and Tabular Combined Notation [13]). However, the work presented here focuses on investigating the idea of applying protocol testing theory to hardware validation. In this context it is preferable to have test cases of a simpler form that eases test execution and analysis.

A natural way to think about test cases for synchronous circuits is: for the given inputs, what should the outputs be? This indicates that test cases can be in the form of trace with alternation of inputs and outputs. For example, test case *JK_t1* can be stored in a file of the form: *J!1; K!0; Q!1; NQ!0; Pass*. In other words, transitions leading to the *Fail* verdict are not explicitly recorded. When implementations have outputs different from the one defined in the test case, a *Fail* verdict should be generated automatically. Moreover, the side-effect of not recording sequences leading to *Fail* easily solves the problem resulting from fixing the order of outputs in specifications.

This method works well with deterministic specifications. However when the specification has non-deterministic behaviour, simply generating traces from a tree raises problems. For example, the test tree of *Sp_t1* cannot be rewritten as *Ip!1; Op!1; Ip!1; Op!0; Pass* and *Ip!1; Op!0; Pass*. If a ⌠ implementation were tested by the first case, it would be given a *Fail* verdict. Conversely, a ⌡ implementation would fail the second test. Actually, both of them might be correct implementations. The problem is that an implementation has to pass all the test cases in a test suite before it is regarded as correct. But for this example, passing only one of the test cases is necessary. This is solved by marking outputs at a contradictory branch to indicate that the corresponding test is inconclusive when the marked outputs are not matched by the IUT.

At some node of a suspension automaton, suppose the test generation program finds that there are two possible output transitions with the same gate offering different values. Both of the outputs should be marked when the corresponding sequences are generated, meaning they are not necessarily matched by the implementation. Coming back to the example above, the tests then become *Ip!1; Op!1⋆; Ip!1; Op!0; Pass* and *Ip!1; Op!0⋆, Pass*. When output *Op!1* from the implementation is compared to the second test case, the ⋆ means this output does not have to be matched and other test outputs should be checked. If this output is matched by another test branch, comparison continues to determine if the subsequent behaviour is satisfied. As digital signals are strictly binary in the DILL model, if test generation produces both traces then no erroneous behaviours of an implementation will be missed.

Test generation is mainly based on traversing suspension automata. If Choice 1 is made, test case generation is complete and a new test case can be begun. Appending an input action to a trace corresponds to selecting Choice 2 in the test generation algorithm. Appending an output event, possibly with a ⋆ mark, equates to Choice 3.

As specifications usually have infinite behaviour, especially if they involve iterations, a test case can hardly be a complete trace unless the circuit has a deadlock state. Therefore a test suite can never cover all the behaviour of a specification. How to generate a test suite with good coverage is an important theme for testing theory based on LTSs, and is expected to be addressed at a later stage of the work presented here. In this paper, when to terminate a test case and test suite selection are mainly based on heuristics.

If covering all behaviour is not achievable, then covering all transitions might be a second-best choice. A suspension automaton is a directed graph. Generating a sequence that visits every edge in the graph at least once is the **Chinese postman problem** [5] that generates a **transition tour**. A single transition tour exists only for a strongly connected graph in which every node in the graph has a path to every other node. Otherwise, more than one tour is needed to cover all the transitions. As suspension automata may not be strongly connected, it is not possible to make direct use of transition tour generation algorithms (e.g. [9]) that guarantee the shortest tour for a strongly connected graphs. In the work presented here, the approach of [8] is adopted because it is suitable for all kinds of directed graph. In this method, depth-first search (DFS) is used whenever possible as it naturally records the transitions traversed. When an unvisited edge cannot reach by DFS, breadth-first search (BFS) is exploited to find a state that has an unvisited edge; DFS then continues from this state. The whole procedure repeats until this is no unvisited edge in a graph.

The tool set CADP supports an application programming interface that allows user-written programs to manipulate the state space of a given LOTOS specification. The programming interface is used to apply the test generation algorithm to synchronous circuits. For example, Figure 5 shows a test case that is generated for the JK Flip-Flop. Note that the test cases are influenced by the order in which suspension automaton edges are stored. This order is adjustable by changing parameters passed to CADP. If more coverage is required, the test generator can be re-run by using different parameters for more test cases.

Cycle	J	K	Q	NQ	Cycle	J	K	Q	NQ
1	1	1	1	0	2	1	1	0	1
3	0	1	0	1	4	1	0	1	0
5	0	0	1	0	6	1	1	0	1
7	0	0	0	1					

Figure 5   Part of the Test Suite Generated for the JK Flip-Flop

Each tour generated in this way is a test case and is saved in a test file. The accumulated test cases are passed to a VHDL simulator that handles a lower-level implementation of the circuit. A VHDL testbench is designed to allows the test cases to be applied and executed against the VHDL description of the circuit. The testbench mainly consists of two processes that are executed concurrently. The first process generates clock signals for the circuit under test. The second process reads the test suite file and generates signal stimuli according to the inputs of each test case. It also compares the outputs generated by the VHDL simulator with the output values specified by test case, giving a fail verdict and aborting the simulation if they are not the same. The testbench also has to determine when to apply the input stimuli and to check the output result. This needs some knowledge of the circuit realisation, such as the propagation delays of components in the circuit. Special care should be given to those outputs which are marked $\star$ as discussed earlier. Between two test cases, a reset signal is generated by the testbench to re-initialise the circuit under test. The assumption is made that a circuit can always be correctly reset. The LOTOS specifications discussed previously do not specify reset behaviour, so a test need not be generated to ensure that reset is correctly achieved.

There is a peculiar situation for which the test generation program is not so suitable, namely when a circuit implementation may have non-deterministic behaviour. For such specifications, most test cases become inconclusive for many correct and incorrect implementations. This makes the test suite less meaningful. Fortunately, this is not so problematic since digital circuits are rarely allowed to be non-deterministic. Although the Single Pulser specification is non-deterministic, both of its implementations have deterministic behaviour.

4.   CASE STUDIES

In this section, test generation and execution are applied to two standard benchmark circuits. [22] gives informal descriptions of these, along with circuit diagrams and timing diagrams. The Single Pulser has already been introduced. Black-Jack Dealer is a device that plays the dealer's hand of the card game also known as Pontoon or Vingt-et-Un. Although the Single Pulser circuit is relatively small, the Black-Jack Dealer has significant complexity.

4.1   SINGLE PULSER

The formal specification of the Single Pulser was given in Section 3.1, along with its automata forms in Figures 2 and 3. The deterministic design given in the benchmark documentation issues an output after a positive transition of the input. The test generation program produces test cases such as:

Test Case 1: *Ip!0; Op!0; Ip!0, Op!0; Ip!1; Op!0⋆, Ip!0; Op!1; Ip!0; Op!0; Ip!1; Op!0; Ip!1; Op!0; Pass*

Test Case 2: *Ip!1; Op!1⋆; Ip!0; Op!0; Ip!0; Op!0; Ip!1; Op!1; Ip!1; Op!0;*
Pass

When the VHDL testbench is used to execute the first test case, simulation is interrupted after the third clock cycle when an input of 1 has been supplied. The circuit outputs 1 but the expected test output is 0. As this output is marked with ⋆, the simulator concludes that the test is inconclusive for this design. The second test, however, successfully passes the simulation, indicating that the implementation can be regarded as correct. Note that the single pulser does not have a reset input, so simulation has to be restarted when the circuit is checked against the second test case.

4.2   BLACK-JACK DEALER

The Black-Jack dealer inputs are *Card_Ready* and *Card_Value* (Ace..King, Clubs..Spades). Its outputs have boolean values: *Hit* (card needed), *Stand* (stay with current cards) and *Broke* (total exceeds 21). The *Card_Ready* and *Hit* signals are used for a handshake with a human operator. Aces have value 1 or 11 at the choice of the player. Numbered cards have values from 2 to 10. Jack, Queen and King also count as 10. The Black-Jack dealer is repeatedly presented with cards. It must assert *Stand* (when its score reaches 17) or *Broke* (when its score exceeds 21). In either case the next card starts a new game.

In the LOTOS specification of the Black-Jack dealer, a new data type *Value* is defined to represent the card value. Although the LOTOS standard data type *NaturalNumber* might appear suitable, CADP cannot generate the corresponding LTS for an infinite data type like this. The key point in the specification is how to handle the ambiguous value of an Ace. To solve the problem, the specification uses the method given by [27]. Specification behaviour occupies about 80 lines including comments. (The circuit diagram is also about a page.)

Using CADP and the test generator program written by the authors, a test suite for the Black-Jack Dealer was derived. The test suite is able to test 181 different hands of cards that a dealer may hold. The VHDL implementation given in [27] was evaluated against this test suite.

Although the circuit was expected to pass the test suite, a *Fail* verdict was recorded after the dealer was given the following cards: 5, 5, 3, 2, 1, 10. In this case the dealer should be *Broke* because the sum of the cards is 26. However the circuit outputs neither *Stand* nor *Broke* since it considers the total to be just 16. Other card combinations including an Ace and causing *Broke* exhibited the same problem. This showed the problem was related to processing an Ace.

The circuit should initially take an Ace as 11. It should be re-valued as 1 (subtracting 10 from the sum) the first time the result would be *Broke*. If the following cards would make the sum exceed 21, no re-valuation should be done as no Ace is 11. But the given benchmark design still re-values the Ace card, so the circuit is not *Broke* in this case. Carefully simulating the circuit discovered a problem with one of the flag registers (*Ace11Flag* in [22]). This indicates if there is an Ace to be 11. It is not reset to zero properly because the effective duration of the signal used to reset it (*ClearAce11Flag*) is too short. By slightly modifying the circuit to remove the cause of this short duration, the circuit was able to successfully pass the test suite.

5.   CONCLUSION AND FUTURE WORK

The framework of formal methods in protocol testing has been used for testing digital circuits. The protocol testing implementation relation *ioconf* has been justified as suitable for testing synchronous hardware circuits. A prototype tool has been developed to generate and execute test cases automatically. The approach has been validated on standard hardware verification benchmarks. It is noteworthy that it could find an error in a published circuit design for the Black-Jack Dealer.

Future work will include applying test selection techniques while generating test cases. Test cases are guaranteed to cover all the transitions of the specification state space, but no further coverage information can be provided. To progress further will involve defining a suitable coverage function or exploiting some existing selection methodologies (e.g. [1, 3]). Applying the method presented in this paper to higher-specification of digital circuits will also be interesting. In the examples of this paper, specification is relatively closer to real implementation. For example, signals are in one-to-one correspondence between both levels. This makes the task of mapping the test cases to the actual implementation very easy. But when specification is more abstract, greater effort will be needed to make this correspondence.

References

[1] J. Alilovic-Curgus and S. T. Vuong. A metric based theory of test selection and coverage. In A. A. S. Danthine, G. Leduc, and P. Wolper, editors, *Proc. Protocol Specification, Testing and Verification XIII*, pages 289–304. North-Holland, 1993.

[2] E. Brinksma. A theory for the derivation of tests. In S. Aggarwal and K. K. Sabnani, editors, *Proc. Protocol Specification, Testing and Verification VIII*. North-Holland, Amsterdam, Netherlands, June 1988.

[3] O. Charles and R. Groz. Basing test coverage on a formalization of test hypotheses. In M. Kim, S. Kang, and K. Hong, editors, *International Workshop on Testing Communicating Systems X*, pages 109–124. Chapman-Hall, London, UK, 1997.

[4] R. De Nicola and M. C. B. Hennessy. Testing equivalences for processes. *Theory of Computer Science*, pages 83–133, 1984.

[5] J. Edmonds and E. L. Johnson. Matching, Euler tours and the Chinese postman. *Mathematical Programming*, 5:88–124, 1972.

[6] J.-C. Fernández, H. Garavel, A. Kerbrat, R. Mateescu, L. Mounier, and M. Sighireanu. CADP (CÆSAR/ALDÉBARAN Development Package): A protocol validation and verification toolbox. In R. Alur and T. A. Henzinger, editors, *Proc. 8th. Conference on Computer-Aided Verification*, number 1102 in Lecture Notes in Computer Science, pages 437–440. Springer-Verlag, Berlin, Germany, Aug. 1996.

[7] J. C. Fernandez, C. Jard, T. Jéron, and C. Viho. Using on-the-fly verification techniques for the generation of test suites. In R. Alur and T. A. Henzinger, editors, *Computer Aided Verification'96*, volume 1102 of *Lecture Notes in Computer Science*, pages 348–359. Springer-Verlag, Berlin, Germany, 1996.

[8] R. C. Ho, C. H. Yang, M. A. Horowitz, and D. L. Dill. Architecture validation for processors. In *Proc. 22nd. Annual International Synposium on Computer Architecture*, 1995.

[9] G. J. Holzmann. *Design and Validation of Computer Protocols*. Prentice-Hall, Englewood Cliffs, New Jersey, USA, 1991.

[10] IEEE. *VHSIC Hardware Design Language*. IEEE 1076. Institution of Electrical and Electronic Engineers Press, New York, USA, 1993.

[11] IEEE. *IEEE Standard Hardware Design Language based on the Verilog Hardware Description Language*. IEEE 1364. Institution of Electrical and Electronic Engineers Press, New York, USA, 1995.

[12] ISO/IEC. *Information Processing Systems – Open Systems Interconnection – LOTOS – A Formal Description Technique based on the Temporal Ordering of Observational Behaviour*. ISO/IEC 8807. International Organization for Standardization, Geneva, Switzerland, 1989.

[13] ISO/IEC. *Information Processing Systems – Open Systems Interconnection – Conformance Testing Methodology and Framework – Part 3: The Tree and Tabular Combined Notation (TTCN)*. ISO/IEC 9646-3. International Organization for Standardization, Geneva, Switzerland, 1991.

[14] Ji He and K. J. Turner. Extended DILL: Digital logic with LOTOS. Technical Report CSM-142, Department of Computing Science and Mathematics, University of Stirling, UK, Nov. 1997.

[15] Ji He and K. J. Turner. Timed DILL: Digital logic with LOTOS. Technical Report CSM-145, Department of Computing Science and Mathematics, University of Stirling, UK, Apr. 1998.

[16] Ji He and K. J. Turner. Modelling and verifying synchronous circuits in DILL. Technical Report CSM-152, Department of Computing Science and Mathematics, University of Stirling, UK, Feb. 1999.

[17] G. Leduc. A framework based on implementation relations for implementing LOTOS specifications. *Computer Networks and ISDN Systems*, 25(1):23–41, Aug. 1992.

[18] D. Moundanos, A. Abraham, and Y. V. Hoskote. Abstraction techniques for validation coverage analysis and test generation. *IEEE Transactions on Computers*, 47:2–14, 1998.

[19] R. D. Nicola. Extensional equivalences for transition systems. *Acta Informatica*, 24:211–237, 1987.

[20] D. H. Pitt and D. Freestone. The derivation of conformance tests from LOTOS specifications. *IEEE Transactions on Software Engineering*, 16(12):1337–1343, Dec. 1990.

[21] J. M. T. Romijn, O. Sies, and J. R. Moonen. A two-level approach to automated conformance testing of VHDL designs. *Testing of Communicating Systems*, 10:432–447, 1997.

[22] J. Staunstrup and T. Kropf. IFIP WG10.5 benchmark circuits. http://goethe.ira.uka.de/hvg/benchmarks.html, July 1996.

[23] J. Tretmans. Conformance testing with labelled transition systems: Implementation relations and test generation. *Computer Networks and ISDN Systems*, 29:25–59, 1996.

[24] J. Tretmans. Test generation with inputs, outputs and repetitive quiescence. *Software Concepts and Tools*, 17:103–120, 1996.

[25] K. J. Turner and R. O. Sinnott. DILL: Specifying digital logic in LOTOS. In R. L. Tenney, P. D. Amer, and M. Ü. Uyar, editors, *Proc. Formal Description Techniques VI*, pages 71–86. North-Holland, Amsterdam, Netherlands, 1994.

[26] F. Vemuri and R. Kalyanaraman. Generation of design verification tests from behavioral VHDL programs using path enumeration and constraint programming. *IEEE Transactions on Very Large Scale Integration Systems*, 3:201–214, 1995.

[27] D. Winkel and F. Prosser. *The Art of Digital Design*. Prentice-Hall, Englewood Cliffs, New Jersey, USA, 1980.

10

AUTOMATED TEST OF
TCP CONGESTION CONTROL ALGORITHMS

Roland Gecse, Péter Krémer

Conformance Center, Ericsson Ltd.

H-1037 Budapest, Laborc u. 1., Hungary

E-mail: { Roland.Gecse, Peter.Kremer } @eth.ericsson.se

Abstract   This paper introduces a new uniform, overall well applicable method for testing TCP implementations fully automatic. Our primary intention was to give an Abstract Test Suite (ATS) for TCP congestion control to provide the Means of Testing (MoT), and to demonstrate it on some of today's implementations. First, we give a short introduction to TCP, and a brief section presents the former trials in automating TCP testing. Then we explain three test cases in detail, and propound the results of their execution on four different platforms. Some interesting faults are also presented. Finally, we point out the benefits of the introduced approach.

Keywords: Conformance test, TCP, TTCN, Internet

1.   INTRODUCTION

Nowadays, it is not easy to predict the way that evolution of communication technology goes. Current tendency shows the convergence of telecom and datacom world as Internet technology gets widespread use in commercial telephony applications. It is no question that this will have great influence on both of them. Everybody knows the reliability and robustness of telecom solutions, which are partly achieved by serious testing procedures. Internet, on the contrary, does not put high demands on conformity of its protocols and applications. While it is common in telecom world to black box test communication equipment against standards, conformance testing of Internet protocols has no roots. Although products undergo interoperability testing[1] in order to prove the ability to interwork with other vendors' devices, this does not seem to be enough. Certain protocol implementations contain sophisticated features, which can be hardly tested when erroneous. This holds especially for communication procedures of Transmission Control Protocol (TCP) [Postel, 1981]. The Network Working Group of Internet Engineering Task Force (IETF) is writing an Internet draft that documents some known TCP implementation problems [Paxson, 1997]. The outline of this draft closely resembles to the structure of a test purpose. It describes desired procedures and impacts to the communication together with traces of both correct and faulty behaviour. But there is no sufficient instruction for testers on how to perform the tests besides connecting a network analyser and observing long traces. On one hand it can be very difficult to trigger the problem, on the other hand it takes a lot of skilled work to find it in the enormous amount of data. Inspired by this draft, we tried to get another perspective and use our knowledge in Conformance Testing Methodology and Framework (CTMF) [ISO 9646, 1997] to develop automated TCP tests.

After a short summary of TCP we give a state of the art overview on Internet testing and describe some previous work. Then, we present some Test Cases (TCs) in the area of TCP's congestion control algorithms. The rest of the paper summarises our test results and draws a conclusion on our work.

2.   TRANSMISSION CONTROL PROTOCOL (TCP)

TCP provides a connection-oriented, reliable, byte stream transport service in the TCP/IP stack. Today's most popular applications, like World Wide Web (WWW) and e-mail, are built upon it, which makes TCP the most used transport protocol of Internet.

The term connection-oriented denotes that every TCP communication takes place through connections. A pair of source and destination side IP addresses and TCP port numbers constitute a connection, which provides a full duplex communication between two endpoints. A connection must be set up before two entities can exchange messages, and it has to be released once the communication having finished. The number of simultaneous connections is limited by the capacity of the underlying media and the number of available ports. This concept makes it possible for more applications to share the same TCP service.

TCP's reliability is expressed by procedures, which make it possible to exchange data over the unreliable Internet Protocol (IP). The application data

---

[1] The term interoperability testing has a different meaning in Internet. It means examining whether two or more different implementations can interwork, rather than the comparision with a reference implementation. In this paper we keep on using interoperability testing in this context.

is broken into best-sized chunks (i.e. Maximum Segment Size – MSS) and put into the payload of IP packets. Each data byte has a sequence number assigned that is used for keeping track of state (sent/received/lost). Whenever a packet is sent, an acknowledgement (ACK) is expected. That means the other end "acks" the received data. If this ACK does not arrive in a given time, TCP considers the packet as lost, and retransmits data bytes beginning with the last ACKed sequence number. The receiver TCP is required to delay the ACK for sending it piggybacked with data if possible.

Since IP is unreliable, packets can be modified, lost, duplicated, or can arrive out of order. TCP detects damaged packets by maintaining header and data checksum. Lost packets are retransmitted; duplicated packets are deleted; above sequence data is buffered within reasonable bounds.

Unlike other transport layer entities, TCP connections do not occupy a prede-fined, constant bandwidth. Neither do they give quality of service guaranties. TCP is said to provide a best effort service, which makes it able to operate over networks having a wide range of transmission capacity. The ability to use as much bandwidth as possible is achieved by flow and congestion con-trol algorithms, such as slow start, congestion avoidance, fast retransmit - fast recovery.

According to [Braden, 1989], all these algorithms must be present in each TCP. Although they did not belong to the original recommendation [Postel, 1981], they became part of it during the evolution of the protocol. All three algorithms operate in the ESTABLISHED state of the TCP state transition diagram. Most problems concerning these algorithms do not occur in interop-erability workshops, because – from the user point of view – TCP seems to work well, even if these procedures are ignored. Indeed, a faulty TCP, which lacks their implementation, usually performs better than a correct one! But this is done, of course, at the expense of suppressing parallel connections, which are competing for resources in vain. In the worst case scenario a faulty implemen-tation could cause serious problems, such as congestion collapse [Jacobson, 1988].

The congestion control algorithms introduce an interesting non-deterministic feature of TCP, that is, we cannot predict how the message flow of a given connection will look like, even if we are aware of all sender and receiver parameters of the network. This means, unlike many other protocols, we cannot define one particular Message Sequence Chart – MSC for a given operation because there are numerous existing good alternatives.

3.   FORMER TRIALS IN TCP TESTING

Many papers deal with TCP test both from telecom and Internet points of view. [Kato, 1997] shows the development of a TCP protocol monitor, which was later used for HTTP monitoring [Ogishi, 1998]. [Hintelmann, 1997] describes TCP's congestion control algorithms in SDL, and compares several different implementations.

On the Internet side, the latest efforts include [Paxson, 1997], which tried to give a general solution to this problem. There are some more TCP tester applications described in [Parker-Schmechel, 1998]. But they are mainly trace analysers and performance monitors. However, none of them managed to give a uniform, overall well applicable method.

3.1   TESTING VS. MONITORING

Current Internet test procedures include packet capturing in real, congested environment, and then they examine the obtained traces. Measurements are triggered either manually or using special applications to cause a particular action. Some of them also requires modifications in Implementation Under Test (IUT), which is unacceptable for us because we focus sharply on black box testing. In addition, most vendors do not provide the source code along with their product.

We see the most significant problem in monitoring is that whenever a packet is lost, we cannot know whether it was because of network overload or measurement error. It is described in [Paxson, 1997; Ogishi, 1998] that packet filters often include ambiguity in the measurement. This could be due to packet drops, additions, resequencing, timing, or even the placement of packet filters. The authors tried to give a workaround by introducing resampling and prediction. In this case the analyser must take numerous possibilities into consideration. Then some features of IUT have to be included in the analyser as well. This has a great disadvantage of losing the general applicability since analyser gets dependent on IUT. And the original idea of testing disappears behind.

All referenced papers are based on monitoring. This could be because of several reasons. First, since all platforms provide some kind of packet capturing functions, it is easy to acquire and filter packets. The obtained traces can be evaluated easily off-line. Second, they have not thought about testing conformity against recommendation. Monitoring is a good solution for interoperability testing, but not for conformance testing.

Third, emulating a TCP is not easy. The difficulty arises from the architecture of TCP implementations. Almost all of today's TCP/IP stacks are written as a whole unit and hidden inside the kernel. Application Programming Interfaces (APIs) do not allow routines to reroute TCP packets to them.

4.   TEST METHODS AND CONFIGURATION

We tried another approach. We use black box method; we do not make any modification to IUT. We use methods and notations of CTMF because we would like to give a general, widely applicable solution. We test the system in laboratory environment, where we do not have to cope with packet loss because of network overload or other unpredictable events. This prevents us from uncertainty coming from measurement errors described earlier. Instead of processing traces documenting former message exchange on a given connection we emulate the peer communicating entity, and control and observe other end's behaviour remotely. Doing so we can focus entirely on the problem given.

For all of our tests we use the remote test method. The UT role is played by an HTTP server. We use the Web server for bulk data transfer. For test purposes we defined a simple WWW site with some documents having different length and data.

5.   MEANS OF TESTING

For the execution of our tests we use the Ericsson proprietary System Certification System (SCS). SCS consists of a set of tools: TTCN Translator translates MP files into EXTEL format, which is then used by Test Component Executor (TCE) that interprets and executes test cases selected from TTCN Manager. The most interesting feature of SCS is the Test Port concept. All Abstract Service Primitives (ASPs) of a given Point of Control and Observation (PCO) have to be mapped onto the Service Primitives (SPs) of the underlying service provider in order to exchange data with IUT. Since different protocols can be based on different services, which use different types of connections. This mapping needs serious programming skills. In order to simplify the task, SCS provides the so called Test Ports. A Test Port is a piece of software that holds routines, which are called after an ASP is received from or before ASPs are sent to TCE. Thus, we get the full control over the SPs.

We developed a new test port for TCP tests. It provides two ASPs for communication with the socket interface: "recv" can only be received, and "send" that can only be sent. The TCP header length and checksum are calculated automatically in the test port, so we do not have to waste time on defining them in the constraints.

6.   TEST CASES

We derived an ATS consisting of 12 test cases (TCs). Altogether these provide almost full coverage of TCP congestion control. Our work was, in part, inspired by problems described in [Paxson, 1998], thus some of our TCs are based on scenarios described therein. The ATS has four groups; the first three groups belong to one of the introduced congestion control algorithms. The fourth group contains all the other test cases. For the sake of this paper we describe the three most important congestion control algorithms: slow start,

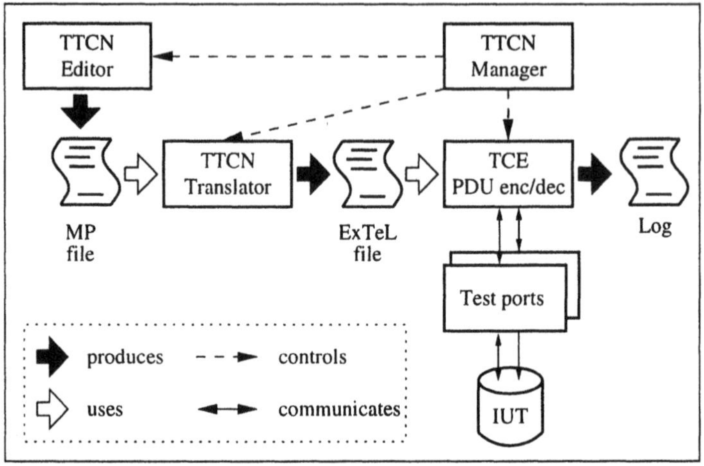

Figure 1   The structure of SCS

congestion avoidance, fast recovery - fast retransmit. But first we should get acquainted with the general ideas of the ATS.

We use Test Case Variables (TCVs) for saving and keeping track of TCP state variables in both sides of the connection. We try to use consistent naming and use the variable names of original TCP recommendation [Postel, 1981] where it is possible. The direction of data flow is tracked from the tester's point of view. Some examples: rcv_mss, for instance, contains the IUT's offered maximum segment size, while snd_mss is the tester's one. The rcv_cwnd holds the value of the current size of IUT's congestion window in bytes, rcv_nxt holds the sequence number of the next data expected from IUT. On the other side, snd_wnd represents the tester's advertised receive window size.

Each receive statement checks whether the received frame has the expected sequence number with the correct source and destination port addresses. The payload of segments is also tested when it is necessary.

Most time we emulate the receiving side of the connection. To fulfill the requirements of [Braden, 1989] we must send acknowledgement messages (ACKs) for incoming data in order to signal the other end that its data has been accepted. The rules for sending ACK are the following: The receiver must send ACK:

    1 after the receipt of two full-sized segments or
    2 on ACK timer (we call it T_ACK in our ATS) expiry or
    3 if there is any data to be sent on the connection.

We must violate the first rule because of our TTCN executor. If we send ACK for every second full-sized segment, we increase IUT's $cwnd$ too aggressively.

But the executor cannot process such amount of data, so sooner or later IUT's *RTO* timer will expire. We try to avoid this situation since most of the time we examine IUT's transmission (and not retransmission) behaviour. To compensate this deficiency our T_ACK timer is not set to the usual $200\,ms$, but only to $10\,ms$. Our experiences show that TCP implementations send all the segments to be transmitted at once. It means that the delay between the first and the last segment is below $1\,ms$. The third situation rarely takes place since we are mostly receiving data.

We implemented active open and both active and passive close. That means, all kinds of connection set-up and release procedures – including simultaneous and half-close – can be performed regardless of TCP options used.

6.1   SLOW START

The slow start and congestion avoidance [Stevens, 1997] were the first TCP congestion control algorithms. Slow start limits the number of packets that TCP can inject into the network. The algorithm introduces a new TCP state variable called congestion window ($cwnd$). This and the receiver side's advertised window (rcv_wnd) give an upper bound to the number of outstanding data bytes per connection. The unit of $cwnd$ is measured in bytes or in sender's maximum segment size (snd_mss). Its initial value is snd_mss, which is increased by at most snd_mss bytes for every incoming non-duplicate ACK.

Slow start is used in almost every TC we have written as a preamble, thus we made great effort to implement it as well as possible. But of course, we have to ensure if it is working properly before using it as a test step.

In our test case we observe whether $cwnd$ is initialised to 1 snd_mss. Nowadays' tendency shows that more and more TCP implementations set this value to 2 snd_mss or even higher. Since [Allman, 1999] permits the use of an initial $cwnd$ with values less then 2 snd_mss but no more than 2 segments, we are prepared to accept initial values, which satisfy these conditions.

Besides the initial value, we also test IUT's rate of $cwnd$ increments. The remaining part of this section describes the behaviour from the tester's point of view. After initialising TCVs we do an active open. Having finished the connection set-up we send an HTTP request to the WWW server located on IUT. This request starts downloading a fairly large document. Then we get into a test step performing slow start (Figure 2), till all data is received.

In this test step, we are waiting for the IUT to send the requested data segments while keeping track of its $cwnd$. First, it should send only one full sized segment and wait for acknowledgement. Receiving this ACK, it should increase its $cwnd$ by one snd_mss. Next, it should send two packets and wait for ACKs and so on. If the IUT transmits more frames than its $cwnd$ value, then it does not implement slow start algorithm well, and our TC will fail. We let

Test Step Dynamic Behaviour					
Test Step Name : SLOW_START_UNLIMITED (transmitted_bytes:INTEGER)					
Group :					
Objective :					
Default : Default_TMAX					
Comments :					
Nr	**Label**	**Behaviour Description**	**Constraints Ref**	**Verdict**	**Comments**
1	label1	+RECV_DATAACK			
2		(rcv_data := packet_length – (4 * header.hlen), rcv_nxt := rcv_nxt + rcv_data, snd_una := header.ack, overall_data := overall_data + rcv_data)			
3		[rcv_data <= 0]			
4		GOTO label1			
5		[rcv_data > 0]			
6		(rcv_cwnd := rcv_cwnd + snd_mss)			
7		[rcv_nxt – rcv_una <= rcv_cwnd]			
8		GOTO label1			
9		[rcv_nxt – rcv_una > rcv_cwnd]			
10		IP ! send	send_param (tcphdr_rst_s)	F	
11		? TIMEOUT T_ACK			
12		[rcv_cwnd >= 2 * snd_mss]			
13		IP ! send (rcv_una := rcv_nxt)	send_param (tcphdr_ack_s)		
14		GOTO label1			
15		[rcv_cwnd < 2 * snd_mss]			
16		GOTO label1			
17		+RECV_FIN_OR_DATAFIN			
18		(rcv_data := packet_length – (4 * header.hlen), rcv_nxt := rcv_nxt + rcv_data + 1, snd_una := header.ack, overall_data := overall_data + rcv_data)			
19		[overall_data = transmitted_bytes]		(P)	
20		+PASSIVE_CLOSE			
21		[overall_data <> transmitted_bytes]		(I)	
22		+PASSIVE_CLOSE			
23		+RECV_ANYACK			
24		IP ! send (rcv_una := rcv_nxt)	send_param (tcphdr_ack_s)		
25		? OTHERWISE		(F)	
26		+ACTIVE_CLOSE			
Detailed Comments :					

Figure 2   The test step of slow start

IUT increase its $cwnd$ constantly, since the data size to be received (100 $kbytes$) is not so much that we should count on packet loss because exceeding the LAN capacity. Moreover, we have to set snd_wnd to an appropriate large value (64 $kbytes$) so that IUT's send rate is limited only by its $cwnd$.

Though we do not expect packet loss in this case, the TC can handle it because our TTCN executor is rather slow. When we transmit very large files (more than 1 $Mbytes$), the delay caused by the executor gets the RTO timer expired and IUT will retransmit the appropriate segments.

We run the algorithm until all data has arrived. Then we either do active or passive close and set the verdict to pass. Figure 3 shows the trace of a normal execution from the Tester's point of view. The diagram is based on a trace, which was taken using tcpdump on IUT's side.

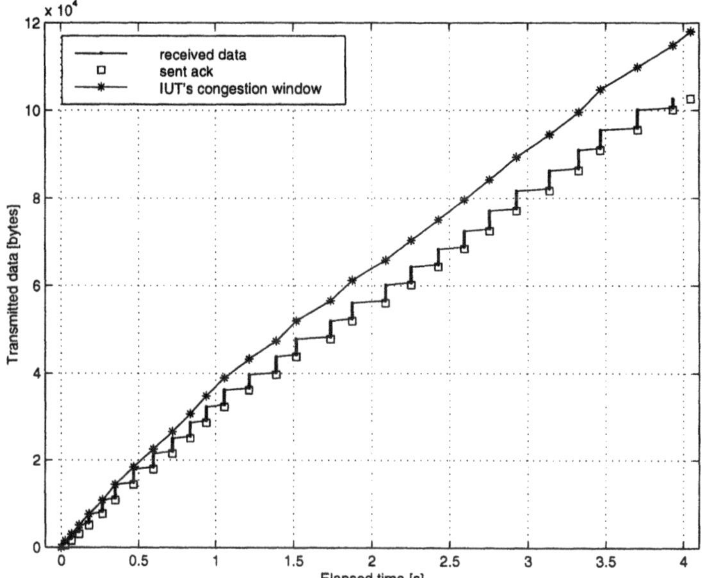

Figure 3   Initial slow start

The dotted line marks the received data (sent by IUT), the squares show the places where the Tester sent acknowledgements (to IUT), and the line with asterisks indicates IUT's congestion window. It can be seen that the network's delay is negligible. The elapsed times between respective ACKs are caused by the executor, e.g. at $3.7\,s$ we receive 8 segments, but send ACK for them at $3.9\,s$, it means that our delay is about $200\,ms$. Now let us see the most exciting part of the figure, which is IUT's congestion window. It is clear that IUT did not send more data than its congestion window, since the dotted line does not exceed the line with asterisks. In the beginning *cwnd* moves away from data, which means that IUT does not send more data than it is permitted by its *cwnd*.

6.2   CONGESTION AVOIDANCE

The congestion avoidance algorithm takes over the control whenever the value of *cwnd* exceeds *ssthresh*. This algorithm slows down the rate of the growth of *cwnd* by increasing it at most by one segment size per *RTT*. The congestion avoidance is performed until *cwnd* congestion is detected.

Current implementations use two methods for counting the actual size of *cwnd* during congestion avoidance. Those implementations, counting the size of *cwnd* in maximum-sized segments calculate the current value of *cwnd* with the formula $snd\_mss^2/cwnd$. The *cwnd* is updated on each new incoming

ACK. Some TCPs, on the other hand, that maintain the value of *cwnd* in bytes increase *cwnd* if the number of ACKed data bytes exceeds *cwnd*. However, the value of *cwnd* must not be increased by more than *snd_mss* bytes.

Some implementations do not follow these rules. They, incorrectly, increase *cwnd* more aggressively than it was described above by a small additive constant. This can diminish performance as described in [Paxson, 1998]. The value of this additive constant is usually *snd_mss*/8.

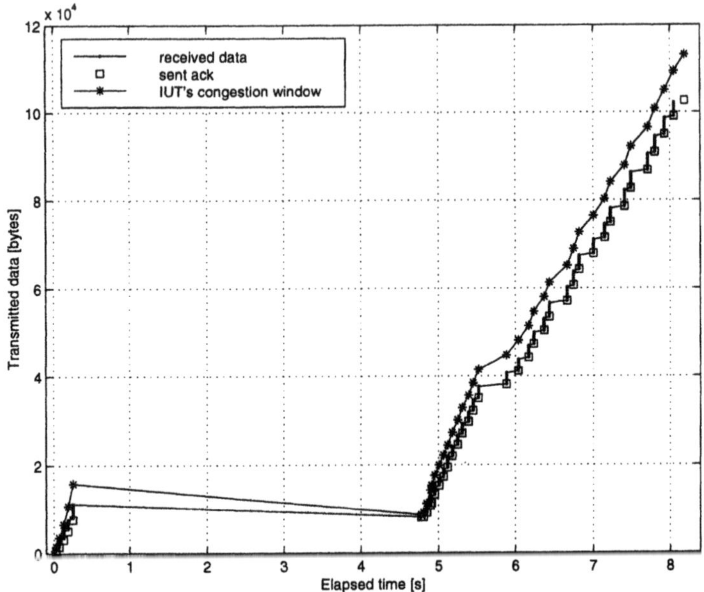

Figure 4   Congestion avoidance

The problem can be hardly found in traces because both *cwnd* and *rcv_wnd* values have to be large. Its identification requires large amount of data to be sent on a connection. In our TC, we first assist IUT to open its congestion window to 16, then we emulate packet loss without sending three duplicate ACKs until IUT's *RTO* timer expires. That makes it set *ssthresh* to 8, *cwnd* to 1 and engage in slow start till *cwnd* does not exceed 8. After that, we take care whether it sends an extra segment after the first 8 full-sized segments. We preset the size of *rcv_wnd* to 64 *kbytes* again so that it does not limit the amount of data sent by IUT. Figure 4 shows the correct behaviour.

6.3   FAST RETRANSMIT - FAST RECOVERY

We describe these two algorithms in one section, since they are closely related and usually implemented together (just like slow start and congestion

control). The fast retransmit - fast recovery is described in [Stevens, 1997].
The idea behind these is to avoid the expensive slow start after retransmission
timeout. After the receipt of the third duplicate ACK we can suppose with large
probability that the packet has been lost, and retransmit it immediately. The
fast recovery allows sending of more segments upon receiving more duplicate
ACKs until rcv_wnd or $cwnd$ allows. But as soon as new data is ACKed, the
sender must retain its $cwnd$ to the half of its original value and continue with
congestion avoidance.

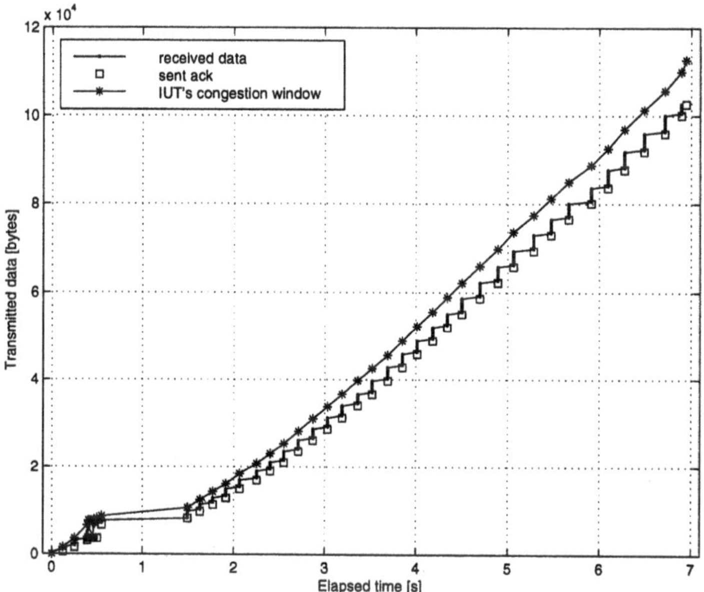

Figure 5   Fast retransmit - fast recovery

The chosen TC was proposed in [Paxson, 1998] as failure of window de-
flation after loss recovery. This error is quite sophisticated. It occurs after
recovery when three subsequent ACKs before RTO time expiry cause IUT's
TCP to retransmit the missing segment according to the fast recovery algorithm.

To trigger the problem, we run slow start till IUT's $cwnd$ is 8. That means
up to 8 segments of unacknowledged data can be out on the network, if the
receiver's advertised window does not limit it. We wait for "some" more frames
to arrive and then send 3 subsequent ACKs to a former but still unacknowledged
segment. After receiving these ACKs, IUT should set $ssthresh$ to $cwnd/2$,
retransmit the missing segment, and increase $cwnd$ by 3. Each time another
duplicate ACK arrives it should increase $cwnd$ by one and transmit a packet
if its $cwnd$ or our advertised receive window allows. However, if new data is

acknowledged, IUT must set its *cwnd* to the half of its original value, which is stored in *ssthresh*, and continue with congestion avoidance.

Figure 5 shows that IUT applies fast retransmit - fast recovery algorithms well. The 1*s* delay is caused by the TC; we ensure this way that IUT does not send more data than allowed.

7.   TEST RESULTS

We executed the 12 TCs on four different platforms. Although most test cases are passed we experienced some minor bugs or non-conformities. Implementation A, for example, failed the TC 'Retransmission sends doubled data' and too low initial RTO was observed on B, C and D platforms. The next subsections discuss the arisen problems in detail. The summary of the results is shown in Table 1.

Test Cases	Platforms	A	B	C	D
No initial slow start		P	P	P	P
No initial slow start with large window		P	P	P	P
No slow start after retransmission timeout		P	P	P	I
No congestion avoidance		P	P	P	P
Uninitialised cwnd		P	P	P	P
Failure to retain above sequence data		P	P	P	P
Initial RTO too low		P	F	F	F*
Failure of window deflation after loss recovery		P	P	P	P
Failure to back off retransmission timeout		P	P	P	P
Strecth ACK violation		P	P	P	P
Failure to send FIN notification promptly		P	P	P	P
Retransmission sends doubled data		F	P	P	F

F: Fail, I: Inconclusive, P: Pass

Table 1   Results of Test Cases

7.1   RETRANSMISSION SENDS DOUBLED DATA

This TC examines the IUT's retransmission behaviour. We send one ACK after every retransmitted segment, and do not acknowledge the retransmitted segment but an older one. Since this ACK goes for an older data, it does not increase the congestion window, thus IUT must ignore it. The TCP implementation on platform A goes another way as the following trace (made by tcpdump) shows:

```
1. 11:19:11.045884 IUT > Tester:  . 3073:3585(512) ack 33 win 9216 (DF)
2. 11:19:11.045952 IUT > Tester:  . 3585:4097(512) ack 33 win 9216 (DF)
3. 11:19:11.046022 IUT > Tester: P 4097:4609(512) ack 33 win 9216 (DF)
4. 11:19:11.066923 Tester > IUT:  . ack 3585 win 2048 (DF)
5. 11:19:11.067005 IUT > Tester:  . 4609:5121(512) ack 33 win 9216 (DF)
6. 11:19:11.067050 IUT > Tester: P 5121:5633(512) ack 33 win 9216 (DF)
```

```
 7. 11:19:15.756634 IUT > Tester:  . 3585:4097(512) ack 33 win 9216  (DF)
 8. 11:19:15.769998 Tester > IUT:  . ack 3585 win 4096  (DF)
 9. 11:19:15.770056 IUT > Tester:  . 3585:4097(512) ack 33 win 9216  (DF)
10. 11:19:15.782244 Tester > IUT:  . ack 3585 win 4096  (DF)
11. 11:19:25.216864 IUT > Tester:  . 3585:4097(512) ack 33 win 9216  (DF)
12. 11:19:25.230479 Tester > IUT:  . ack 3585 win 4096  (DF)
13. 11:19:44.117747 IUT > Tester:  . 3585:4097(512) ack 33 win 9216  (DF)
14. 11:19:44.131287 Tester > IUT:  . ack 3585 win 4096  (DF)
15. 11:20:21.927916 IUT > Tester:  . 3585:4097(512) ack 33 win 9216  (DF)
16. 11:20:21.943187 Tester > IUT:  . ack 3585 win 4096  (DF)
```

In row 7, IUT resends segment 3585 : 4097 for the first time. We reply with an ACK for the former segment and IUT retransmits segment 3585 : 4097 because of our ACK. Note that the elapsed time between the last two retransmissions (row 7 and 9) is only $13.422\,ms$. It is also remarkable that implementation A performs this "doubled retransmission" only after the first retransmission. This inconsistency causes only one single segment to be transmitted in vain, and this single packet also increases the load of the network unnecessarily.

Implementation D also exhibits the similar problem.

7.2   INITIAL RTO TOO LOW

It is quite surprising that almost all examined implementation failed this TC. This phenomenon arises from early BSD TCP implementations [Wright-Stevens, 1995]. It measures the elapsed time using two different timers. One of them is set to expire every $200\,ms$, the other every $500\,ms$. The $500\,ms$ timer is used for RTO measurement. For instance the initial RTO is measured by waiting for this timer to expire six times. It means that actual delay could fall between 2500 and $3000\,ms$. Although the TCs failed, we do not consider this clock skew as a serious failure since – as the following dumps show – this technique is used in almost all of today's TCP implementations.

In implementation B, the measured RTO value is $2976.8\,ms$, which is fairly close to the recommended $3s$:

```
1. 14:15:58.007950 IUT > Tester:  . 513:1025(512) ack 33 win 16384  (DF)
2. 14:16:00.984750 IUT > Tester:  . 1:513(512) ack 33 win 16384  (DF)
```

On the C platform, the initial value of the RTO timer is probably $3s$. The $2994\,ms$ measured RTO by the test case was less then the real value ($2999.204\,ms$), but it is still below the desired $3s$:

```
1. 13:20:36.836817 IUT > Tester: P 1:513(512) ack 33 win 32736  (DF)
2. 13:20:39.836021 IUT > Tester: P 1:513(512) ack 33 win 32736  (DF)
```

We also observed this deficiency on our fourth tested platform. The retransmission timer of implementation D is initially set above $3s$. The TTCN executor measured the RTO to be $2998\,ms$, but the real value was $3001.525\,ms$. The cause of this problem is probably the time-variant feature both of the network and the executor or the delay in the test case. In the test case we can only start to measure the elapsed time after processing the received data. This inherent delay may consumes the missing $\approx 2\,ms$:

```
1. 18:25:50.534270 IUT > Tester: . 1:513(512) ack 33 win 8160 (DF)
2. 18:25:53.535795 IUT > Tester: . 1:513(512) ack 33 win 8160 (DF)
```

7.3   NO SLOW START AFTER RETRANSMISSION TIMEOUT

Implementation D has a major fault in its TCP implementation. The problem occurs when we send ACK to a retransmitted segment. In row 8 IUT performs the second retransmission (the first was in row 6), we send ACK to an older data and IUT replies with a new segment (row 10). When the retransmission timer expires again, IUT replies to our ACK with two new segments (row 15 and 16). Since we send ACKs for older data, these ACKs do not allow increasing IUT's congestion window. It means that under some circumstances IUT does not perform slow start after retransmission timeout! But if the first retransmission is successful, IUT works well. Although the mentioned test case passed we had to modify the final result to inconclusive, because of this problem.

```
 1. 08:38:27.504181 IUT > Tester: . 3073:3585(512) ack 33 win 8160 (DF)
 2. 08:38:27.504655 IUT > Tester: P 3585:4097(512) ack 33 win 8160 (DF)
 3. 08:38:27.539684 Tester > IUT: . ack 3073 win 2048 (DF)
 4. 08:38:27.540300 IUT > Tester: . 4097:4609(512) ack 33 win 8160 (DF)
 5. 08:38:27.540777 IUT > Tester: . 4609:5121(512) ack 33 win 8160 (DF)
 6. 08:38:29.559949 IUT > Tester: . 3073:3585(512) ack 33 win 8160 (DF)
 7. 08:38:29.573549 Tester > IUT: . ack 3073 win 4096 (DF)
 8. 08:38:33.766114 IUT > Tester: . 3073:3585(512) ack 33 win 8160 (DF)
 9. 08:38:33.779216 Tester > IUT: . ack 3073 win 4096 (DF)
10. 08:38:33.779834 IUT > Tester: P 3585:4097(512) ack 33 win 8160 (DF)
11. 08:38:42.178411 IUT > Tester: . 3073:3585(512) ack 33 win 8160 (DF)
12. 08:38:42.191494 Tester > IUT: . ack 3073 win 4096 (DF)
13. 08:38:42.192108 IUT > Tester: . 3073:3585(512) ack 33 win 8160 (DF)
14. 08:38:42.205000 Tester > IUT: . ack 3073 win 4096 (DF)
15. 08:38:42.205620 IUT > Tester: P 3585:4097(512) ack 33 win 8160 (DF)
16. 08:38:42.206096 IUT > Tester: . 4097:4609(512) ack 33 win 8160 (DF)
```

8.   CONCLUSION

Although, in this article, we addressed only a subset of possible problems in TCP, we think that we managed to take the first step towards automated TCP tests. We pointed out the major drawbacks of monitoring techniques, which are in use for current Internet interoperability tests. Then, we introduced how conformance tests could be carried out in TCP implementations. However not excepted, we found some errors while performing our tests on today's most widespreadly used TCPs. This fact justified our thought that interoperability test, as it is used in Internet, does not cover all protocol features. One could think that 12 test cases are not enough to cover the field of congestion control, but remember that these algorithms could have several good outcomes. Our test cases verify all the correct possibilities, not only one. Hopefully, we could show that there is a high need for testing Internet protocols for conformance.

References

M. Allman (editor): TCP congestion control, <draft-eitf-tcpimpl-cong-control-05.txt>, TCP Implementation Working Group, February 1999.

B. Baumgarten, A. Giessler: OSI conformance testing methodology and TTCN, North holland, 1994.

R. Braden (editor): Requirements for Internet hosts – communication layers, RFC 1122, IETF Network Working Group, October 1989.

R. Gecse: Conformance testing methodology of Internet protocols, Testing of Communicating Systems, Tomsk, Russia, September 1998.

J. Hintelmann, R. Westerfeld: Performance analysis of TCP's flow control mechanisms using queueing SDL, SDL '97: TIME FOR TESTING - SDL, MSC and Trends, Elsevier Science B. V., 1997.

V. Jacobson: Congestion avoidance and control, ACM SIGCOMM '88, Stanford, California, August 1988.

T. Kato, T. Ogishi, A. Idoue and K. Suzuki: Design of protocol monitor emulating behaviours of TCP/IP protocols, Testing of Communicating Systems, Cheju Island, Korea, September 1997.

T. Ogishi, A. Idoue, T. Kato and K. Suzuki: Intelligent protocol analyzer for WWW server accesses with exception handling function, Testing of Communicating Systems, Tomsk, Russia, September 1998.

OSI - Open System Interconnection, Conformance testing methodology and framework, ISO/IEC 9646, 1997.

S. Parker, C. Schmechel: Some testing tools for TCP implementors, RFC 2398, IETF Network Working Group, August 1998.

V. Paxson: Automated packet trace analysis of TCP implementations, ACM SIGCOMM'97, Cannes, France, September 1997.

V. Paxson (editor): Known TCP implementation problems, <draft-ietf-tcpimpl-prob-05.txt>, IETF Network Working Group, November 1998.

J. Postel (editor): Transmission Control Protocol, RFC 793, September 1981.

W. R. Stevens: TCP/IP Illustrated, Volume 1, The Protocols, Addison-Wesley, 1994.

W. R. Stevens: TCP slow start, congestion avoidance, fast retransmit, and fast recovery algorithms, RFC 2001, IETF Network Working Group, January 1997.

G. R. Wright, W. R. Stevens: TCP/IP Illustrated, Volume 2, The Implementation, Addison-Wesley, 1995.

V
TEST GENERATION METHODS

11

TEST TEMPLATES FOR TEST GENERATION

Marco Hollenberg

Philips Research Laboratories Eindhoven

hollenbe@natlab.research.philips.com

Abstract   We describe the use of *test templates* in test generation from deterministic Finite State Machines. A test template is an expression in a formal language that describes a test or group of tests. Such templates can be used to guide the test generation process.

Keywords: Conformance testing, TTCN, Finite State Machines.

1.   INTRODUCTION

Within Philips, we have developed PHACT, the Philips Automated Conformance Tester ([3]). From deterministic Finite State Machines (FSMs) TTCN ([4]) test suites are generated. This is powered by the Conformance Kit ([1]), which provides a number of fixed strategies for test generation, principal amongst these a UIO-method ([5]), referred to as the partitioned tour method. PHACT also has test execution environments for different platforms (pSOS, Windows) such that these suites can be executed. For this we of course need to write a translation function, linking the abstract events of the TTCN test suite to concrete events that the System Under Test (SUT) understands.

This paper is the result of research targeting the test generation module of PHACT (i.e, the Conformance Kit). We were dissatisfied with the fact that only a fixed number of strategies were given to us. The ability to guide the generation process was considered desirable. The need for this occurs both when one deals with large and with relatively small specifications.

If a specification is large the strategies given are too expensive. Calculating UIO-sequences for thousands of states is not always feasible. Even if it was, executing the resulting test suite could be too expensive. In large specifications, we would like to steer the test generation process so that it at least covers certain areas of the specification that we find interesting.

It may also occur that your specification is rather small. We have encountered the need for small specifications in a number of our projects. This does not necessarily mean that the SUTs have a small statespace. Typically, we do not model the behaviour of the entire system in one specification. Instead, we produce a number of specifications, each dealing with a different aspects of the system. This is because our SUTs are not always purely protocol oriented, but their behaviour may also be governed by a quickly changing environment, such as an audio/video bitstream, which may not be under adequate control of the tester. In such cases it may be necessary to have different specifications, taking into account different environments. The result is that we get a number of possibly quite small specifications.

If we use the strategies supplied by the Conformance Kit for test generation, this may not yield a very large test suite. In PHACT, once a test suite generated from a particular FSM can be executed (the translation function between abstract and concrete events has been written, the test environment has been set up), everything is in place to execute other test suites that are generated from the same FSM. Thus, if *more* tests could be generated, above and beyond the ones generated using the supplied strategies, these could be executed almost for free. A greater number of tests provides the possibility of more exhaustive testing which will yield a better coverage and hence a greater confidence in correctness of the implementation. Of course, we do not just want test suites with randomly generated tests: we want some control, we want to guide the generation of these extra test suites.

To provide more guidance to the test generation process in PHACT, we came up with the notion of *test templates*. A test template is an expression in a formal language (akin to that of regular expressions) that allows you to state the structure of a test, what building blocks should go into the test and how they may be combined. The idea is to generate from such a test template a test suite made up out of tests that satisfy the template. In this way, we can specify a greater number of strategies than we had available before, and guide the generation process towards test suites that we find desirable.

2.   TRACES AND TESTS

In this section we provide the background for the definition of test templates. We begin by defining what kind of specifications we consider and what constitutes a test whether an SUT conforms to the specification.

Definition 2..1

A *Finite State Machine* (FSM) is a tuple $\mathcal{F} = (S, L_I, L_O, \rightarrow, s_0)$, where:

- S is a nonempty finite set of *states*;

- L_I is a finite set of *input actions*;

- L_O is a finite set of *output actions*;

- $\to\, \subseteq S \times L_I \times L_O \times S$ is the set of *transitions*.

- $s_0 \in S$ is the *initial state*.

$\square$

There is no requirement that L_I is disjoint from L_O, as sometimes happens in the definition of FSMs, although requiring such a disjointness would not change anything in this paper.

Let us fix some notation that is convenient when discussing FSMs:

- $s \xrightarrow{a?/b!} t$ denotes: $(s, a, b, t) \in\, \to$.

- $s \xrightarrow{a_1?/b_1!...a_n?/b_n} t$ denotes: there are $s_1,\ldots,s_n \in S$ such that $s \xrightarrow{a_1?/b_1!} s_1 \ldots \xrightarrow{a_n?/b_n!} s_n = t$.

- If in the notation $s \xrightarrow{a_1?/b_1!...a_n?/b_n!} t$ we omit the outputs $b_1,\ldots,b_n$ or the state t we mean that these can be found such that the statement becomes true. For instance: $s \xrightarrow{a_1?...a_n?}$ means that there are outputs $b_1,\ldots,b_n \in L_O$ and there is a state $t \in S$ such that: $s \xrightarrow{a_1?/b_1!...a_n?/b_n!} t$.

Next, we list some properties of FSMs. The FSMs that we will consider in this paper all have these properties.

- An FSM is *deterministic* if for every $s \in S$ and every $a \in L_I$ there is at most one pair (b, t) such that $s \xrightarrow{a?/b!} t$.

- An FSM is *synchronizing* if there is a sequence $a_1,\ldots,a_n \in L_I$ such that $s \xrightarrow{a_1?...a_n?} s_0$ for every state $s \in S$. This means that we can use that sequence to take us from any state to the initial state. The sequence $a_1 \ldots a_n$ is the called a synchronizing sequence.

- An FSM is *connected* if for every $s \in S$ there is some sequence $a_1 \ldots a_n \in L_I$ such that $s_0 \xrightarrow{a_1?...a_n?} s$. I.e., every state can be reached from the initial state.

Throughout this paper we only consider deterministic synchronizing connected FSMs.

Definition 2..2

A *trace* through an FSM $\mathcal{F}$ is a pair $(s, a_1 \ldots a_n)$ with each $a_i \in L_I$ such that $s \xrightarrow{a_1?...a_n?}$.

Because we assume FSMs are deterministic we can define the *output sequence* of a trace $(s, a_1 \ldots a_n)$ as the unique sequence $b_1, \ldots, b_n \in L_O$ such that $s \xrightarrow{a_1?/b_1!\ldots a_n?/b_n!}$ (such a sequence must exist, if $(s, a_1 \ldots a_n)$ is in fact a trace as defined above).

□

A *test* of a deterministic FSM is defined to be a trace that begins in the initial state of the FSM. An implementation passes a test $(s_0, a_1 \ldots a_n)$ with output sequence $b_1, \ldots, b_n$ if after bringing the system into its initial state (by means of a synchronizing sequence) we successively supply the stimuli $a_1, \ldots, a_n$, it responds with, respectively, $b_1, \ldots, b_n$. Otherwise it fails the test. From a test specified as a trace it is possible to create a TTCN test case. Section 5. contains an example of such a test case (table 1).

3.   TEST TEMPLATES

Now that everything is in place, it is time to define the notion of test templates. As tradition dictates we first define the syntax, then the semantics.

3.1   SYNTAX

Given a set of S of FSM-states the set of *guards over S* is defined as follows:

$$\phi ::= s \mid \top \mid \phi \vee \phi \mid \neg\phi$$

where $s \in S$. In other words: the set of guards over S is simply the set of boolean expressions over S. A guard will be interpreted (formally defined in the next section) as a set of FSM-states.

From the same set S of states and a set L_I of input-actions we can define the set of *test templates over S and L_I* as follows:

$$\pi ::= a \mid . \mid [\phi] \mid \text{TS} \mid \text{SIOS} \mid \pi + \pi \mid \pi; \pi \mid \pi^*$$

where $a \in L_I$ and ϕ is a guard over S.

Test templates are in essence regular expressions. They will be interpreted as sets of traces. An example of a test template is:

$$\text{TS}; x; ([s \vee t]; y)^*; [s \vee t] \tag{1}$$

A trace satisfies this template if it starts with a *transfer sequence* to an arbitrary state (a sequence of inputs that brings us from the intial state to this arbitrary state, calculated by some efficient function), then does an x-step so that we end up in either state s or state t, and then does an arbitrary number of y-steps so that the states s or t are not left. This could be used to guide the test generation process towards tests that perform y-actions in states s or t, preceded by an x-action. All of this will be formally defined in the next section.

If an FSM $\mathcal{F}$ is clear from the context, a test template is understood to be a test template over its set of states and its set of input-actions and similarly for guards.

3.2   SEMANTICS

Test templates are to be interpreted as sets of traces. Of ultimate interest are only the *tests* among these: the ones that start in the initial state. We give a formal definition in this section. Assume fixed an FSM $\mathcal{F} = (S, L_I, L_O, \rightarrow, s_0)$.

Guards are interpreted as subsets of S. If ϕ is a guard, its interpretation $[\![\phi]\!]$ is defined as follows:

$$
\begin{aligned}
[\![s]\!] &= \{s\} && \text{if } s \in S \\
[\![\top]\!] &= S \\
[\![\phi \vee \psi]\!] &= [\![\phi]\!] \cup [\![\psi]\!] \\
[\![\neg\phi]\!] &= S \setminus [\![\phi]\!]
\end{aligned}
$$

Test templates are interpreted as sets of traces. They are build up from guards, input-actions and two special building blocks: transfer sequences and Simple Input-Output Sequences (SIOSs). These need to be defined before we can proceed with the semantics of test templates.

A *transfer sequence* to a state $s \in S$ is a trace $(s_0, a_1 \ldots a_n)$ such that $s_0 \xrightarrow{a_1?\ldots a_n?} s$. We assume present a function `makets` that, for every $s \in S$ produces a transfer sequence to s. Such a function can exist because we assume FSMs to be connected.

A SIOS for a state s is a trace $(s, a_1 \ldots a_n)$ such that for every $t \in S$: $(t, a_1 \ldots a_n)$ is a trace and its output sequence differs from that of $(s, a_1 \ldots a_n)$. A SIOS need not exist for every state. SIOSs can be used to gain confidence in the idea that at some point of a test the SUT has reached a particular state and not some other. We assume present a function `makesios` that for any $s \in S$ produces a SIOS for s if it exists, and yields the trace (s, ε) otherwise (where ε is the empty sequence).

Now we can define the semantics of test templates:

- $[\![a]\!] = \{(s, a) \mid s \in S \ \& \ s \xrightarrow{a} \}$.

- $[\![.]\!] = \{(s, a) \mid s \in S \ \& \ a \in L_I \ \& \ s \xrightarrow{a} \}$.

- $[\![[\phi]]\!] = \{(s, \varepsilon) \mid s \in [\![\phi]\!]\}$.

- $[\![\text{TS}]\!] = \{\texttt{makets}(s) \mid s \in S\}$.

- $[\![\text{SIOS}]\!] = \{\texttt{makesios}(s) \mid s \in S\}$.

- $[\![\pi_1 + \pi_2]\!] = [\![\pi_1]\!] \cup [\![\pi_2]\!].$

- $[\![\pi_1; \pi_2]\!] = [\![\pi_1]\!] \circ [\![\pi_2]\!]$ where $\circ$ is defined on sets of traces as follows:
  $T_1 \circ T_2 =$

$$
\left\{ (s, a_1 \ldots a_n a_{n+1} \ldots a_{n+m}) \;\middle|\; \exists t \in S. \left(\begin{array}{l} (s, a_1, \ldots, a_n) \in T_1 \;\&\; \\ s \xrightarrow{a_1?\ldots a_n?} t \;\&\; \\ (t, a_{n+1} \ldots a_{n+m}) \in T_2 \end{array} \right) \right\}
$$

- Using the above notion of composition we can also define trace *iteration*. If T is a set of traces we define:

$$
\begin{array}{lcl}
T^0 & = & \{(s, \varepsilon) \mid s \in S\} \\
T^{n+1} & = & T^n \circ T \\
T^* & = & \bigcup_{n \geq 0} T^n
\end{array}
$$

Now we can define: $[\![\pi^*]\!] = [\![\pi]\!]^*.$

Consider again the test template 1. The interpretation of this will consist of all traces $(s_0, a_1 \ldots a_n x y_1 \ldots y_m)$ such that for some state u, $\mathtt{makets}(u) = (s_0, a_1 \ldots a_n)$ and for some states $v_0, \ldots, v_m \in \{s, t\}$, $s_0 \xrightarrow{a_1?\ldots a_n?} u \xrightarrow{x?} v_0 \xrightarrow{y_1?} v_1 \ldots \xrightarrow{y_m?} v_m$ and $y_1 = \ldots = y_m = y$. It depends completely on the FSM if such traces exist or not.

4.   THE RESULTING TEST SUITE

If we have an FSM and a test template π, what tests do we want in our test suite? We can not always include all tests that satisfy π. For a start, if π uses iteration, we would get an infinite suite. But even if this is not the case, the suite might get too large. A simple solution is to let the user specify a range of lengths of tests $[n, m]$, where $0 \leq n \leq m$. The length of a trace $(s, a_1 \ldots a_n)$ is the number of inputs in the trace: n in this case.

The test suite can be pruned some by removing redundancy. A test that tests what happens if we supply an input a_1 and then an input a_2 (from the initial state) subsumes a test that only tests a_1. Thus *prefixes* of tests can be removed from the suite.

If it is not always feasible for a test generation tool to generate a test suite containing all tests satisfying the template, even if we remove all redundant tests, we need to define what test suites a tool is allowed to generate. We give such a definition here:

Definition 4..1
Assume given a synchronizing, connected, deterministic FSM $\mathcal{F}$. Let π be a test template, and let $[n, m]$ be the specified range ($0 \leq n \leq m$). Then a test suite T *satisfying* π and $[n, m]$ is a set such that:

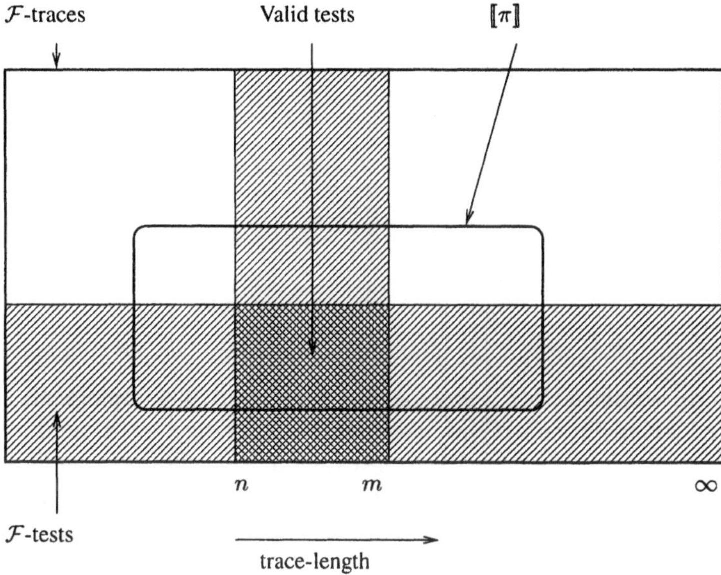

Figure 1   Tests generated from FSM, test template and specified range

- T satisfies the test template π: $T \subseteq [\![\pi]\!]$.

- T contains only tests: if $(s, a_1 \ldots a_k) \in T$ then $s = s_0$.

- T satisfies the length-requirement: if $(s_0, a_1 \ldots a_k) \in T$ then $n \leq k \leq m$.

- T contains no redundancies: if $(s_0, a_1 \ldots a_k) \in T$ then for no $l < k$: $(s_0, a_1 \ldots a_l) \in T$. A test subsumes all of its proper prefixes.

$\square$

There can be many test suites that satisfy the same template and range. Some of these will be less informative (i.e. test less) than others. This notion is easy to define. Read $\sqsubseteq$ as "is less informative than, or just as informative". First $(s_0, a_1 \ldots a_n) \sqsubseteq (s_0, c_1 \ldots c_m)$ iff $a_1 \ldots a_n$ is a prefix of $c_1 \ldots c_m$ (or they are equal). Then for two test suites T_1 and T_2: $T_1 \sqsubseteq T_2$ iff for each $t \in T_1$ there is a $u \in T_2$ such that $t \sqsubseteq u$. Now it is easy to see that for each test template and each range there is a finite test suite that satisfies the above requirements and is maximal with respect to $\sqsubseteq$ among all those that satisfy the requirements. Such a test suite we call *maximal* with respect to the template and the range.

Figure 1 depicts where the tests that go into satisfying a template and a range are located.

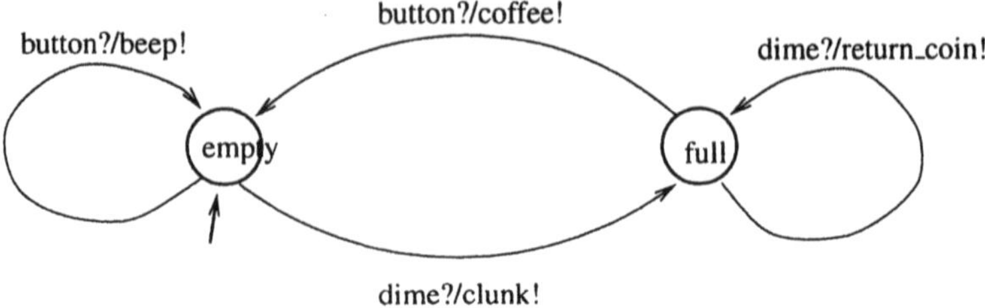

Figure 2 A coffee machine FSM

We have written a Unix prototype tool, `tp2ttcn`, that implements the ideas presented so far. From an FSM, a test template, a range, a specified maximum number of tests it produces a TTCN test suite that satisfies the template and the range. We will not go into the algorithm of the tool here. Suffice it to say the test template is used to build a tree whose nodes represent tests and where some nodes are labelled to indicate that they represent tests that satisfy the template.

5.   AN EXAMPLE

In this section we work out a few examples of test templates, to get a better feel. Our running example is the FSM in figure 2. It is easily verified to be deterministic, synchronizing and connected. The synchronizing sequence is `button`. The FSM represents a simple coffee machine. Coffee costs a dime. If a dime has been inserted a button can be pressed to obtain coffee. Pressing the button too soon produces a warning and inserting too much money causes the excess to be returned to you.

A decent test template would be the *partitioned tour*:

$$\text{TS}; .; \text{SIOS}$$

This should produce a test suite such that each test case starts with a transfer sequence to some state, proceeds with some arbitrary step and finally does a SIOS to check that it really is in the state it expects to be in.

Suppose that the functions `makets` and `makesios` are defined as follows:

$$\texttt{makets}(s) = \begin{cases} (\texttt{empty}, \varepsilon) & \text{if } s = \texttt{empty} \\ (\texttt{empty}, \texttt{dime}) & \text{if } s = \texttt{full} \end{cases}$$

$$\texttt{makesios}(s) = (s, \texttt{dime})$$

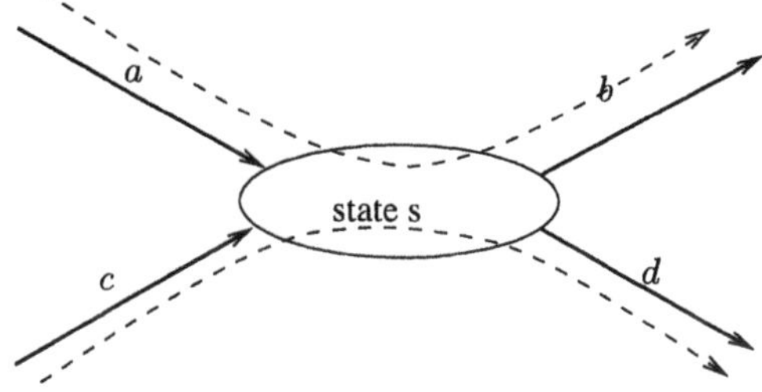

Figure 3   A traversal that does not test all pairs of transitions

Then $[\![\text{TS}; .; \text{SIOS}]\!] =$

$$\left\{ \begin{array}{l} (\texttt{empty}, \texttt{dime} \cdot \texttt{dime}), \\ (\texttt{empty}, \texttt{button} \cdot \texttt{dime}), \\ (\texttt{empty}, \texttt{dime} \cdot \texttt{dime} \cdot \texttt{dime}), \\ (\texttt{empty}, \texttt{dime} \cdot \texttt{button} \cdot \texttt{dime}) \end{array} \right\}$$

Because the template started with TS it is no coincidence that this set contains only tests: every trace that satisfies TS starts in the initial state. The maximal test suite for this template and a range of, say $[0, 3]$ consists of $[\![\text{TS}; .; \text{SIOS}]\!]$ without $(\texttt{empty}, \texttt{dime} \cdot \texttt{dime})$, which is redundant because of the presence of $(\texttt{empty}, \texttt{dime} \cdot \texttt{dime} \cdot \texttt{dime})$. The tool creates TTCN from such traces. A module of the Conformance Kit is used for this. Table 1 contains a few of the tables of the resulting TTCN, as generated by the Conformance Kit.

The partitioned tour happens to be one of the strategies that the Conformance Kit supplies. For us, it is just a template among others though. The partitioned tour produces tests that visit all transitions at least once. What if we are worried about *combinations* of transitions? Consider the partial state machine presented in figure 3. The dotted path show a traversal of the state machine by a test suite. All transitions are covered by the transitions. But note that we do not find errors in this way that occur when an a-transition is followed by a d-transition, or a c-transition is followed by a b-transition.

In the test template language we can specify an alternative to the partitioned tour that tests all transitions and all pairs of transitions:

$$\text{TS}; (. + (.;.)); \text{SIOS}$$

The maximal test suite for this template (taking a sufficiently broad range) contains six tests, even though the interpretation of the template (without removing redundancy) contains ten. If larger examples are considered the amount of redundancy could be a real bottleneck to test generation from templates, if we did not remove it at its source. The tool `tp2ttcn` does precisely that.

Table 1   TTCN tables

Test Case Dynamic Behaviour			
Test Case Name: test_1			
Test Group: coffeeMachine/coffee/			
Purpose: Elementary sequence "test_1" (1) from state "empty"			
Defaults Reference: general_default			
Nr	**Behaviour Description**	**CRef**	**Verdict**
1	+ ss		
2	pco ! button		
3	pco ? beep		
4	pco ! dime		
5	pco ? clunk		PASS
Detailed Comments:			
line 1: *force IUT in start state "empty"*			

Test Step Dynamic Behaviour			
Test Step Name: ss			
Test Group: coffeeMachine/steps/coffee/			
Objective: Synchronize to start state "empty"			
Defaults Reference: general_default			
Nr	**Behaviour Description**	**CRef**	**Verdict**
1	pco ! button		
2	**START** no_output_timer		
3	+ gobble_one		
4	CANCEL no_output_timer		
	gobble_one		
5	pco **? OTHERWISE**		
6	**? TIMEOUT** no_output_timer		

Default Dynamic Behaviour			
Default Name: general_default			
Test Group: coffeeMachine/defaults/coffee/			
Objective: On any other event: fail			
Nr	**Behaviour Description**	**CRef**	**Verdict**
1	pco **? OTHERWISE**		FAIL

Our final example is of a template with an infinity of tests that satisfy it. It is here where the need for specifying a range is clear. Suppose we wanted to create a large test suite containing tests that check whether the coffee machine does not give out free coffee, i.e. bring it to the empty state and press the button. A suitable test template for this purpose would be

```
.*;[empty];button
```

We could now say that we want all tests that satisfy this template but have a length ≤ 10: this would produce a maximal test suite of 2^9 test cases (if we remove redundancy). Note that such a test suite is no longer purely functional: it borders on a stress test.

6.   CONCLUSIONS AND FURTHER RESEARCH

This paper has described test templates. They are useful for guiding the test generation process in relatively small specifications, where the requirement that every transition is covered by a test is easily met, but one still wants more. For large specifications they are also useful: covering every transition may not be feasible so one wants to guide the test generation process towards a test suite that covers all the *interesting* parts (whatever those may be).

We conclude with a list of possible extensions. First, our tool uses FSMs as its specification language. One could consider using slightly more advanced languages. For instance: EFSMs (Extended FSMs), FSMs extended with variables. One would then need to extend the test template language as well. An example of an extended test template would be:

$$\text{TS}; [\texttt{value} = 0 \lor \texttt{value} = 50 \lor \texttt{value} = 100]; .; \text{SIOS}$$

which denotes a partitioned tour that only considers steps from states where the `value`-variable has a boundary value (0 or 100) or a chosen medium value (50).

In our test template language we consider two functions, one for calculating transfer sequences, one for calculating SIOSs. This was done because these are available in the Conformance Kit. One could of course consider other functions, perhaps to implement other UIO-strategies, and add the corresponding building blocks to the language.

A final extension would be to consider *nondeterministic* FSMs. This would require a major overhaul of the above work. In nondeterministic FSMs, tests can no longer be equated with lists of input sequences. It makes much more sense to equate tests with *trees* labeled with both inputs and outputs. When tests are traces, it makes sense to choose a regular expression-like language for test templates. When tests are trees, this is no longer a good choice. A modal or temporal language, such as CTL ([2]), is more suitable. A drawback for us would be that the Conformance Kit could not be reused: this tool does not support nondeterministic FSMs.

References

[1] S.P. van de Burgt, J. Kroon, E. Kwast, H.J. Wilts. The RNL Conformance Kit. In: J. de Meer, L. Mackert and W. Effelsberg, eds., *Proc. of the 2nd*

International Workshop on Protocol Test Systems, pp. 279-294. North-Holland, 1989.

[2] E.M. Clarke, E.A. Emerson, A.P. Sistla. Automatic verification of finite-state concurrent systems using temporal logic specification. In: *ACM Transactions in Programming Languages and Systems*, Vol. 8, No. 2, 1986, pp. 244-263.

[3] L.M.G. Feijs, F.A.C. Meijs, J.R. Moonen, J.J. van Wamel. Conformance Testing of a Multimedia System Using PHACT. In: A. Petrenko, N. Yevtushenko, *Testing of Communicating Systems*, Proceedings of IWTCS'98, pp. 193–210, Kluwer, 1998.

[4] ISO. *Information Technology – Open Systems Interconnection, Conformance Testing Methodology and Framework – Part 3: Tree and Tabular Combined Notation.* International Standard IS)/IEC 9646-3, 1991.

[5] H. Ural. Formal methods for test sequence generation. In: *Computer Communications*, Vol 15, No. 5, 1992, pp. 311-325.

12

FORMAL TEST AUTOMATION:
A SIMPLE EXPERIMENT*

Axel Belinfante[†], Jan Feenstra, René G. de Vries, Jan Tretmans
University of Twente

Nicolae Goga, Loe Feijs, Sjouke Mauw
Eindhoven University of Technology

Lex Heerink
Philips Research Laboratories

Abstract   In this paper we study the automation of test derivation and execution in the area of conformance testing. The test scenarios are derived from multiple specification languages: LOTOS, PROMELA and SDL. A central theme of this study is the usability of batch-oriented and on-the-fly testing approaches. To facilitate the derivation from multiple formal description techniques and the different test execution approaches, an open, generic environment called TORX is introduced. TORX enables plugging in existing or dedicated tools. We have carried out several experiments in testing a conference protocol, resulting in requirements on automated testing and benchmarking criteria.

1   INTRODUCTION

Conformance testing is an important activity in the development of reactive systems. Its aim is to gain confidence in the correctness of the system, by means of experimenting with the system implementation. To judge whether an implementation is a valid

*This research is supported by the Dutch Technology Foundation STW under project STW TIF.4111: *Côte de Resyste* – COnformance TEsting of REactive SYSTEms. URL: http://fmt.cs.utwente.nl/CdR
[†]Corresponding author: Axel Belinfante, University of Twente, Formal Methods and Tools research group, Faculty of Computer Science, P.O. Box 217, NL-7500 AE Enschede, The Netherlands, email: Axel.Belinfante@cs.utwente.nl

realization of the specification, and to compare tests and test results, we need a precise notion of correctness. Using formal methods we can achive this. Another benefit from the use of formal techniques is that it allows to automate the conformance testing process. This is important since test derivation is an error-prone and time-intensive process. In [8] a framework for conformance testing based on formal methods (FMCT) is described. Previously, we showed in [11] the testing of a relatively simple example, the conference protocol, in order to assess the feasibility of the FMCT model. One of the major conclusions was that FMCT provides a sound basis for conformance testing based on formal methods, but automation is necessary.

In this paper, we study this automation of conformance testing, by comparing different approaches of test execution, different formalisms as the basis for test derivation, and different supporting tool sets. The aim is to get insight in the strengths and weaknesses of the different approaches, to identify shortcomings, and to identify comparison criteria, required computational effort and means to accommodate automation. These results are an inspiration for an elaborated study of benchmarking existing automated testing methods and tools. The research reported in this paper is part of the *Côte-de-Resyste* project (a Dutch joint venture of the Universities of Eindhoven and Twente and the industrial partners Philips Research and KPN Research). The project aims, among other goals, at comparing existing automated test methods and developing open testing tools together with the underlying formal theory.

The results of this paper are obtained from the same case study as in [11], i.e. testing the *conference protocol*. The difference is that now we do it automatically. Tests are derived from multiple *formal description techniques* (FDTs): LOTOS, PROMELA and SDL. The "simplicity" of our "experiment" merely refers to this case study, not to the testing theory or tools that we use. The conference protocol is a rather simple, almost toy-like protocol.

In this paper, the conference protocol is automatically tested in two ways: on-the-fly and batch-wise. Our on-the-fly testing experiments are based on LOTOS and PROMELA. The batch testing experiment is based on the TAU tool [1] AUTOLINK which is based on the work in [10], i.e. the derivation of TTCN test cases from SDL in a semi-automated way. Test derivation and test execution are facilitated by *Côte de Resyste* tools available in the TORX environment. This is a tool architecture giving means to link different kinds of test tools within a test derivation/execution site, without reengineering of the whole test site, thus facilitating an open generic environment. For instance, for this study we can plug in several modules for the support of the FDTs. Also TTCN is supported, which can be plugged into the execution part of TORX.

This paper is structured as follows. Section 2 gives an overview of test derivation and test execution methods. The TORX environment will be introduced. Section 3 will explain the conference protocol case study: informally, the formal specifications in LOTOS, PROMELA and SDL, and the implementations. Section 4 deals with the test architecture for testing conference protocol entity implementations. Section 5 reports about the test activities we have carried out. We give an evaluation of the results and directions for future work in the final section.

2   AUTOMATED TESTING

In system development we build an implementation i based on a specification s. Formally, i is said to be a correct implementation of s if i **imp** s, where **imp** is an implementation relation, i.e. the notion of correctness. An implementation which is assessed on its correctness by testing is called an *Implementation under Test (IUT)*. During test execution a set of test experiments, called a *test suite*, is carried out on this IUT, resulting in a verdict of either *pass* or *fail*. A test suite is exhaustive if we can conclude from the verdict *pass* that the implementation is correct; it is sound if the verdict *fail* never occurs with a correct implementation. Exhaustive testing, i.e. showing the absence of errors, usually requires an infinite test suite and is therefore not feasible in practice. A minimal requirement on a test suite is that it is sound [8]. The *test derivation* process aims at deriving a test suite from the specification, given **imp**.

It follows from the above discussion that there are two main phases in the testing process: *test derivation*, i.e. obtaining a test suite, and *test execution*, i.e. applying the test suite to the IUT. Both phases can be automated. This can be done in two ways: as two separate phases, or in an integrated manner.

In the first approach, in the first phase a test suite is derived and stored in some representation, usually TTCN. In the second phase this test suite is executed along with the IUT. This principle is called *batch testing*. Batch test derivation is computationally expensive and suffers from the state space explosion problem. This complexity can be reduced by user guidance and on-the-fly derivation techniques [4].

The second approach is called *on-the-fly testing*. As opposed to batch testing, test derivation and test execution occur simultaneously. Instead of deriving a complete *test case* (one test scenario in a test suite), the test derivation process derives *test primitives* from the specification. Test primitives are actions that are immediately executed in the test run. While executing a test case, only the necessary part of a test case is considered: the test case is derived *lazily* (cf. lazy evaluation of functional languages). Using observations during the test execution we can reduce the effort in deriving test information from the specification compared with batch derivation; see also [13].

Now that we have introduced the two test methodologies, batch testing and on-the-fly testing, we make a few remarks about their respective qualities. Firstly, the batch-wise approach is better suited for manual test case preparation and for semi-automatic test case preparation. Humans are good at test selection, but they are not fast enough to do it at run time, except perhaps for very slow protocols. This was also one of the traditional ideas behind TTCN. But now we are moving towards a further automation of the process, and therefore this advantage of batch-wise approach counts as less significant.

The second remark concerns the system dependent PIXIT software, sometimes called mapping software, glue software, encoding/decoding, or interfacing. For batch-wise test derivation it is possible to compile the abstract test cases into concrete test cases which have all the mapping details encoded. For the on-the-fly approaches the encodings and decodings have to be done by run-time facilities. In this paper we will show that this is feasible.

The third remark is that in case of on-the-fly testing all computations have to be done at run-time, whereas batch-wise testing allows some of the work to be moved to compile-time. So, the batch-wise approach has an advantage which makes it easier to satisfy the IUT's real-time requirements. But the price to pay for this is that many test-steps which do not happen at run-time are pre-computed, just because the system happens to choose another branch. This leads to test-suites of an enormous size, and the amount of pre-computation work and the storage demands involved may well undo the advantage.

Test Tool Architecture   A test tool architecture was devised which allows on-the-fly testing, batch test derivation and batch test execution for different specification formalisms. This architecture was baptized TORX. The main characteristics of TORX are its flexibility and openness. Flexibility is obtained by requiring a modular architecture with well-defined interfaces between the components – this allows easy replacement of a component by an improved version, or by a component implementing another specification language or implementation relation. Openness is acquired by choosing, when possible, existing (industry standard) interfaces to link the components – this enables integration of 'third party' components that implement these interfaces, in our tool environment. Later we will show how this general architecture was instantiated for on-the-fly testing with LOTOS and PROMELA and for batch testing with SDL. We now discuss the TORX architecture (Figure 1) in terms of its components, the interfaces between them, and the currently available component implementations.

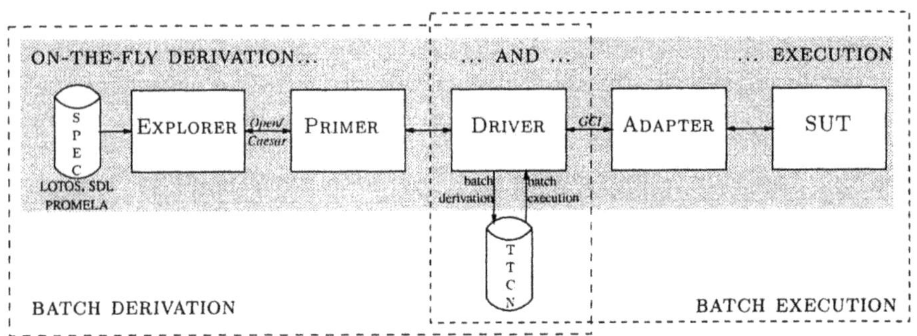

Figure 1   TORX tool architecture

TORX consists of the following components (modules): EXPLORER, PRIMER, DRIVER, ADAPTER, and TTCN storage. Figure 1 depicts how these components are linked for batch derivation (with EXPLORER, PRIMER DRIVER and TTCN storage) batch execution (with TTCN storage, DRIVER, ADAPTER), and on-the-fly derivation and execution (involving all components without storage of TTCN). The SUT is the *System under Test*. It is the IUT together with its test context, i.e. its surrounding environment (see Section 4).

EXPLORER.   The EXPLORER is a specification language-specific component that offers functions to explore the transition-graph of a specification and to provide, for a given state, the set of transitions that are enabled in this state. For the interface between EXPLORER and PRIMER we use the Open/Caesar interface [6] which is a C API that provides exactly such state and transition functions for labeled transition systems. This implies that we can use existing tools to implement the EXPLORER. Currently, we use CAESAR as our LOTOS EXPLORER. An SDL explorer also exist (but, to our knowledge, is not publically available).

PRIMER. The PRIMER uses the functions provided by the EXPLORER to implement the test derivation algorithm, for which it keeps track of the set of states that the specification might be in. It offers functions to generate inputs (stimuli) for the implementation and to check outputs (observations) from the implementation. Our current implementation of the PRIMER implements the test derivation algorithm for the implementation relation **ioco** [12]. Since this algorithm does not contain test data selection criteria, test selection is currently implemented by making random choices using a random number generator. The seed of this generator is a parameter of the PRIMER.

DRIVER. The DRIVER is the central component of the tool architecture; it controls the progress of the testing process. It decides whether to do an input action or to observe and check an output action from the implementation. The DRIVER uses the PRIMER to obtain an input and to check whether the output of the implementation is correct. It uses the ADAPTER to execute the selected inputs by sending these inputs to the IUT, and to observe outputs that are generated by the IUT. For batch testing the derived tests are first stored in a filesystem (indicated as "TTCN" in Figure 1). To execute tests in batch mode the test events are obtained from storage rather than from the PRIMER.

The PRIMER-DRIVER interface is only used by the on-the-fly tester, where PRIMER and DRIVER are connected by pipes that the DRIVER uses to write (textual) commands to and receive (textual) responses from the PRIMER.

ADAPTER. The ADAPTER provides the connection with the SUT. It is responsible for sending inputs to and receiving outputs from the SUT on request of the DRIVER. The ADAPTER is also responsible for encoding and decoding of abstract actions to concrete bits and bytes, and vice versa. This also involves mapping of the *quiescent action δ* onto time-outs, see [12].

We currently use two interfaces between DRIVER and ADAPTER. In our current on-the-fly tester we use a simple (ad hoc) interface based on calling conventions for the (TCL) functions that implement the en/decoding functions. We are in the process of replacing this interface by the Generic Compiler/Interpreter Interface (GCI) [2] which we already use for batch execution of TTCN test-suites derived by TAU. The GCI has been developed in the INTOOL project for TTCN-based testers as an interface between the TTCN-dependent part (code generated by a TTCN compiler, or a TTCN interpreter) and the other parts (manager, responsible for configuration and logging, and adapter, responsible for encoding and decoding and access to the SUT). The GCI is defined in a language-independent way; a C language binding has been defined, which we use.

Test Approaches in TORX   As Figure 1 shows, the components can be put together for the three different ways of testing: on-the-fly testing, batch test derivation and batch test execution. We now describe how these configurations were used for testing based on LOTOS, PROMELA and SDL specifications. We will pay special attention to the differences and similarities in the use of the basic building blocks.

On-the-fly testing.   Testing for LOTOS and PROMELA is performed on-the-fly based on the implementation relation **ioco** [12]. They use the same DRIVER and major parts of the ADAPTER modules; they differ in the EXPLORER and PRIMER modules.

The DRIVER sends commands to the PRIMER to request a menu of possible input or output events, or the 'execution' of a specific transition from such a menu. Even though it can be configured to fully automatically select a trace to test, the random choices that it makes are 'parameterized' by the seed of its random number, which can be chosen by the user. The test trace that is derived and executed can be logged and replayed during a subsequent test run to guide the DRIVER. The DRIVER is implemented using the scripting language EXPECT because the high-level constructs offered by EXPECT allow rapid prototype development and easy interaction with external programs.

The ADAPTER contains the en/decoding routines and the SUT connection programs, and the function that maps abstract actions onto PCOs (*Points of Control and Observation*, see Section 4). The en/decoding routines and the mapping function are specification language dependent, because they depend on the representation of the abstract values that they have to en/decode, respectively map, which will vary between languages. The SUT connection programs are specification language independent; separate programs are used that handle specific protocols (like TCP, UDP) and can be controlled via standard input and output.

For both LOTOS and PROMELA on-the-fly testing, the mapping of TORX components on tool-implementation programs is the same: the DRIVER forms the main program together with the en/decoding part of the ADAPTER; EXPLORER and PRIMER are integrated in a second program (a.k.a. the *specification module*), and for the SUT connection part of the ADAPTER we use separate protocol-specific *connection programs*.

LOTOS.   The specification-dependent EXPLORER module can be automatically generated from a LOTOS specification using the CADP tool set. This EXPLORER module is linked with the **ioco**-PRIMER using the Open/Caesar interface, which gives us a program that has to be configured with lists of input and output gates, and (optionally) with the seed of its random number generator. The PRIMER module is independent of the specification, and, is in principle even specification language independent (for all specification languages for which there is a compiler that compiles to the Open/Caesar interface). In practice this will only be true if the labels generated by the compiler sufficiently resemble LOTOS events (our current PRIMER requires that). The labels generated by the CADP generated EXPLORER do not contain free variables, because CADP expands free variables by enumerating the values in the domains of the variables after which it generates a label for each possible combination of these values.

PROMELA. For PROMELA on-the-fly testing, we can automatically generate a single module that implements both the EXPLORER and PRIMER. This is done by the the TROJKA tool which is described in detail in [13]. The resulting specification module can be configured in the same way as the LOTOS one, but the configuration parameters are different, for example because input and output channel operations are already identified in a PROMELA specification and need not be given by the user. If the tool SPIN were able to supply an Open/Caesar interface for PROMELA, we would have been able to use the same PRIMER as for LOTOS. Such an extension is left for future work.

Batch SDL testing.   For SDL we use batch testing to derive and execute test suites in TTCN. For this purpose we use the TAU tool set [1]. The functionality of the EXPLORER, PRIMER and DRIVER is covered by TAU's AUTOLINK test derivation tool [10]. It generates the constraints and dynamic parts of TTCN test suites, guided by *message sequence charts* (MSCs) that have to be provided by the user. These MSCs can be derived by hand from the SDL specification using the SDL simulator that is integrated in TAU. The TTCN test suite derived by AUTOLINK has then to be completed with declarations (e.g. for types and PCOs) using a program that is automatically generated from the SDL specification by TAU's LINK TOOL; the result is a complete TTCN test suite.

The batch test execution DRIVER module is automatically generated from the complete TTCN suites by the TTCN compiler of TAU. This DRIVER is linked with an ADAPTER module using the GCI interface; the result is a single program that can execute the test suites. The ADAPTER differs from the one for the on-the-fly testers in the following aspects. Firstly, the ADAPTER uses the GCI interface, and it does not have to provide a function to map labels to PCOs because the TTCN test suite already explicitly refers to PCOs. Secondly, the ADAPTER does not have to use external programs to provide the connection to the SUT, because support for several connection types (protocols) has been built in. Finally, the en/decoding routines are implemented in C instead of TCL, and they translate using an intermediate representation.

An important difference between the on-the-fly testers used for LOTOS and PROMELA on the one hand, and the batch testing provided using TAU on the other hand, lies in the level of automation offered for test derivation: the on-the-fly testers are able to automatically derive and execute tests without human intervention or guidance, whereas the TAU batch test deriver cannot do its work without a (manually derived) MSC that represents the test purpose.

3   THE CONFERENCE PROTOCOL

The example protocol that is used as the case study for the test experiments is the Conference Protocol [5, 11]. Some aspects of it are highlighted here; an elaborate description together with the complete formal specifications and the set of implementations can be found in [9].

Informal description.   The conference service provides a multicast service, resembling a 'chatbox', to users participating in a conference. A conference is a group of users that can exchange messages with all conference partners in that conference.

Messages are exchanged using the service primitives *datareq* and *dataind*. The partners in a conference can change dynamically because the conference service allows its users to *join* and *leave* a conference. Different conferences can exist at the same time, but each user can only participate in at most one conference at a time.

The underlying service, used by the conference protocol, is a point-to-point connectionless and unreliable service provided by the *User Datagram Protocol* (UDP), i.e. data packets may get lost or duplicated or be delivered out of sequence but are never corrupted or misdelivered.

The object of our experiments is testing a *Conference Protocol Entity* (CPE). The CPEs send and receive *Protocol Data Units* (PDUs) via the underlying service provide the conference service. The CPE has four PDUs: *join-PDU, answer-PDU, data-PDU* and *leave-PDU*, which can be sent and received according to a number of rules, of which the details are omitted here. Moreover, every CPE is responsible for the administration of two sets, the *potential conference partners* and the *conference partners*. The first is static and contains all users who are allowed to participate in a conference, and the second is dynamic and contains all conference partners (in the form of names and UDP-addresses) that currently participate in the same conference.

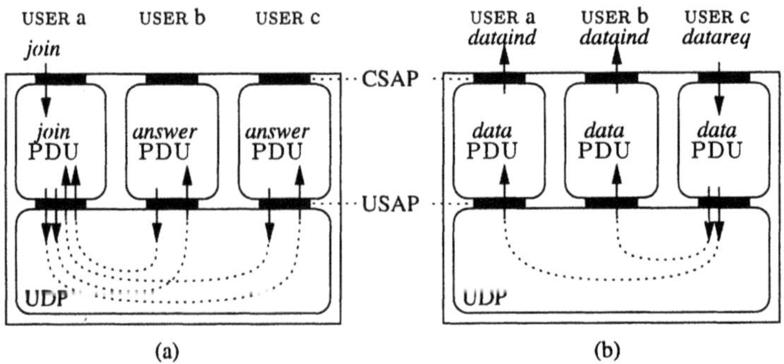

Figure 2   The conference protocol

Figure 2 gives two example instances of behaviour: in (*a*) a *join* service primitive results in sending a *join-PDU*, which is acknowledged by an *answer-PDU*; in (*b*) a *datareq* service primitive leads to a *data-PDU* being sent to all conference partners, which, in turn, invoke a *dataind* primitive.

Formal specifications.   Three formal specifications were developed for the Conference Protocol using LOTOS, PROMELA and SDL.

LOTOS   The LOTOS specification was mainly taken from [11]. The core of the specification is a (state-oriented) description of the conference protocol entity behaviour. The CPE behaviour is parameterized with the set of potential conference partners and its CSAP and USAP addresses, and is constrained by the local behaviour at CSAP and USAP. The instantiation of the CPE with concrete values for these parameters is part of the specification.

PROMELA   Communication between conference partners has been modelled by a set of processes, one for each potential receiver, to 'allow' all possible interleavings between the several sendings of multicast PDUs. Instantiating the specification with three potential conference users, a PROMELA model for testing is generated which consists of 122 states and 5 processes. For model checking and simulation purposes, the user needs not only the behaviour of the system itself but also the behaviour of the system environment. For testing this is not required, see [13]. Only some channels have to be marked as observable, viz. the ones where observable actions may occur.

SDL   For SDL, several specifications of the conference protocol were produced. In the TAU tool set it is possible to use different kinds of test formalisms, e.g., interoperability and conformance testing. For each of these formalisms different specifications are required. The conference protocol model is specified in a natural way by decomposing the problem into three subprocesses: one for reception and translation of incoming messages, one for managing the conference data structures, and one for composing and broadcasting outgoing PDUs.

Conference Protocol Implementations.   The conference protocol has been implemented on SUN SPARC workstations using a UNIX-like (SOLARIS) operating system, and it was programmed using the ANSI-C programming language. Furthermore, we used only standard UNIX inter-process and inter-machine communication facilities, such as uni-directional pipes and sockets.

A conference protocol implementation consists of the actual CPE which implements the protocol behaviour and a user-interface on top of it. We require that the user-interface is separated (loosely coupled) from the CPE to isolate the protocol entity; only the CPE is the object of testing. This is realistic because user interfaces are often implemented using dedicated software.

The conference protocol implementation has two interfaces: the CSAP and the USAP. The CSAP interface allows communication between the two UNIX processes, the user-interface and the CPE, and is implemented by two uni-directional pipes. The USAP interface allows communication between the CPE and the underlaying layer UDP, and is implemented by sockets.

In order to guarantee that a conference protocol entity has knowledge about the potential conference partners the conference protocol entity reads a *configuration file* during the initialization phase.

Error seeding.   For our experiment with automatic testing we developed 28 different conference protocol implementations. One of these implementations is correct (at least, to our knowledge), whereas in 27 of them a single error was injected deliberately. The erroneous implementations can be categorized in three different groups: *No outputs*, *No internal checks* and *No internal updates*. The group *No outputs* contains implementations that forget to send output when they are required to do so. The group *No internal checks* contains implementations that do not check whether the implementations are allowed to participate in the same conference according to the set of potential conference partners and the set of conference partners. The group *No internal updates* contains implementations that do not correctly administrate the set of conference partners.

4   TEST ARCHITECTURE

For testing a conference protocol entity (CPE) implementation, knowledge about the environment in which it is tested, i.e. the *test architecture*, is essential. A test architecture can (abstractly) be described in terms of a tester, an *Implementation Under Test* (IUT) (in our case the CPE), a test context, *Points of Control and Observation* (PCOs), and *Implementation Access Points* (IAPs) [8]. The test context is the environment in which the IUT is embedded and that is present during testing, but that is not the aim of conformance testing. The communication interfaces between the IUT and the test context are defined by IAPs, and the communication interfaces between the test context and the tester are defined by PCOs. The SUT (*System Under Test*) consists of the IUT embedded in its test context. Figure 3(a) depicts an abstract test architecture.

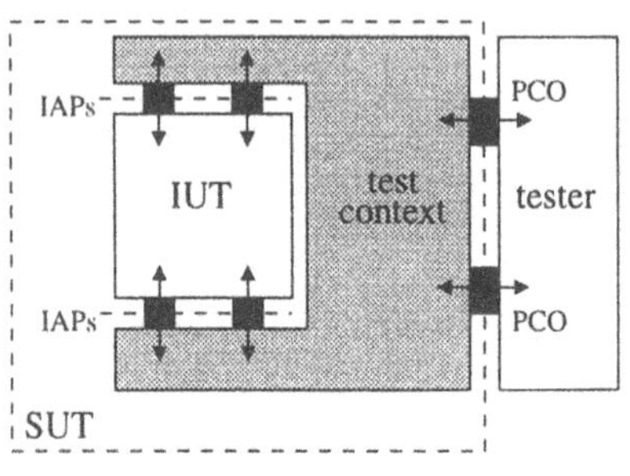

(a) Abstract test architecture

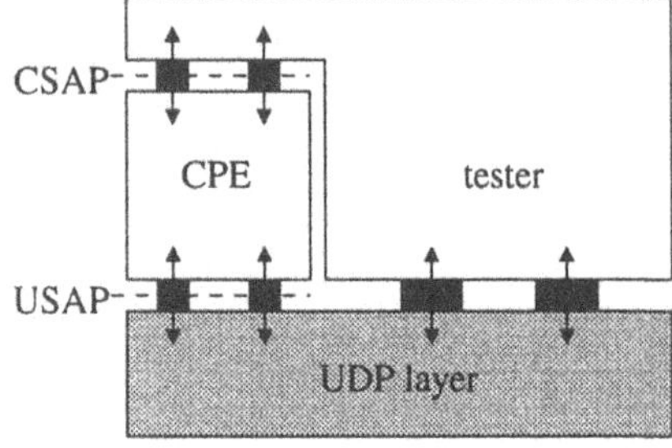

(b) Test architecture for conference protocol entities

Figure 3   Test architecture

Ideally, the tester accesses the CPE directly at its IAPs, both at the CSAP and the USAP level. In our test architecture, which is the same as in [11], this is not the case. The tester communicates with the CPE at the USAP via the underlying UDP layer; this UDP layer acts as the test context. Since UDP behaves as an unreliable channel, this complicates the testing process. To avoid this complication we make the assumption that communication via UDP is reliable and that messages are delivered in sequence. This assumption is realistic if we require that the tester and the CPE reside on the same host machine, so that messages exchanged via UDP do not have to travel through the protocol layers below IP but 'bounce back' at IP.

With respect to the IAP at the CSAP interface we already assumed in the previous section that the user interface can be separated from the core CPE. Since the CSAP interface is implemented by means of pipes the tester therefore has to access the CSAP interface via the pipe mechanism.

Figure 3(b) depicts the concrete test architecture. The SUT consists of the CPE together with the reliable UDP service provider. The tester accesses the IAPs at the CSAP level directly, and the IAPs at USAP level via the UDP layer.

Formal model of the test architecture   For formal test derivation, a realistic model of the behavioural properties of the complete SUT is required, i.e. the CPE and the test context, as well as the communication interfaces (IAPs and PCOs). The formal model of the CPE is based on the formal protocol specifications in Section 3. Using our assumption that the tester and the CPE reside on the same host, the test context (i.e. the UDP layer) acts as a reliable channel that provides in-sequence delivery. This can be modelled by two unbounded first-in/first-out (FIFO) queues, one for message transfer from tester to CPE, and one vice versa. The CSAP interface is implemented by means of pipes, which essentially behave like bounded first-in/first-out (FIFO) buffers. Under the assumption that a pipe is never 'overloaded', this can also be modelled as an unbounded FIFO queue. The USAP interface is implemented by means of sockets. Sockets can also be modelled, just as pipes, by unbounded FIFO queues. Finally, the number of communicating peer entities of the CPE, i.e. the set of potential conference partners, has been fixed in the test architecture to two. Figure 4 visualizes the complete formal model of the SUT.

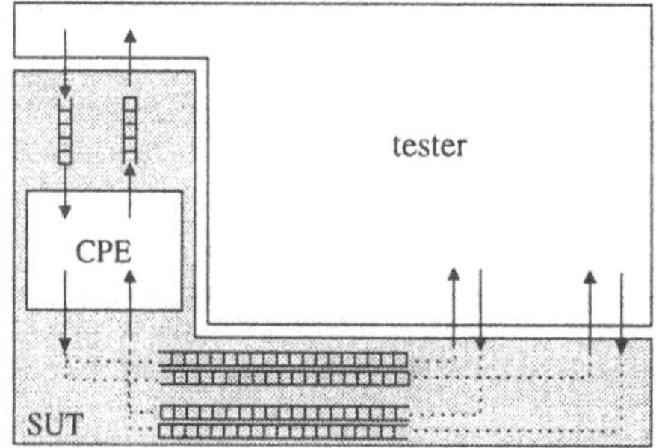

Figure 4   Formal model of the SUT

To compare test derivation and test execution based on LOTOS, SDL and PROMELA, we have built models of the SUT in these three languages. In the LOTOS description of the SUT the behaviour description of the CPE is extended with queues that model the underlying UDP service, the pipes and the sockets. As two consecutive unbounded queues behave exactly equivalent as a single unbounded queue, an optimization was made with respect to such consecutive queues. A consequence of the introduction of queues is that the hidden synchronizations between the queues and the CPE at CSAP and USAP lead to internal steps. In Section 5 we will see the impact that these internal steps have on run-time and memory consumption of the on-the-fly tester.

Similar to LOTOS, a complete specification of the SUT, including the underlying UDP layer and the interfaces, has been constructed in SDL. For automatic derivation of TTCN the use of certain SDL constructs had to be restricted, e.g. the built-in type PId for address construction had to be converted.

In PROMELA, also a model of the SUT has been constructed. In contrast to the LOTOS and SDL specifications, an optimization has been made in PROMELA that allows removal of the queues without changing the observable behaviour of the protocol.

5   TESTING ACTIVITIES

This section describes our testing activities. After summarizing the overall results we will elaborate on the test activities for each of the specification languages. We used the LOTOS and PROMELA specifications for on-the-fly test derivation and execution, using the correctness criterion **ioco** [12]. We used the SDL specification to derive and execute TTCN in batch mode.

For each of the specification languages we started with initial experiments to identify errors in the specifications and to test the (specification language specific) en/decoding functions of the tester. Once we had sufficient confidence in the specification and test tools, we tested the (assumed to be) correct implementation, after which the 27 erroneous mutants were tested by people who did not know which errors had been introduced in these mutants.

In the following table we summarize the results of these experiments. Each implementation was tested several times: on-the-fly with different seeds, and batch-wise with 13 different test cases. For each implementation and each FDT we indicate the verdict, and the minimal and maximal number of execution steps, i.e. LOTOS, PROMELA or TTCN test events that were taken to reach the verdict. If all tests led to a **pass** verdict, we indicate **pass**. For tests that have a **pass** verdict we do not indicate the number of steps. If at least one test led to a **fail** verdict we indicate **fail**. If no test led to a **fail** verdict, but at least one test led to an **inconclusive** verdict because of a timeout, we indicate **timeout** (this only applies to the SDL-based tests). A **timeout** implies a deadlock, and therefore it must be viewed as a serious warning.

LOTOS   To instantiate our test architecture with LOTOS specific components, we only needed to produce a specification-dependent EXPLORER component and specific en/decoding routines for the ADAPTER component; the remaining components of our test tool architecture could be reused. The EXPLORER module was automatically generated using the CADP tool set, see section 2. The ADAPTER en/decoding routines were hand-written. The resulting tester is parameterized with the seed of its random number generator and the target depth.

We started by repeatedly running the tester in automatic mode, each time with a different seed for the random number generator, until either a depth of 500 steps was reached or an inconsistency between tester and implementation was detected (i.e. **fail**, usually after some 30 to 70 steps). This uncovered some errors in both the implementation and the specification, which were repaired. In addition we have run the tester in user-guided, manual mode to explore specific scenarios and to study failures that were found in fully automatic mode.

Once we had sufficient confidence in the quality of the specification and implementation we repeated the previous 'automatic mode' experiment, but now with a (target) depth of $1,000,000$ steps. This depth has not been reached, because the LOTOS EXPLORER-PRIMER module runs out of memory (without discovering an inconsistency) after a few thousand steps – the longest trace consisted of $27,803$ steps and took 1.4 Gb of memory. For this trace the computation of a single step took from 3 seconds (99.9% of the steps) up to 43 minutes of CPU time on a 296 MHz Sun UltraSPARC-II processor. The main reason for the huge memory consumption is that

mu-tant nr.	LOTOS			Promela			SDL	
	verdict	steps min	steps max	verdict	steps min	steps max	verdict	steps min
				'correct' implementation				
0	pass	-	-	pass	-	-	pass	-
			Incorrect Implementations – No outputs					
1	fail	37	66	fail	9	51	pass	-
2	fail	21	37	fail	6	116	timeout	7
3	fail	63	78	fail	24	498	timeout	7
4	fail	65	68	fail	20	83	timeout	7
5	fail	11	17	fail	2	10	timeout	7
6	fail	31	192	fail	14	81	timeout	7
			Incorrect Implementations – No internal checks					
7	fail	57	126	fail	31	392	timeout	12
8	fail	31	37	fail	38	200	pass	-
9	pass	-	-	pass	-	-	timeout	12
10	pass	-	-	pass	-	-	pass	-
			Incorrect Implementations – No internal updates					
11	fail	26	126	fail	29	143	timeout	12
12	fail	21	44	fail	6	127	timeout	7
13	fail	21	45	fail	6	19	timeout	7
14	fail	57	76	fail	28	146	fail	7
15	fail	207	304	fail	19	142	fail	17
16	fail	40	208	fail	25	83	fail	25
17	fail	35	198	fail	9	46	timeout	8
18	fail	31	238	fail	12	121	timeout	7
19	fail	29	467	fail	9	165	pass	-
20	fail	57	166	fail	33	142	timeout	7
21	fail	63	178	fail	15	219	fail	7
22	fail	57	166	fail	31	144	timeout	7
23	fail	21	35	fail	5	33	fail	7
24	fail	69	126	fail	31	127	pass	-
25	fail	37	55	fail	7	51	timeout	7
26	fail	66	91	fail	24	235	pass	-
27	fail	46	210	fail	23	139	fail	17

there are (known) memory leaks in the EXPLORER, while we could not use the garbage collector supplied in CADP because it does not cooperate with our PRIMER (this is being fixed). The main reason for the long computation time lies in the internal steps in the model of the CPE extended with test context (see Section 4), because our PRIMER has to explore large numbers of states for traces that contain long sequences of internal steps.

To test the error-detection capabilities of our tester we repeatedly ran the tester in automatic mode for a depth of 500 steps, each time with a different seed for the random number generator, on the 27 mutants. The tester was able to detect 25 of them.

The two mutants that could not be detected accept PDUs from any source – they do not check whether an incoming PDU comes from a potential conference parter. This is not explicitly modeled in our LOTOS specification, and therefore these mutants are **ioco**-correct with respect to the LOTOS specification, which is why we can not detect them.

Promela   To instantiate our tester with PROMELA-specific components, we generated a single module, which implements both the EXPLORER and PRIMER modules, automatically from the PROMELA specification using the TROJKA tool [13]. TROJKA is based on the SPIN tool for PROMELA [7]. We could reuse most of the ADAPTER developed for LOTOS; changes were only necessary in the parts that depend on the specification labels.

With the PROMELA based tester we repeated the experiments that we did with the LOTOS based one. We received the same results, but in a much shorter time (on average about 1.1 steps per second), and we were able to reach greater depths (450,000 steps), using less memory (400Mb). The differences with the LOTOS based tester can be explained by the use of memory-efficient internal data representations and the use of hashing techniques to remember the results of unfoldings. These techniques were inherited by TROJKA from SPIN. The PROMELA based tester was able to detect the same 25 of the 27 mutants as the LOTOS based tester (which is no surprise because both testers check for the same correctness criterion **ioco**).

SDL   We used the TAU tool kit for batch derivation and execution of test suites in TTCN from SDL [1]. The functionality of the EXPLORER, PRIMER and (batch derivation) DRIVER were covered by TAU's AUTOLINK test derivation tool, which generates TTCN test suites from the SDL specification, guided by MSCs that were derived manually from the SDL specification using TAU's SDL simulator.

Before running the derived TTCN test suite against our implementations, we ran it against the original SDL specification to validate the TTCN test suite, by running the SDL simulator and TAU's TTCN simulator in connection. This uncovered some problems in AUTOLINK. Another problem encountered was that for some MSCs no TTCN suites could be derived, although the MSC could be successfully verified (this depends on the memory of our computers). An interesting aspect is that the partial temporal ordering of the events in the MSC is sometimes not respected in the TTCN code derived by AUTOLINK. Finally, we have experienced some minor problems and bugs. Some of the abovementioned problems may have been solved already in recent releases of the software involved.

The batch execution DRIVER module was automatically generated from the complete TTCN suites using the TTCN compiler of TAU. This DRIVER is linked with an ADAPTER module using the GCI interface; the result is a single program that can execute the test suites (see Section 2). Some parts of the en/decoding functions of the ADAPTER could be generated using KIMWITU [3]; the rest consists of hand-written C code.

We have derived 15 test cases (MSCs) from the SDL specification. The time to build an MSC by means of simulation is 3 minutes for an MSC with 7 events which includes the time to split it into parts (two parts in this case). This was the shortest time. The longest time was 45 minutes for 44 events (15 parts). The average time was 12 minutes for an average of 20 events and 5 parts. The test derivation time consists of the abovementioned times plus the AUTOLINK generation step's time which was less than 8 seconds. The preceding splitting turned out to be essential for most cases (e.g. 14 minutes without splitting becomes 7 seconds *with* splitting). For two of these MSCs no TTCN could not be derived, not even after splitting, because AUTOLINK ran out of memory.

We will try to sketch our informal strategy for defining test purposes in the next few lines. Most of the test purposes are concerned with a single conference. Various arbitrary interleavings of join actions, data transfer and leave actions give rise to one test purpose each. The other test purposes check the absence of interference between two simultaneous conferences (for 3 users it makes no sense to have more than 2 conferences).

The test execution time, running the TTCN which was derived in a batch-wise way against the implementation, took from 2 to 5 seconds.

The detection of errors was done by repeating all the 13 test cases for which TTCN could be derived for the 27 mutants. Six **fail** verdicts were obtained, next to 15 **inconclusive** verdicts that were the effect of a timeout. We felt that such **inconclusive** verdicts ought to be viewed a serious warning, because they indicate a deadlock. Six errors went undetected, although some of them could have been found by a larger test suite.

6   EVALUATION AND CONCLUSIONS

Conclusions.   In this paper we have studied the feasibility of automatic test derivation and execution from a number of formal specifications. To conduct this study, a protocol has been modelled in three formal specification languages: LOTOS, PROMELA and SDL. Also, a set of concrete implementations has been constructed, some of which were injected with faults that were unknown to the person that performed the testing. To test these implementations based on the formal specifications an open test architecture has been defined that was successfully instantiated for on-the-fly testing with LOTOS and PROMELA and batch-wise testing with SDL. The results have been compared with respect to the number of erroneous implementations that could be detected for each of the specifications, and the time and effort that it took to test the implementations.

The tool architecture that was used to conduct the experiment supports both on-the-fly testing and batch testing. Several existing tool sets, such as CADP and TAU, were used as plug-ins for the tool architecture, thereby illustrating its openness.

In the on-the-fly approach, tests were fully automatically derived (in a random way) and executed. In the batch approach the construction of tests needed manual assistance. Execution in the batch approach was done automatically. Both the on-the-fly approach and the batch approach were able to detect many erroneous implementations. Using the on-the-fly techniques all erroneous implementations could be detected, except for those that contained errors that simply could not be detected due to modelling choices in the specification and the choice of implementation relation (hence, were formally no errors). Using batch testing based on SDL fewer erroneous implementations were detected. On the one hand this is caused by the occurrence of timeouts (which should be considered as indications of potential errors such as deadlocks), and on the other hand by the fact that less tests were executed due to the fact that manual assistance during test derivation was needed. By deriving more test cases in the batch approach it will be possible to increase the error detecting capability. Although with batch SDL testing certainly less erroneous implementations were detected, the current experiments are too restricted to deduce general conclusions from them. More experiments are needed where also more care is taken that the starting points are comparible.

In the on-the-fly approach, tests were derived using random selection of inputs. How many steps it takes to trigger errors depends very much on the random choices that the tester makes, and the seed of the random choice generator. However, the results in this paper support the assumption that if the tester runs 'sufficiently long' then eventually all errors will be found. In the batch approach more human assistance is needed. Consequently, an error can often be found in less steps using the batch approach than in the on-the-fly approach. We found that for long traces that lead to errors, it is difficult to analyse the exact conditions under which such an error was triggered. Especially in the on-the-fly approach, error analysis was more difficult because no support was available to diagnose the trace that led to the error. Analysis could be made easier if the traces that lead to an error could be transformed into a form in which they can be studied in the existing development environments for the FDTs.

Further work.   The experimental results that are presented in this paper are based on a case study of a single protocol and a limited number of implementations. To obtain more valuable results the number of cases studies and the number of experiments per case study should be increased. To enable a rigorous comparison of test derivation and test execution tools by different vendors, one (or more) case studies containing specifications and sets of implementations should be made publically available so that they can be used by several tool vendors to compare their test derivation/execution tools. The case study in this paper can be seen as one of the first initiatives towards such a *test tool benchmarking* activity. To promote this kind of benchmarking, the description of our case study together with the formal specifications and all implementations are available on the Web [9]. Everybody is thus invited to conduct and publish analogous experiments with his or her favourite test tools.

The on-the-fly approach for test derivation is currently implemented using a random strategy. Therefore, it is difficult to steer the test that is being derived and executed. In practice, more advanced and user controlled strategies that allow for the derivation of tests that are targeted towards specific test purposes, or that avoid deriving the same tests more than once, are needed. This requires research in the field of test purpose oriented test derivation, and adaptive test derivation techniques. The batch approach is currently limited by the fact that human intervention is needed to derive tests, and due to the fact that sometimes inconclusive verdicts are reached as a result of timeouts. More advanced batch test derivation techniques might overcome these deficiencies.

Most of the work to instantiate the on-the-fly tester goes into the making and validation of the formal specification. After this the specification module can be generated fully automatically. For the batch approach the most laborious task is to derive tests. Also, a laborious task is to produce the ADAPTER. Creation of the ADAPTER can be eased by developing a general framework in which the method to connect to the IUT is orthogonal to the data en/decoding, and to allow easy reuse of connection modules for common protocols. In addition, it is worth studying whether it is possible to develop languages and tools to specify the data en/decoding functions in a compact syntax and generate automatically code from that.

References

[1] Telelogic AB. Telelogic TAU Documentation, 1998.

[2] F. Brady and R.M. Barker. Infrastructural Tools for Information Technology and Telecommunications Conformance Testing, INTOOL/GCI, Generic Compiler/Interpreter (GCI) Interface Specification, Version 2.2, 1996. INTOOL doc. nr. GCI/NPL038v2.

[3] P. van Eijk, A. Belinfante, H. Eertink, and H. Alblas. The Term Processor KIMWITU. In E. Brinksma, editor, *TACAS'97*, pages 96–111. LNCS 1217, Springer-Verlag, 1997.

[4] J.-C. Fernandez, C. Jard, T. Jéron, and C. Viho. Using On-the-Fly Verification Techniques for the generation of test suites. In R. Alur et al., editor, *CAV'96*. LNCS 1102, Springer-Verlag, 1996.

[5] L. Ferreira Pires. Protocol Implementation: Manual for Practical Exercises 1995/1996. Lecture notes, University of Twente, The Netherlands, 1995.

[6] H. Garavel. OPEN/CÆSAR: An Open Software Architecture for Verification, Simulation, and Testing. In B. Steffen, editor, *TACAS'98*, pages 68–84. LNCS 1384, Springer-Verlag, 1998.

[7] G. J. Holzmann. *Design and Validation of Computer Protocols*. Prentice-Hall Inc., 1991.

[8] ISO/IEC JTC1/SC21 WG7, ITU-T SG 10/Q.8. *Framework: Formal Methods in Conformance Testing*. CD 13245-1, ITU-T Z.500. ISO – ITU-T, Geneve, 1996.

[9] Project Consortium Côte de Resyste. Conference Protocol Case Study. URL: `http://fmt.cs.utwente.nl/ConfCase`.

[10] M. Schmitt, A. Ek, B. Koch, J. Grabowski, and D. Hogrefe. – AUTOLINK – Putting SDL-based Test Generation into Practice. In A. Petrenko et al., editor, *IWTCS'98*, pages 227–243. Kluwer Academic Publishers, 1998.

[11] R. Terpstra, L. Ferreira Pires, L. Heerink, and J. Tretmans. Testing theory in practice: A simple experiment. In T. Kapus et al., editor, *COST 247 Workshop on Applied Formal Methods in System Design*, pages 168–183. University of Maribor, Slovenia, 1996.

[12] J. Tretmans. Test generation with inputs, outputs and repetitive quiescence. *Software—Concepts and Tools*, 17(3):103–120, 1996.

[13] R.G. de Vries and J. Tretmans. On-the-Fly Conformance Testing using SPIN. In G. Holzmann et al., editor, *Fourth* SPIN *Workshop*, ENST 98 S 002, pages

115–128. Ecole Nationale Supérieure des Télécommunications, Paris, France, November 2 1998.

13

GENERATING TEST CASES FOR A TIMED I/O AUTOMATON MODEL

Teruo Higashino[†], Akio Nakata[††], Kenichi Taniguchi[†] and Ana R. Cavalli[†††]

† : *Dept. of Informatics and Mathematical Science, Graduate School of Engineering Science, Osaka University, Toyonaka, Osaka 560-8531, Japan*

Tel : +81-6-6850-6590   Fax : +81-6-6850-6594

Email : higashino@ics.es.osaka-u.ac.jp

†† : *Dept. of Computer Science, Hiroshima City University, Japan*

††† : *Institut National des Télécommunications, Evry France*

Abstract   Recently various real-time communication protocols have been proposed. In this paper, first, we propose a timed I/O automaton model so that we can simply specify such real-time protocols. The proposed model can handle not only time but also data values. Then, we propose a conformance testing method for the model. In order to trace a test sequence (I/O sequence) on the timed I/O automaton model, we need to execute each I/O action in the test sequence at an adequate execution timing which satisfies all timing constraints in the test sequence. However, since outputs are given from IUTs and uncontrollable, we cannot designate their output timing in advance. Also their output timing affects the executable timing for the succeeding I/O actions in the test sequence. Therefore, in general, the executable timing of each input action in a test sequence can be specified by a function of the execution time of the preceding I/O actions. In this paper, we propose an algorithm to decide efficiently whether a given test sequence is executable. We also give an algorithm to derive such a function from an executable test sequence automatically using a technique for solving linear programming problems, and propose a conformance testing method using those algorithms.

Keywords: real-time protocols, timed automata, conformance testing, traceability

1.   INTRODUCTION

Conformance testing is one of methods to improve the reliability of communication protocols [3, 11]. However, conformance testing for models with the notion of time has not sufficiently studied. Recently, some conformance testing methods have been proposed in [4, 5, 13, 15] for real-time models such as timed automata [1].

Alur's timed automata model [1] is a simple and nice model for specifying real-time systems, and a lot of verification techniques have been proposed [2]. However, basically it does not treat data values used in communication protocols. Especially, in multimedia communication protocols such as QoS control for video streams, sometimes one may want to specify different timeout intervals depending on the size of data to be transmitted. Also, it is possible that the timing of each action of the process may depend on the type and/or size of transmitted data. For the above purposes, we need models that can treat not only *time* but also *data values*. It is also desirable that such models have efficient verification and/or testing methods. Here, we need a model which combines timed automata with EFSMs.

In testing EFSMs or real-time systems, there are some problems to be solved. First, a given test sequence is not always executable. In order to execute the test sequence, we must find some appropriate input values or execution timing which satisfy its transition conditions. In contrast to the case of EFSMs, in the case of real-time systems, the tester can designate the input timing. However, in general, the output timing is not controlled by the tester and it is decided by each IUT itself. Moreover, the executable timing of some I/O action may depend on the execution time of its preceding I/O actions. It is desirable that whenever the preceding output actions are executed, there always exists some adequate input timing such that its succeeding sequence is executable. Thus, in this paper we propose a *timed I/O automaton model* for specifying real-time protocols and a conformance testing method for the model which handles the problems described above.

In our timed I/O automaton model, each transition is either input or output action. Moreover, in order to describe timing constraints among actions, we introduce some *variables* and one special *global time variable* which always holds the current time. The *variables* can hold not only time values but also values expressed as linear expressions of the time values and input data. Each *transition condition* can be specified by a *logical conjunction* of linear inequalities of those two types of variables. Note that any Alur's time automata can be specified in this model.

We define two kinds of executability (traceability) of test sequences, *must-traceability* and *may-traceability*. A must-traceable test sequence can be always executed if we specify some appropriate input timing for its input actions, no matter when its output actions are executed. A may-traceable test sequence can be executed only when the execution time of its output actions belongs to the sub-ranges which make the succeeding actions executable. In this paper, we present an efficient algorithm for checking the must/may-traceability of given test sequences and obtaining the upper and lower bounds for each input action as functions of the execution time of its preceding I/O actions.

Based on UIOv-method [16], we propose a conformance testing method for our model. Our method can be used for improving the reliability of a given IUT. In our method, we assume that the errors in IUTs fall into some specific types, and under the assumption we check the correctness of IUTs by checking (1) the traceability of the derived test sequences and (2) the executability of each transition condition at some specified boundary time where the boundary time denotes the moments when the truth value of the transition condition changes.

2. TIMED I/O AUTOMATON MODEL

2.1 DEFINITIONS

Definition 1 *A* timed I/O automaton *is a 10-tuple* $M = \, <S, A, I/Otype, t, V, Pred, Def, \delta, s_{init}, \{x_{1init}, x_{2init}, \ldots, x_{kinit}\} >$, *where*

- $S = \{s_0, s_1, \ldots, s_n\}$ *is a finite set of states.*

- *A is a finite set of I/O actions.*

- $I/Otype = \{!, ?\} \cup \{?_{\mathbf{v}} | \mathbf{v}$ *is an input variable*$\}$ *is a set of I/O types, where the symbols ? and ! represent input and output, respectively. The symbol* $?_{\mathbf{v}}$ *represents that the input value is assigned to the variable* $\mathbf{v}$ *whose value may be used in the transition conditions of its succeeding actions. Each variable* $\mathbf{v}$ *can hold a rational number.*

- *t denotes the global clock variable which always holds the current time as a rational number.*

- $V = \{x_1, x_2, \ldots, x_k\}$ *is a finite set of variables which can hold rational numbers.*

- *Pred is a set of linear inequalities* $P[t, x_1, x_2, \ldots, x_k]$ *on rational numbers and their logical* conjunctions.

- *Def is a set of* assignments. *An assignment is a function which maps variables* $x_i \in V$ *to a linear expression* $f(t, \mathbf{v}, x_1, x_2, \ldots, x_k)$, *denoted by* $x_i \leftarrow f(t, \mathbf{v}, x_1, x_2, \ldots, x_k)$.

- $\delta \subseteq S \times A \times I/Otype \times Pred \times Def \times S$ *is a transition relation.*

- $s_{init} \in S$ *is the initial state of M.*

- $\{x_{1init}, x_{2init}, \ldots, x_{kinit}\}$ *is a set of the initial values for variables* $x_1, x_2, \ldots, x_k \in V$.

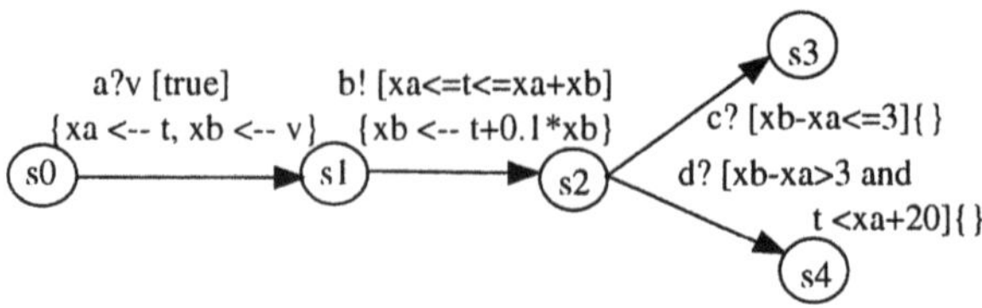

Figure 1   Semantics of the proposed model

- *There is no state from which there are two outgoing transitions with the same I/O actions.*

Note that, from the last constraint in the above definition, the timed I/O automaton is deterministic. Output values are omitted in this model since if they affect succeeding transition conditions, such conditions can be specified using the variables in V, input variables and global clock variable t. An element of a transition relation $(s, a, \$, P, D, s') \in \delta$ of M is denoted by $s \xrightarrow{a\$[P]D}_M s'$. If M is understood from the context, we simply write $s \xrightarrow{a\$[P]D} s'$.

Intuitively, the semantics of our timed I/O automaton model is as follows. For instance, at state s_0 in Fig. 1, the input action $a?_\mathbf{v}$ is executable. If $a?_\mathbf{v}$ has been executed, the state moves into s_1 and the execution time of $a?_\mathbf{v}$ is assigned to x_a according to the assignment $\{x_a \leftarrow t\}$. Also, the input value $\mathbf{v}$ is assigned to x_b according to the assignment $\{x_b \leftarrow \mathbf{v}\}$. If the execution time of $a?_\mathbf{v}$ is 5 and the input value $\mathbf{v}$ is 3, then the values of x_a and x_b become 5 and 3, respectively. So the output action $b!$ will be executed on time t such that the transition condition $5 \leq t \leq 5+3$ holds, that is, within time 8. If the execution time of $b!$ is 6, the value of $t + 0.1 * x_b$, which is equal to $6 + 0.1 * 3 = 6.3$, is assigned to variable x_b and the state moves into s_2. At state s_2, there are two choices. If $x_b - x_a \leq 3$ holds, then the input action $c?$ is executable. For the above case, since $x_b - x_a = 6.3 - 5 = 1.3 \leq 3$ holds, $c?$ is executable. In the case that $x_b - x_a > 3$ holds, another input action $d?$ is executable for 20 seconds after $a?$ is executed.

Relation to Timed Automata.   Our model can simulate any timed automaton in the original version of Alur's timed automata [1]. Let's consider the example in Fig. 2. Since the global clock variable t has the current time in our model, by replacing each clock variable t_1 in Alur's timed automaton into $t - t_1$, we can obtain the same time constraints in our model. The $reset(t_1)$ operation can be simulated by assigning the current time t to the clock variable t_1 ($\{t_1 \leftarrow t\}$). If Alur's model has several clock variables, then the corresponding our model also has (at most) the same number of variables.

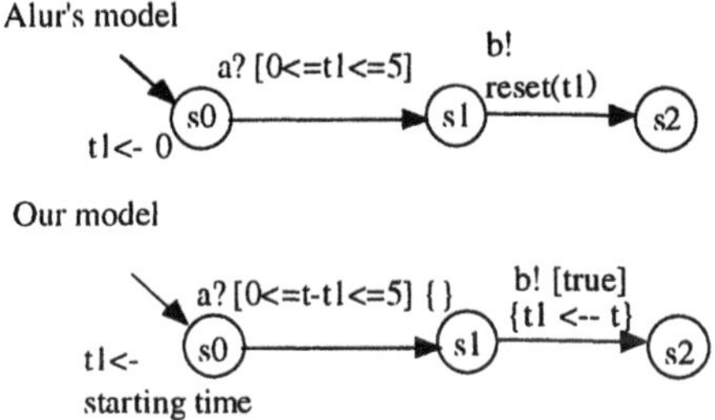

Figure 2   Simulation of Alur's automata

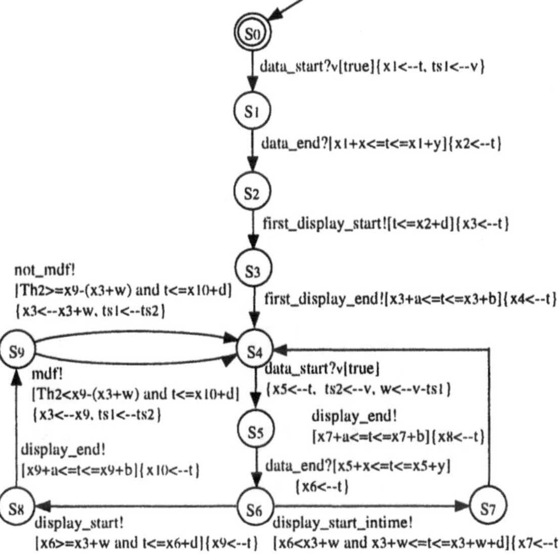

Figure 3   Media synchronization protocol

2.2   EXAMPLE

As an example of the timed I/O automata, we present a slightly modified specification of a receiving node of a media synchronization protocol[8] in Fig. 3. The media synchronization protocol specifies how to synchronize real-time continuous media such as video stream when the transfer rate quickly changes.

In this system, the sending node tries to send data continuously at the fixed rate to the receiving node. First, the sending node sends to the receiving node a video packet with a time-stamp representing its sending time. The receiving node starts to receive the data at time x_1, and assigns its time-stamp to ts_1.

Depending on its network propagation delay (minimum x, maximum y), it finishes to receive the data at time x_2 ($x_1 + x \leq x_2 \leq x_1 + y$). Next, it outputs a signal which instructs to start playing the video packet before time $x_2 + d$ where d denotes the maximum time necessary for preparation to play video. Then, it outputs a signal which instructs to finish playing the video packet after the duration necessary to finish playing (minimum a, maximum b), which depends on the decoding time of the data such as MPEG encoded data. After that, the receiving node carries out the above work continuously. Moreover, it synchronizes the playing rate with the sending node. To do so, it sets the objective starting time for playing the next video packet to time $x_3 + w$, where x_3 is the starting time for playing the previous video packet and w is a difference between two time-stamps of the most recent video packet and its previous one. If the actual completion time of receiving data is earlier than the objective one ($x_6 < x_3 + w$), the system waits until the objective time and then outputs the signal to start playing. Otherwise, it immediately (at time x_9) outputs the signal ("quick recovery of synchronization errors"). However, the error is within the specified tolerance T_{h2} ($T_{h2} \geq x_9 - (x_3 + w)$), it considers that the playing has started in time (*not_mdf*!), and that the system does not adjust the synchronization interval. Otherwise ($T_{h2} \leq x_9 - (x_3 + w)$), the actual starting time x_9 of playing is assigned to x_3, and the system adjusts the synchronization interval (*mdf*!). Then, it repeats the behaviour described above. Note that a, b, d, x, y and T_{h2} in Fig. 3 are constant parameters specified by designers.

3.   EXECUTABILITY OF TRANSITION SEQUENCES

3.1   MUST/MAY TRACEABILITY

A *transition sequence* of a timed I/O automaton M is viewed as an execution path of the transition graph of M. However, in a timed I/O automaton, the value of each variable may change by executing each transition. In order to decide whether a given transition sequence is executable, we must consider how to change the values of those variables on execution of the actions in the transition sequence. To reflect such a change, we apply the technique inspired by the general symbolic execution technique (e.g. [5, 10]). First, we assign an unique name for each occurrence of variables in the transition sequence. Then we replace each occurrence of variables with an expression containing the initial values of the variables and the uniquely named variables.

Definition 1 *Let α denote a transition sequence* $s_0 \xrightarrow{a_1 \$_1 [P_1] D_1} s_1 \xrightarrow{a_2 \$_2 [P_2] D_2}$ *$\ldots \xrightarrow{a_n \$_n [P_n] D_n} s_n$ of a timed I/O automaton M where s_0 is the initial state s_{init} of M. Let $t_1, \ldots, t_n$ denote n different variables which represent the execution times of $a_1, \ldots, a_n$, respectively. Similarly, let $\mathbf{v}_1, \ldots, \mathbf{v}_m$ denote* *m different variables which represent the input values of the corresponding m occurrences of data-input actions in α. Let $x_1^{(i)}, \ldots, x_k^{(i)}$ represent the values of variables $x_1, \ldots, x_k \in V$ on i-th state s_i of α. Then, using the following algorithm, we express $x_1^{(i)}, \ldots, x_k^{(i)}$ by expressions containing only variables in $\{t_1, \ldots, t_i, x_{1init}, \ldots, x_{kinit}, \mathbf{v}_1, \ldots, \mathbf{v}_m\}$.*

- *$j := 0$, for $p = 1$ to k do $x_p^{(0)} := x_{pinit}$*

- *for $i = 1$ to n do*

    - *if $\$_i =?_{\mathbf{v}}$ then $j := j + 1$; $\$_i :=?_{\mathbf{v}_j}$*
    - *for $p = 1$ to k do*

        * *if $x_p \leftarrow f(t, \mathbf{v}, x_1, \ldots, x_k) \in D_i$*
          $$x_p^{(i)} := f(t_i, \mathbf{v}_j, x_1^{(i-1)}, \ldots, x_k^{(i-1)})$$
        * *else (there are no assignments to x_p in D_i)*
          $$x_p^{(i)} := x_p^{(i-1)}$$

For each transition condition $P_i[t, x_1, \ldots, x_k]$ of α, let $\widehat{P_i}$ be a condition defined by

$$\widehat{P_1} \stackrel{\text{def}}{=} P_1[t_1/t, x_{1init}/x_1, \ldots, x_{kinit}/x_k]$$

$$\widehat{P_i} \stackrel{\text{def}}{=} P_i[t_i/t, x_1^{(i-1)}/x_1, \ldots, x_k^{(i-1)}/x_k] \wedge (t_{i-1} \leq t_i)$$

where each $x_k^{(i)}$ is the expression obtained by the above algorithm, and $P_i[t_i/t, u_1/x_1, \ldots, u_k/x_k]$ means the expression obtained from $P_i[t, x_1, \ldots, x_k]$ by substituting $t_i, u_1, \ldots, u_k$ into $t, x_1, \ldots, x_k$, respectively. Each t_i represents the execution time of action a_i. We call $w \stackrel{\text{def}}{=} (a_1\$_1, t_1, \widehat{P_1}) \ldots (a_n\$_n, t_n, \widehat{P_n})$ as a symbolic trace for α.

Example 1 *The symbolic trace w for the transition sequence (which contains a loop) $s_0 \xrightarrow{a![t \leq x_b + 7]\{x_a \leftarrow t\}} s_1 \xrightarrow{b?_{\mathbf{v}}[t \leq x_a + 3]\{x_b \leftarrow t, x_a \leftarrow x_a + v\}} s_2 \xrightarrow{c?[t \leq x_b + 5 \wedge x_a + 15 \leq t]\{\}}$ $s_0 \xrightarrow{a![t \leq x_b + 7]\{x_a \leftarrow t\}} s_1$ of a timed I/O automaton M is obtained as $w = (a!, t_a, t_a \leq x_{binit} + 7) \, (b?_{\mathbf{v}_1}, t_b, t_b \leq t_a + 3 \wedge t_a \leq t_b) \, (c?, t_c, t_c \leq t_b + 5 \wedge t_a + \mathbf{v}_1 + 15 \leq t_c \wedge t_b \leq t_c) \, (a!, t_a', t_a' \leq t_b + 7 \wedge t_c \leq t_a').$*

Definition 2 *We say that a symbolic trace w is* must-traceable, *if whenever each output action is executed, there always exists some input timing for each input action such that the rest of the sequence can be executed. On the other hand, we say that a symbolic trace w is* may-traceable, *if for some output timing there exists some input timing such that the rest of the sequence can be executed.*

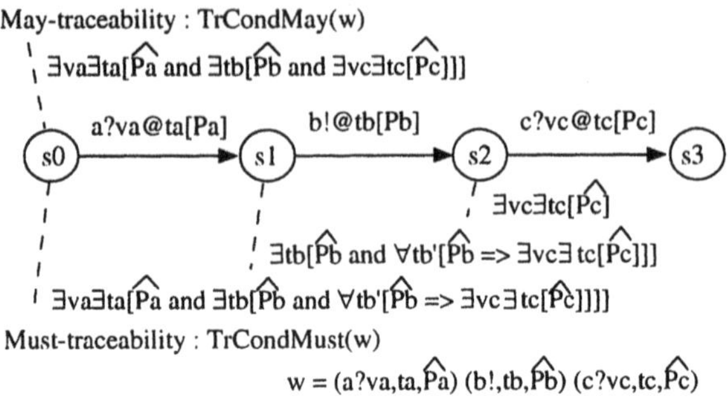

Figure 4   Must/may traceability

For a symbolic trace w, let $TrCondMust(w)$ and $TrCondMay(w)$ denote the expressions representing whether w is must/may traceable or not. That is, w is *must-traceable* (or *may-traceable*) if and only if $TrCondMust(w)$ (or $TrCondMay(w)$) is true. $TrCondMust(w)$ and $TrCondMay(w)$ are calculated recursively like the expressions in Fig. 4. For $TrCondMust(w)$, the sub-expression $\exists t_b[\widehat{P_b}]$ denotes that there exists a timing such that the output action $b!$ is executable. And, the sub-expression $\forall t_{b'}[\widehat{P_b} \Rightarrow \exists v_c \exists t_c[\widehat{P_c}]]$ denotes that there exists an input timing t_c executing $c?$ and an adequate input value v_c for any output timing in which $b!$ is executed.

Example 2 *A symbolic trace* $w = (a!, t_a, 1 \leq t_a \leq 5)(b?_\mathbf{v}, t_b, t_b \leq t_a + \mathbf{v} \wedge t_b \leq 5 \wedge t_a \leq t_b)$ *is must-traceable because the following expression is true.*

$$\exists t_a[1 \leq t_a \leq 5 \wedge \forall t_a'[1 \leq t_a' \leq 5 \Rightarrow \exists \mathbf{v} \exists t_b[t_b \leq t_a' + \mathbf{v} \wedge t_b \leq 5 \wedge t_a' \leq t_b]]]$$

A symbolic trace $w' = (a!, t_a, 1 \leq t_a \leq 6)(b?_\mathbf{v}, t_b, t_b \leq t_a + \mathbf{v} \wedge t_b \leq 5 \wedge t_a \leq t_b)$ *is not must-traceable, since the input action* $b?_\mathbf{v}$ *is not executable if the output action* $a!$ *is executed at time* t_a *such that* $5 < t_a \leq 6$. *However,* w' *is may-traceable since* $b?_\mathbf{v}$ *is executable if* $a!$ *is executed at time* t_a *such that* $1 \leq t_a \leq 5$. *That is,*

$$\exists t_a[1 \leq t_a \leq 6 \wedge$$
$$\forall t_a'[1 \leq t_a' \leq 6 \Rightarrow \exists \mathbf{v} \exists t_b[t_b \leq t_a' + \mathbf{v} \wedge t_b \leq 5 \wedge t_a' \leq t_b]]]$$
$$\equiv false$$

$$\exists t_a[1 \leq t_a \leq 6 \wedge \exists \mathbf{v} \exists t_b[t_b \leq t_a + \mathbf{v} \wedge t_b \leq 5 \wedge t_a \leq t_b]] \equiv true$$

Decision of must/may traceability.   In general, $TrCondMust(w)$ and $TrCondMay(w)$ become rational Presburger sentences and it is known that there exists a decision procedure for the general class [7]. However, the decision problem is NP-hard in the general class. Here, we only use inequalities on rational numbers and their logical conjunctions. Hereafter, for this restricted class, we will propose an efficient decision algorithm and decide the must/may traceability efficiently.

3.2   EFFICIENT DECISION OF MUST/MAY TRACEABILITY

For simplicity, we regard a linear inequality $f(t, x_1, x_2, \ldots) < t$ as $f(t, x_1, x_2, \ldots) + \rho \leq t_n$ for a sufficiently small positive rational number ρ and consider only two inequality relations $\leq$ and $\geq$. We may omit the ρ in the rest of the paper.

First, we will consider must-traceability. Intuitively, the proposed algorithm for checking must-traceability of a symbolic trace $w = (a_1\$_1, t_1, \widehat{P_1}) \ldots (a_n\$_n, t_n, \widehat{P_n})$ works as follows.

Since the last action a_n has no succeeding actions, the executable time t_n of a_n is a solution of the constraint P_n. In our model, we can transform the constraint into the conjunctions of the following three types of linear inequalities by appropriate transposition: (1) $\{f_i(t_1, \ldots, t_{n-1}) \leq t_n | i \in I\}$, (2) $\{t_n \leq g_j(t_1, \ldots, t_{n-1}) \,|\, j \in J\}$, and (3) a conjunction $R_n(t_1, \ldots, t_{n-1})$ of inequalities which contain no t_n's. Therefore, we can obtain the lower bound $t_n^{inf} = max\{f_i(t_1, \ldots, t_{n-1}) | i \in I\}$ and the upper bound $t_n^{sup} = min\{g_j(t_1, \ldots, t_{n-1}) | j \in J\}$ as the interval of the executable time t_n. In order that there exists an executable time t_n of a_n, the expression [$t_n^{inf} \leq t_n^{sup}$ and $R_n(t_1, \ldots, t_{n-1})$] must be true. Thus, let $TrCondMust_n(w)$ denote [$t_n^{inf} \leq t_n^{sup} \wedge R_n(t_1, \ldots, t_{n-1})$]. Since we can also express $t_n^{inf} \leq t_n^{sup}$ by $\bigwedge_{i \in I} \bigwedge_{j \in J} [f_i(t_1, \ldots, t_{n-1}) \leq g_j(t_1, \ldots, t_{n-1})]$, $TrCondMust_n(w)$ can be obtained as a conjunction of linear inequalities which contain only variables $t_1, \ldots, t_{n-1}$.[1] If $TrCondMust_n(w)$ is true, then the executable time t_n of a_n belongs to the interval $[t_n^{inf}, t_n^{sup}]$.

Now we will treat the general case. For each input action a_k such that $\$_k =?$ and $k < n$, the executable time t_k of a_k satisfies the constraint $P_k \wedge TrCondMust_{k+1}(w)$. From this constraint, we can obtain t_k^{sup}, t_k^{inf} and $TrCondMust_k(w)$ similarly. Here, $TrCondMust_k(w)$ represents the condition to make the transition sequence $a_k, a_{k+1}, \ldots, a_n$ in w must-traceable.

For each output action $a_{k'}$ such that $\$_{k'} =!$ and $k' < n$, we can obtain $TrCondMust_{k'}(w)$ as follows. Unlike input actions, the output action $a_{k'}$ can be executed at any output timing $t_{k'}$ satisfying $P_{k'}$, since the output timing is uncontrollable. Thus, first, from the constraint $P_{k'}$, we obtain $t_{k'}^{inf} = l_{P_{k'}}(t_1, \ldots, t_{k'-1})$, $t_{k'}^{sup} = u_{P_{k'}}(t_1, \ldots, t_{k'-1})$ and the conjunction of remain-

---

[1] If a_n is a data-input action $a_n?_\mathbf{v}$, $TrCondMust_n(w)$ may also contain the variable $\mathbf{v}$.

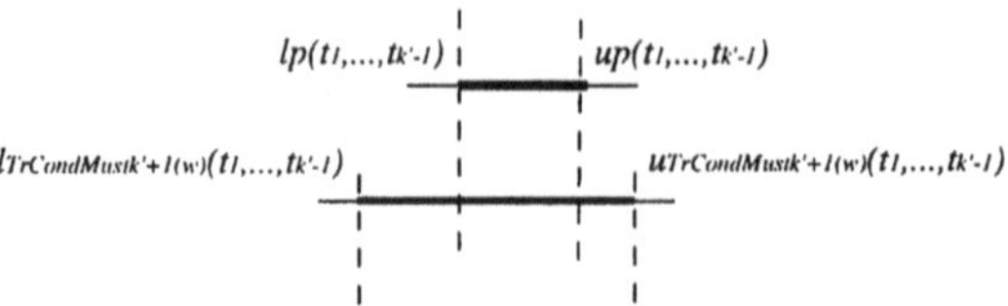

Figure 5   Must traceability for output actions

der clauses $R_{k'}(t_1, \ldots, t_{k'-1})$ which does not contain $t_{k'}$. Here, $[t_{k'}^{inf}, t_{k'}^{sup}]$ denotes the interval of executable output timing for $a_{k'}$. Next, we transform $TrCondMust_{k'+1}(w)$ into the conjunction of the following three forms : (1) $\{f_i(t_1, \ldots, t_{k'-1}) \leq t_{k'} | i \in I'\}$, (2) $\{t_{k'} \leq g_j(t_1, \ldots, t_{k'-1}) \, | j \in J'\}$, and (3) a conjunction of inequalities $R'_{k'}(t_1, \ldots, t_{k'-1})$ which does not contain $t_{k'}$. Then, we obtain the lower bound $l_{TrCondMust_{k'+1}(w)}(t_1, \ldots, t_{k'-1}) = max\{f_i(t_1, \ldots, t_{k'-1}) | i \in I'\}$ and the upper bound $u_{TrCondMust_{k'+1}(w)}(t_1, \ldots, t_{k'-1}) = min\{t_{k'} \leq g_j(t_1, \ldots, t_{k'-1}) \, | j \in J'\}$ as the interval of executable time $t_{k'}$ which satisfies $TrCondMust_{k'+1}(w)$. The remainder $R'_{k'}(t_1, \ldots, t_{k'-1})$ for $TrCondMust_{k'+1}(w)$ is also obtained as an expression with variables $t_1, \ldots, t_{k'-1}$.

Then, $a_{k'}$ is executable and the succeeding sequence is also executable for any output timing $t_{k'}$ of $a_{k'}$ if and only if $t_1, \ldots, t_{k'-1}$ satisfy the following three conditions :

- $l_P(t_1, \ldots, t_{k'-1}) \leq u_P(t_1, \ldots, t_{k'-1})$ holds (i.e. $a_{k'}$ is executable),

- as shown in Fig. 5, the interval $[l_P(t_1, \ldots, t_{k'-1}), u_P(t_1, \ldots, t_{k'-1})]$ is included in the interval $[l_{TrCondMust_{k'+1}(w)}(t_1, \ldots, t_{k'-1}),$ $u_{TrCondMust_{k'+1}(w)}(t_1, \ldots, t_{k'-1})]$ (that is, $l_{TrCondMust_{k'+1}(w)}(t_1, \ldots, t_{k'-1}) \leq l_P(t_1, \ldots, t_{k'-1}) \wedge u_P(t_1, \ldots, t_{k'-1}) \leq u_{TrCondMust_{k'+1}(w)}(t_1, \ldots, t_{k'-1})$ holds), and

- $R_{k'}(t_1, \ldots, t_{k'-1}) \wedge R'_{k'}(t_1, \ldots, t_{k'-1})$ holds.

We use the logical conjunction of these conditions as the condition $TrCond$ $Must_{k'}(w)$ and carry it over to the preceding actions.

By applying this method recursively, we can finally obtain $TrCondMust_1(w)$. In general, $TrCondMust_1(w)$ is a logical combination of linear inequalities which may contain the input variables $\mathbf{v}_1, \ldots, \mathbf{v}_m$ and variables in V. By assigning the initial values to the variables in V, we obtain a formula $D(\mathbf{v}_1, \ldots, \mathbf{v}_m)$ which contains $\mathbf{v}_1, \ldots, \mathbf{v}_m$ only. If this formula is satisfiable, we conclude that the given symbolic trace is must-traceable. The formula corresponds to $TrCondMust(w)$. Its satisfiability can be checked using the

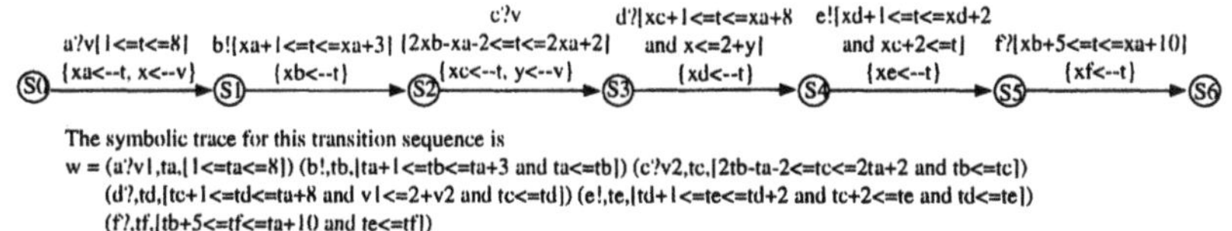

Figure 6   A transition sequence

technique for solving linear programming problems. A solution of $\mathbf{v}_1, \ldots, \mathbf{v}_m$ can be also obtained by the technique.

Sufficient condition for more efficient decision of traceability.   Note that the formula $l_{TrCondMust_{k'+1}(w)}(t_1, \ldots, t_{k'-1}) \leq l_P(t_1, \ldots, t_{k'-1})$, which is generally expressed by $max\{f_i(t_1, \ldots, t_{k-1}) \,|i = 1, ..., p\} \leq max\{f'_j(t_1, \ldots, t_{k-1}) \,|j = 1, ..., q\}$, is actually checked by dividing into q cases, $max\{f_1, \ldots, f_p\} \leq f'_1 \vee max\{f_1, \ldots, f_p\} \leq f'_2 \vee \ldots \vee max\{f_1, \ldots, f_p\} \leq f'_q$. In each case, $max\{f_1, \ldots, f_p\} \leq f'_j$ can be expressed by a logical conjunction of inequalities such as $\bigwedge_{i=1,...,p} (f_i \leq f'_j)$, so the presented algorithm can be applied.

Here, if we replace the condition $max\{f_i(t_1, \ldots, t_{k-1}) \,|i = 1, ..., p\} \leq max\{f'_j(t_1, \ldots, t_{k-1}) \,|j = 1, ..., q\}$ with its sufficient condition $max\{f_i(t_1, \ldots, t_{k-1}) \,|i = 1, ..., p\} \leq min\{f'_j(t_1, \ldots, t_{k-1}) \,|j = 1, ..., q\}$, the $TrCond$ $Must_1$ obtained using the replaced condition is also a sufficient condition that the given symbolic trace is must-traceable. In this case, the replaced condition can be expressed by one logical conjunction of inequalities (without disjunction) such as $\bigwedge_{i=1,...,p} \bigwedge_{j=1,...,q}(f_i(t_1, \ldots, t_{n-1}) \leq f'_j(t_1, \ldots, t_{n-1}))$. So we do not have to divide into cases and can efficiently check must-traceability by applying the presented algorithm (the sufficient condition works well for our example cases in Section 5).

Example 3 *For example, we will apply the algorithm to the symbolic trace in Fig. 6. The result is shown in Fig. 7. According to the result, if we give input values $\mathbf{v}_1$ and $\mathbf{v}_2$ which satisfy $\mathbf{v}_1 \leq 2 + \mathbf{v}_2$, the sequence is must-traceable. As understood from the result, the executable interval for each action is expressed by the function of the execution time of preceding actions. On the actual execution of the sequence, we can choose any timing within the interval $[t_i^{inf}, t_i^{sup}]$ for input action $a_i?$.[2] Once the actual timing t'_i is decided, every occurrence of the variable t_i in the interval expression of every succeeding action is replaced with the actual value t'_i. Then the resulting expression is*

---

[2]For each output action, we observe its output timing and check whether the timing is in the interval.

$$
\begin{aligned}
(t_f^{inf}, t_f^{sup}) &= (max\{t_e, t_b + 5\}, t_a + 10) \\
TrCondMust_f(w) &= [max\{t_e, t_b + 5\} \le t_a + 10] \\
(t_e^{inf}, t_e^{sup}) &= (max\{t_d, t_d + 1, t_c + 2\}, t_d + 2) \\
TrCondMust_e(w) &= [max\{t_d, t_d + 1, t_c + 2\} \le t_d + 2 \wedge t_d + 2 \le t_a + 10 \\
&\quad \wedge t_b + 5 \le t_a + 10] \\
(t_d^{inf}, t_d^{sup}) &= (max\{t_c, t_c + 1\}, min\{t_a + 8, t_a + 10\}) \\
TrCondMust_d(w) &= [max\{t_c, t_c + 1\} \le min\{t_a + 8, t_a + 10\} \wedge t_b + 5 \le t_a + 10 \\
&\quad \wedge \mathbf{v}_1 \le 2 + \mathbf{v}_2] \\
(t_c^{inf}, t_c^{sup}) &= (max\{t_b, 2t_b - t_a - 2\}, min\{2t_a + 2, t_a + 7\}) \\
TrCondMust_c(w) &= [max\{t_b, 2t_b - t_a - 2\} \le min\{2t_a + 2, t_a + 7\} \\
&\quad \wedge t_b + 5 \le t_a + 10 \wedge \mathbf{v}_1 \le 2 + \mathbf{v}_2] \\
(t_b^{inf}, t_b^{sup}) &= (max\{t_a, t_a + 1\}, t_a + 3) \\
TrCondMust_b(w) &= [max\{t_a, t_a + 1\} \le t_a + 3 \\
&\quad \wedge t_a + 3 \le min\{t_a + 5, 2t_a + 2, t_a + 7, 1.5t_a + 2, t_a + 4.5\} \\
&\quad \wedge \mathbf{v}_1 \le 2 + \mathbf{v}_2] \\
(t_a^{inf}, t_a^{sup}) &= (max\{1, 2\}, 8) \\
TrCondMust_a(w) &= [max\{1, 2\} \le 8 \wedge \mathbf{v}_1 \le 2 + \mathbf{v}_2] \\
TrCondMust(w) &= \exists \mathbf{v}_1 \exists \mathbf{v}_2 [\mathbf{v}_1 \le 2 + \mathbf{v}_2]
\end{aligned}
$$

Figure 7　Checking must traceability

partially computed. The actual interval for each succeeding action is gradually fixed by substituting the actual timing for preceding actions.

May-traceability.　The "may-traceability" case is quite similar to the must case with a slight modification. We only treat each output action as an input action and derive the intervals of execution time for those I/O actions. If the execution time of each output action belongs to the derived time interval, which is generally narrow than the original executable time interval for the output action, then its succeeding actions can be executed. Thus, we omit the detail.

4.　CONFORMANCE TESTING FOR TIMED I/O AUTOMATA

In this section, we will propose a conformance testing method. Here, we use UIOv-method[16] where, for each state, a transfer sequence and a succeeding UIO sequence are treated as a test case and we derive executable time interval for each input action in the test case using the algorithm explained above.

4.1   PROPOSED TESTING METHOD

Correctness of implementation of states and transitions.   In the proposed method, we identify all states in the IUT as follows.

1 For a given timed I/O automaton, we consider the corresponding FSM in which all transition conditions are removed. We compose a set of UIO sequences $U = \{u_1, \ldots, u_n\}$ $(i = 0, \ldots, n)$ for each state s_i in the FSM. Also we generate the set of transfer sequences $V = \{v_1, \ldots, v_n\}$, each of which leads M from its initial state to a state s_i.

2 Let $V.U_\circ = \bigcup_i v_i.u_i$ and $V.U_\times = \bigcup_{i \neq j} v_i.u'_j.a_{ij}$, where u'_j denotes the longest executable prefix of u_j after v_i is executed, and a_{ij} denotes the first unexecutable I/O action of u_j. Intuitively, $V.U_\circ$ is a set of sequences which check the *executability* of the UIO sequence for each state, and $V.U_\times$ is a set of sequences which check the *un-executability* of UIO sequences for the different states.

3 We decide the must/may-traceability for all sequences in $V.U_\circ$. If all sequences are must-traceable, or at least may-traceable, we can use those sequences as test cases.[3] If not, we return to step (1) and construct different U and V again.

4 We give the generated test cases $V.U_\circ$ and $V.U_\times$ to the IUT. At that time, we execute each input action of a test sequence in $V.U_\circ$ with an executable timing which is calculated from the function obtained by the must/may-traceability checking algorithm described in Section 3 by substituting the actual execution time of preceding actions into the function. For each output action, we check whether the output action is observed at a time which satisfies the timing constraints described in the specification.For each test sequence $v_i.u'_j.a_{ij}$ in $V.U_\times$, we confirm that it is impossible to execute a_{ij} after execution of $v_i.u'_j$ with various timing for more than F (sufficiently large number) times.[4]

5 If we can observe the same response from the IUT as that of the specification (that is, all the sequences of $V.U_\circ$ are executable and those of $V.U_\times$ are not executable), we conclude that we have identified all states in the IUT.

We also identify all transitions in the IUT like the above method. At that time, we construct $V.A.U_\circ = \{v_i.a_{ij}.u_j | a_{ij}$ is an outgoing transition (action)

---

[3]It is desirable that, for each test sequence $v_i.u_i$, the UIO sequence part u_i is must-traceable even if the transfer sequence part v_i is only may-traceable.

[4]Note that we may confirm that we can execute only a partial sequence of u'_j.

from state s_i }, $V.A_{\times} = \{v_i.a_{ij} | a_{ij}$ is not an outgoing transition (action) from state $s_i\}$ where the set V of transfer sequences is the same one as used in the above state identification. Then we confirm that we can observe the same response from the IUT for $V.A.U_o$ as that of the specification, and that we cannot execute a_{ij} in each test sequence of $V.A_{\times}$ after giving v_i with various timing for more than F (sufficiently large number) times. If we can observe the same response from the IUT as that of the specification, we conclude that we have identified all transitions in the IUT.

When we give $V.U_o$ and $V.U_{\times}$ to the IUT and observe the same response as the specification, we regard that the IUT has n different states, which have the same set of UIO sequences as those of the corresponding states of the specification. For the identification of transitions, we use the same set V of transfer sequences as used for state identification. Similar to the traditional UIOv-method, we also regard that for each state in the specification, the corresponding state in the IUT has the same outgoing transitions. Giving an UIO sequence to the destination state of the outgoing transition, we regard that the destination state corresponds to that of the specification.

Correctness of implementation of transition conditions .   Here, we present a method to detect faults in transition conditions of IUTs if the faults are restricted to some typical ones.

On testing of the correctness of transition conditions, we find transitions from the initial state step by step using the breadth first search and test them. For testing of the last k-th transition of a transition sequence whose length is k, we assume that we have already tested the correctness of all the transitions from the first one to the $(k-1)$-th one, and that they are correctly implemented.

To simplify our explanation, we assume that the timing constraint of the k-th transition on a test sequence is specified as $(A_1 \leq t_k) \wedge (A_2 \leq t_k) \wedge \ldots \wedge (A_n \leq t_k) \wedge (t_k \leq B_1) \wedge (t_k \leq B_2) \wedge \ldots \wedge (t_k \leq B_m)$, and that only one of them, for example, $(A_h \leq t_k)$ may be incorrectly implemented and its fault is either one of the following six types of errors : (1) $(A_h < t_k)$, (2) $(A_h \geq t_k)$, (3) $(A_h > t_k)$, (4) $(A_h = t_k)$, (5) $(A_h + C \leq t_k)$ and (6) $(A_h - C \leq t_k)$ (c: positive constant).

As we mentioned in Section 3, the execution time of k-th transition on a given test sequence is specified as $max(A_1, \ldots, A_n) \leq t_k \leq min(B_1, \ldots, B_m)$. Then if A_h is less than $max(A_1, \ldots, A_n)$, it makes no effects even if $(A_h \leq t_k)$ is implemented as $(A_h < t_k)$ incorrectly. For example, even if the timing constraint $(t_{k-1} + 1 \leq t_k) \wedge (t_{k-1} \leq t_k)$ on a specification is implemented as $(t_{k-1} + 1 \leq t_k) \wedge (t_{k-1} < t_k)$ incorrectly, the incorrect part $(t_{k-1} < t_k)$ makes no effects on the correctness of the whole transition condition because the faulty constraint is equal to the correct one $(t_{k-1} + 1 \leq t_k)$. Because we cannot test the correctness of such a condition, we try to test only the correctness of the transition condition which is executable even if we add the constraint $A_p + \epsilon < A_h$ (ϵ: sufficiently small constant, $\epsilon \ll C$) for all other A_p.

For the execution time of k-th transition, we select three types of execution time, (a) $max(A_1, \ldots, A_n) - \epsilon$, (b) $max(A_1, \ldots, A_n)$, (c) $max(A_1, \ldots, A_n) + \epsilon$. The response from the specification (correct IUT) and wrong IUTs with one fault in (1)–(6) is described as follows (the marks $\circ$ and $\times$ mean "executable" and "unexecutable", respectively).

Response	(a)	(b)	(c)
Spec.	$\times$	$\circ$	$\circ$
(1)	$\times$	$\times$	$\circ$
(2)	$\circ$	$\circ$	$\times$
(3)	$\circ$	$\times$	$\times$
(4)	$\times$	$\circ$	$\times$
(5)	$\times$	$\times$	$\times$
(6)	$\circ$	$\circ$	$\circ$

Clearly as this table, we can detect every fault because those responses are different from the response in the specification. For the above example, we generate a symbolic trace w corresponding to the transition sequence from the first transition to k-th transition. Then, if k-th transition is an input action, we try to check that it is possible to execute k-th transition with the above execution timing (b) and (c), and that it is impossible to execute k-th transition for more than F times with the above execution timing (a) after execution of w with various timings. If we can confirm the above, we conclude that the implementation of k-th transition is correct. While if k-th transition is an output action, we try to check that the output action can be executed with the execution timing (b) and (c), and that it cannot be executed with (a) for many times. If we can confirm them, we regard that the transition condition is correct. In general, there may exist other types of errors than the above (1)–(6). Also there may exist multiple faults. So, we cannot detect all types of errors as easily as we proposed above. However we are sure to be able to detect a significant number of errors by our method because these types of errors are typical.

5.   EXAMPLE

We can apply our testing method to the example in Fig. 3 by constructing the set of UIO sequences U as follows.

$$
\begin{aligned}
U = \{ \quad & data\_start?data\_end?first\_display\_start!, \\
& data\_end?first\_display\_start!, \\
& first\_display\_start!, first\_display\_end! \\
& data\_start?data\_end?display\_start!,
\end{aligned}
$$

$$data\_end? display\_start!, display\_start!,$$
$$display\_start\_intime!, display\_end! not\_mdf!,$$
$$display\_end! data\_start?, not\_mdf!\}$$

The set of shortest paths from the initial state to all states can be used as the set of transfer sequences V. We have confirmed we can generate the set of must-traceable test sequences : (A) $V.U_\circ$ and $V.U_\times$ for identifying states, and (B) $V.A.U_\circ$ and $V.A_\times$ for identifying transitions. For example, for a test sequence $w = data\_start?_{\mathbf{v}_1}\ data\_end?\ first\_display\_start!\ first\_display\_end!\ data\_start?_{\mathbf{v}_2}\ data\_end?\ display\_start!\ display\ \_end!\ not\_mdf!$ for identifying state s_9, the condition to make the sequence w must-traceable is $D(\mathbf{v}_1, \mathbf{v}_2) = [b \leq T_{h2} + \mathbf{v}_2 - \mathbf{v}_1 - x]$. If the constants b, T_{h2} and x are actually $10, 2$ and 10, respectively, any input data $\mathbf{v}_1, \mathbf{v}_2$ which satisfy the inequality $10 \leq 2 + \mathbf{v}_2 - \mathbf{v}_1 - 10$ (that is, $18 \leq \mathbf{v}_2 - \mathbf{v}_1$) can be given in order to make w must-traceable. The executable time interval for each input action in w can be specified as a function of the execution time of its preceding actions and input values $\mathbf{v}_1$ and $\mathbf{v}_2$.

We have developed a tool to decide whether a formula like formulas in Fig. 4 is true or not. By using the tool, we can check the must/may-traceability of test sequences like the above example within a few seconds for most cases (Pentium II 200MHz). In general, if we apply the general algorithms checking the satisfiability of rational Presburger sentences [7], it takes much more time to decide the must/may-traceability of a given test sequence (in some cases, a few hours or more). Also, the sufficient condition introduced in Section 3.3 makes the decision time further short. For the example in Fig. 3, we can prove all the test sequences are must-traceable using the sufficient condition introduced in Section 3.

We have applied our technique to the following examples which can be modeled as our timed I/O automata using our test case derivation tool :

- the multiplexing protocol for multimedia communication like H.223[9] which sends packets consisting of three media, that is, movie (MPEG2 etc.), sound and text data.

- Ethernet protocol[14] and its time-out mechanism.

For the above examples, we have derived must-traceable test sequences for most cases. However, for checking some time-out actions (or interruption), we can only derive may-traceable test sequences since such time-out actions can be executed only when their preceding output actions are executed very late. Even for such cases, we can recognize when the preceding output actions should be executed in order to make a time-out action executable.

6.   CONCLUSION

In this paper, we have proposed a timed I/O automaton model for specifying real-time communication protocols, and a method to generate conformance test cases with execution timing for their I/O actions. In general, from a state, if there is a choice (branch) between two output actions and/or between an input action and an output action, then the tester cannot control which output action should be executed, that is, the decision of such a choice is made by the IUT itself. If an undesirable branching output action is executed during a test run, we must reset the test run and try the same test run again. The paper [12] treats such a problem and has proposed a suitable testing method. The method for treating such uncontrollable output actions is also applicable to our method.

As the future work, we are planning to develop a tool which translates given specifications written in time extended SDL and E-LOTOS into the proposed timed I/O automata, and to derive test sequences from those specifications directly.

References

[1] R. Alur and D.L. Dill : "A Theory of Timed Automata", Theoretical Computer Science, Vol. 126, pp.183-235 (1994).

[2] R. Alur and D.L. Dill : "Automata-Theoretic Verification of Real-Time Systems", In Formal Methods for Real-Time Computing, Trends in Software Series, pp.55-82 (1996).

[3] B. S. Bosik and M. U. Uyar : "Finite State Machine Based Formal Methods in Protocol Conformance Testing", Computer Networks ISDN Systems, 22, pp. 7-33 (1991).

[4] D. Clarke and I. Lee : "Automatic Generation of Tests for Timing Constraints from Requirements", Proc. of 3rd Int. Workshop on Object-Oriented Real-Time Dependable Systems (1997).

[5] L. K. Dillon : "Using Symbolic Execution for Verification of Ada Tasking Programs", ACM Trans. on Programming Languages and Systems, Vol. 12, No. 4, pp.643-669 (1990).

[6] A. En-Nouaary, R. Dssouli, F. Khendek and A. Elqortobi : "Timed Test Cases Generation Based on State Characterization Technique", Proc. of 19th IEEE Real-Time Systems Symposium (RTSS'98) (1998).

[7] J. E. Hopcroft and J. D. Ullman : "Introduction to Automata Theory, Languages, and Computation", Addison-Weslay (1979).

[8] Y. Ishibashi, S. Tasaka and E. Minami : "Performance Measurement of a Stored Media Synchronization Mechanism: Quick Recovery Scheme", Proc. of GLOBECOM'95, pp.811-817 (1995).

[9] ITU-T : "- Multiplexing Protocol for Low Bit Rate Multimedia Communication", Recommendation H.223 (1996).

[10] R. Kneuper : "Symbolic Execution: A Semantic Approach", Science of Computer Programming, Vol. 16, No. 3, pp.207-249 (1991).

[11] D. Lee and M. Yannakakis : "Principles and Methods of Testing Finite State Machines - A survey", Proc. of the IEEE, Vol. 84, No. 8 (1996).

[12] Q. M. Tan and A. Petrenko : "Test Generation for Specifications Modeled by Input/Output Automata", Proc. of 11th IFIP Workshop on Testing of Communicating Systems (IWTCS'98), pp.83-99 (1998).

[13] D. Mandrioli, S. Morasca, A. Morzenti: "Generating Test Cases for Real-Time Systems from Logic Specifications", ACM Trans. on Computer Systems, Vol. 13, No. 4, pp.365-398 (1995).

[14] R. Metcalfe and D. Boggs : "Ethernet: Distributed Packet Switching for Local Computer Networks", Communication of the ACM, Vol. 19, No. 7 (1976).

[15] J. Springintveld, F. Vaandrager, and P. R. D'Argenio : "Testing Timed Automata", Technical Report, CTIT 97-17, University of Twente (1997).

[16] S. T. Vuong, W.L. Chan and M. R. Ito : "The UIOv-Method for Protocol Test Sequence Generation", Proc. of 2nd IFIP Workshop on Protocol Test Systems (IWPTS'89), pp.161-175 (1989).

14

TEST GENERATION DRIVEN BY USER-DEFINED FAULT MODELS

I. Koufareva†, A. Petrenko‡ and N. Yevtushenko†

† Tomsk State University
36 Lenin St
Tomsk, 634050, Russia
phone: +7 (3822) 41-39-64
irene@gpb.tspace.ru, yevtushenko@elefot.tsu.tomsk.su

‡ CRIM, Centre de Recherche Informatique de Montréal
550 Sherbrooke West, Suite 100
Montréal, H3A 1B9, Canada
phone: +1 (514) 840-1234, fax: +1 (514) 840-1244
petrenko@crim.ca

Abstract   In this paper, we consider the problem of test derivation from a specification FSM, assuming that all possible implementation FSMs are submachines of some nondeterministic FSM. The latter represents a restricted class of faults defined by the user. The state number in an implementation may exceed that of the specification. We present a method for test generation that can deliver shorter tests than the other existing methods. The method is also more flexible than the traditional FSM-based methods, which embody a universal fault model defined only by a state number.

Keywords: Conformance testing, test generation, fault models, finite state machines.

1.   INTRODUCTION

Conformance testing is often formalized as the FSMs equivalence problem. One usually assumes that all possible implementation faults for a given specification machine are described by a certain set of "mutant" FSMs, called a fault domain. The test suite is complete w.r.t. a given fault domain if for each mutant FSM, nonequivalent to the specification FSM, the test suite has a sequence detecting this machine.

In cases when the fault domain is defined only by maximal number of states of implementations under test (a universal fault domain) we can use a number of test derivation methods [Vasi73], [Chow78], [VCI89], [YePe90], [Petr91], [FBKA91], [YaLe95]. The total length of a test suite complete in such a domain exponentially depends on the value $(m\text{-}n)$ where m is the maximal number of states of implementation or mutant FSMs, while n is the number of states of the specification FSM. Shorter tests are usually required when the fault domain is further restricted to user-defined faults [GrPe88], [BrJu92], [PeYe92], [PoRe97]. The question is how one can efficiently represent a fault domain that is a proper subset of a universal fault domain.

In cases when it is possible to explicitly enumerate all the FSMs of the fault domain one can determine for each FSM, nonequivalent to the specification FSM, an input sequence distinguishing the two machines to eventually derive a complete test suite [Gill62], [PoMc64]. We do not know, however, whether this technique provides the shortest test suite. In cases, when the number of mutants tends to explode, as it often happens for practical systems, an implicit enumeration of mutants, proposed in [PoRe97], can be attempted. According to this approach, one determines the set of all "smallest" partially specified FSMs and for each machine finds a sequence that distinguishes it from the given specification machine. The approach does not rely on a state cover set and distinguishing sequences of a given specification FSM opposed to classical methods, such as *W, Wp, UIOv, HSI*-methods. By this reason, in the worst case situation, when only the maximal number of states of an implementation FSM is known, total length of tests derived by the method [PoRe97] exponentially depends not on the value $(m\text{-}n)$, but on $(m+n\text{-}1)$.

A general representation of mutant FSMs based on a so-called *fault function* has been proposed in [GrPe88], [PeYe92]. Existing fault models for restricted faults such as output faults [NaTs81], input faults [Kozl81], component faults of a system of communicating FSMs [GrPe88], [PYD94], "black-box" faults [PYB96], and others correspond to particular forms of the fault function. Methods for test derivation are developed in [GrPe88],

[PeYe92] for the cases when no mutant FSM has more states than a specification FSM.

In this paper, we extent the notion of a fault function to the notion of a mutation machine driven by the observation that faults actually may increase the number of states. In fact, a mutation machine is a nondeterministic FSM such that the set of its submachines is the user-defined fault domain. The number of states of a mutation machine may exceed that of the specification FSM. To elaborate a strategy that provides shorter tests than those derived by classical testing methods we generalize the notion of universal traversal set that is the key construction in the methods, to a family of traversal sets and propose an algorithm for deriving a complete test suite.

The rest of the paper is structured as follows. In Section 2, we give necessary basic notions and definitions. Section 3 demonstrates that a test suite derived by *W*-method can be reduced when the user-defined fault domain is smaller than a universal fault domain defined by the state number. In Section 4, we construct a so-called distinguishing automaton to elaborate a test derivation approach for specification and mutation machines. An algorithm for test derivation is proposed in Section 5.

2.   PRELIMINARIES

2.1   Finite State Machines

A *finite state machine* (*FSM*), often simply called a machine throughout this paper, is an initialized (possibly nondeterministic) machine, i. e. 5-tuple $A=(S,X,Y,h,s_0)$, where S is a finite set of n states with $s_0 \in S$ as the initial state, X and Y are finite sets of input and output symbols, respectively, and h is a behavior function, $h: S \times X \rightarrow P(S \times Y)$ where $P(S \times Y)$ is a set of all non-empty subsets of $S \times Y$. The machine A is *deterministic* if $|h(s,x)|=1$ for all $(s,x) \in S \times X$; otherwise, it is *nondeterministic*.

For any finite alphabet X, let X^* denote the set of all finite sequences in X containing the empty sequence ε. As usual, we extend the behavior function h of the machine A to a mapping from the set $S \times X^*$ to the set $P(S \times Y^*)$. The function h has two projections h^1 and h^2, usually called the *next state function* and the *output function*. The machine A is said to be *initially connected* or simply *connected* if each state is reachable from the initial state, i.e. for each state $s \in S$ there exists $\alpha \in X^*$ such that $s \in h^1(s_0, \alpha)$.

Given the machine $A=(S,X,Y,h,s_0)$, a machine $B=(S',X,Y,h',s_0)$ is a *submachine* of A if $S'\subseteq S$ and $h'(s,x)\subseteq h(s,x)$ for all $(s,x)\in S'\times X$. The set of all initially connected deterministic submachines of A is denoted $Sub(A)$.

Given two states, s of FSM $A=(S,X,Y,h,s_0)$ and t of FSM $B=(T,X,Y,g,t_0)$, states s and t are said to be *equivalent*, written $s\cong t$, if $g^2(t,\alpha)=h^2(s,\alpha)$ for all input sequences $\alpha\in X^*$, otherwise, states s and t are *distinguishable*, written $s\not\cong t$. For each pair of distinguishable states (s,t), there exists an input sequence $\alpha\in X^*$ such that $g^2(t,\alpha)\ne h^2(s,\alpha)$. In this case, the sequence α is said to *distinguish* states s and t. The machine is said to be *reduced*, if any pair of its states is distinguishable.

The machines are *equivalent* if their initial states are equivalent; otherwise, they are *distinguishable*. An input sequence is said to *distinguish* two machines if it distinguishes their initial states.

2.2   Fault Model

Let $A=(S,X,Y,h,s_0)$ be a deterministic reduced connected FSM, called the *specification machine*. We assume that all possible implementations machines are represented by the set of all deterministic submachines of the FSM $M=(T,X,Y,F,t_0)$, called a *mutation machine*. A submachine B of M is called a *nonconforming implementation* if it is not equivalent to A, otherwise, it is called a *conforming implementation*. In other words, we consider the fault model $<A,\cong,Sub(M)>$ [PYB96].

Any finite set $E\subseteq X^*$ of input sequences is a *test suite* w.r.t the fault model $<A,\cong,Sub(M)>$. The test suite E *detects* a nonconforming implementation machine $B\in Sub(M)$ if there exists an input sequence $\alpha\in E$ distinguishing machines A and B. The test suite E is said to be *complete* w.r.t the fault model $<A,\cong,Sub(M)>$ if it detects each nonconforming implementation machine $B\in Sub(M)$.

Next section is devoted to the discussion of possible approaches for deriving tests complete w.r.t. the fault model $<A,\cong,Sub(M)>$.

3.   TEST DERIVATION USING EXISTING METHODS

3.1   Universal traversal set

Given an input alphabet X, an output alphabet Y and integer m, there exists a special nondeterministic FSM such that any machine with up to m

states defined over alphabets X and Y is isomorphic to its submachine. In particular, consider a *chaos* machine $Ch_m(X,Y)=(P,X,Y,h,p_0)$, where $|P|=m$ and $h(p,x)=P\times Y$ for all $(p,x)\in P\times X$. The fault model $<A,\cong,Sub(Ch_m(X,Y))>$ is a well known "black-box" model. If a given mutation machine M has m states then a complete test suite w.r.t. $<A,\cong,Sub(Ch_m(X,Y))>$ is also complete w.r.t. $<A,\cong,Sub(M)>$, for any submachine of the mutation machine M is isomorphic to a submachine of $Ch_m(X,Y)$. A number of methods exist for deriving tests complete w.r.t. the fault model $<A,\cong,Sub(Ch_m(X,Y))>$. Below we briefly sketch the W-method [Vasi73], [Chow78].

A finite set W of finite input sequences is called a *characterization set* [Chow78] of a reduced FSM A if for each pair of different states of A, the set W has a sequence, distinguishing the states. Given a finite subset V X^* and an integer k, we further denote VX^k the set obtained by concatenating each sequence $\alpha\in V$ with each sequence in X^* up to length k. Let $n(V)$ be a number of distinct states, where sequences of the set V take the FSM A from the initial state, i.e. $n(V)=|\{\delta(s_0,\alpha)\mid\alpha\in V\}|$. The set $VX^{m-|n(V)|+1}W$ is a complete test suite w.r.t. the fault model $<A,\cong,Sub(Ch_m(X,Y))>$.

A	P	Q	R
x	$R/1$	$Q/1$	$P/1$
y	$P/0$	$P/1$	$Q/1$

Figure 1. FSM A.

Example. Consider the machine A (Figure 1) with three states, P, Q, and R; the initial state is P. It has two inputs, x and y; and two outputs, 0 and 1. We apply the W-method to generate tests, assuming that any implementation has at most four states. A characterization set of A is $W=\{yy\}$. Given a set $V=\{\varepsilon,x,xy\}$ of input sequences, $n(V)=3$. We construct the test suite $VX^2W=\{xxxyy,\ xxyyy,\ xyxxyy,\ xyxyyy,\ xyyxyy,\ xyyyyy,\ yxyy,\ yyyy\}$ complete w.r.t. the fault model $<A,\cong,Sub(Ch_4(X,Y))>$.

The set $X^{m-|n(V)|+1}$ that is a key construction of the W-method is called a *universal traversal set* [YaLe95] and enjoys a nice property.

Proposition 3.1. For each nonconforming implementation FSM $B\in Sub(Ch_m(X,Y))$, at least one of the following conditions holds.
- There exist sequences $\alpha\in V$ and $\beta\in X^{m-|n(V)|+1}$ such that the sequence $\alpha\beta$ distinguishes B from A or visits distinguishable states in the FSM

A from the initial state simultaneously visiting one and the same state in the FSM *B*.

- There exist sequences $\alpha, \gamma \in V$ and $\beta \in X^{m-|n(V)|+1}$ such that the sequences $\alpha\beta$ and γ take *A* from the initial state to distinguishable states while taking *B* from the initial state to one the same state.

Since $Ch_m(X, Y)$ is a chaos machine, for any sequences $\alpha, \gamma \in V$, $\beta \in X^{m-|n(V)|+1}$ and any pair of different states of *A*, there exists $B \in Sub(Ch_m(X, Y))$ such that the conditions of Proposition 3.1 hold. Therefore, to obtain a test suite complete w.r.t. the fault model $<A, \cong, Sub(Ch_m(X, Y))>$ we should concatenate each sequence of the set $VX^{m-|n(V)|+1}$ with a characterization set *W*. As mentioned above, a test suite complete w.r.t. $<A, \cong, Sub(Ch_m(X, Y))>$ is also complete w.r.t. $<A, \cong, Sub(M)>$ for any mutation FSM *M* with at most *m* states. However, in particular cases, when *M* is a proper submachine of $Ch_m(X, Y)$ the test suite may be reduced without loss of its completeness.

3.2   **Test Minimization**

Let *TS* be a test suite complete w.r.t. the fault model $<A, \cong, Sub(Ch_m(X, Y))>$ and *Pref(TS)* be the set of all prefixes of *TS*. There may exist a proper subset of *Pref(TS)* that is complete w.r.t. $<A, \cong, Sub(M)>$. If it is possible to explicitly enumerate all the submachines of the mutation machine *M*, the subset can be determined as a column coverage of a special Boolean matrix Θ. The item Θ_{ij} of the matrix is '1' iff the sequence α_i of the set *Pref(TS)* detects a nonconforming submachine B_j of *M*. The existing methods (see, for example, [John74]) can solve the problem for matrices with thousands rows and columns.

M	1	2	3	4
x	3/1 4/1	3/0,1 4/0,1	1/0,1 2/0,1	4/1
y	1/0,1 4/0,1	4/1	4/1	1/0,1

Figure 2: Mutation FSM *M*.

Example. Consider the specification machine *A* with the initial state *P* shown in Figure 1. Suppose now that all the faulty machines are submachines of the FSM *M* (Figure 2) with the initial state 1. The entries of this table are interpreted as follows. Consider as an example {1/0,1; 4/0,1}. It means that any implementation in response to input *y* will transfer to state 1 or 4. In either case, it will produce output 0 or 1. The fault domain consists of 256 machines. By direct inspection, one can assure the set $\{xyy, xxyy\}$ is a column coverage of the corresponding Boolean matrix, i.e. the test suite $\{xyy, xxyy\}$ is complete w.r.t. the fault model $<A,\cong,Sub(M)>$. Its total length is much less than that of a complete test suite w.r.t. $<A,\cong,Sub(Ch_m(X,Y))>$ obtained in the previous section. The example indicates that a complete test suite obtained by the W-method may well be redundant when a given mutation machine is a proper submachine of the chaos machine.

In cases when elements of the set $Sub(M)$ cannot be explicitly enumerated, we may use a test strategy based on traversal sets and state identifiers implemented in the W or other methods. However, given a mutation machine M, Proposition 3.1 may hold for a proper subset of the set $X^{m-|n(V)|+1}$ when M is a proper submachine of $Ch_m(X,Y)$. As an example, consider a mutation FSM M that has only invalid outputs at the initial state for some input x. In this case, the set $\{x\}$ may serve as a traversal set, for each submachine of M is distinguished from A by x. Thus, for a "non-chaotic" mutation FSM M, the universal traversal set $X^{m-|n(V)|+1}$ may be reduced. Moreover, concatenation of each sequence of the traversal set with a characterization set may also be unnecessary. For some mutation FSMs, as in the above example, we do not need a characterization set at all.

Given a specification FSM A, a mutation FSM M and a finite set V of input sequences, we generalize the notion of universal traversal set to the case when not every FSM with up to m states can be an implementation machine. We extent this notion to a so-called family of M-traversal sets w.r.t. the set V. Necessary properties of traversal sets are established in terms of pairs of states, where an input sequence simultaneously takes specification and implementation machines from their initial states (Proposition 3.1). By this reason, to construct M-traversal sets we use distinguishing automata that are similar to products of FSMs, see e.g. [PoMc64], [ACY95].

4. DISTINGUISHING AUTOMATA

4.1   Finite Automata

A *finite automaton* over the finite alphabet X is a 4-tuple $R=(S,X,\delta,s_0)$, where S is a finite set of states, $s_0 \in S$ is an initial state and δ is a transition function, $\delta\colon S \times X \to P(S)$. The automaton R is *deterministic* if $|\delta(s,x)|=1$ for all states $s \in S$ and symbols $x \in X$, otherwise it is *nondeterministic*.

Given an automaton $R=(S,X,\delta,s_0)$, a sequence $\eta \cdot \beta = s x_1 s_1 x_2 s_2 \ldots x_k s_k$, where $\eta = s s_1 s_2 \ldots s_k \in S^*$, $\beta = x_1 \ldots x_k \in X^*$, is called a *path* of the automaton R at the state s if $s_1 \in \delta(s,x_1)$ and $s_i \in \delta(s_{i-1},x_i)$ for all i, $1 < i\ k$; the sequence β is called *X-projection* of the path $\eta \cdot \beta$. Given a sequence $\beta \in X^*$, any path $\eta \cdot \beta$ at the initial state of R is said to be *created* by the sequence β in the automaton R. As usual, the integer k is called *length* of the path, while states s and s_k are called the *head* and the *tail* states of the path. A path at the initial state of the automaton is simply called a *path of the automaton* throughout this paper. When we want to indicate that the head state of the path $\eta \cdot \beta$ is s we rewrite it as $s\eta \cdot \beta$. For each state s of the automaton R, there always exists a special path $ss \cdot \varepsilon$, written $s \cdot \varepsilon$ for short, where ε is the empty sequence. By definition, this path has zero length and its head and tail states coincide. A path $\eta \cdot \beta$ is said to *traverse* state $s \in S$ if the sequence η includes state s . State s of the automaton R is called *reachable* if there exists a path of R traversing the state s.

Given the automaton $R=(S,X,\delta,s_0)$, an automaton $R\ =(S\ ,X,\delta\ ,s_0)$ is a *sub-automaton* of R if $S\ \subseteq S$ and $\delta\ (s,x)\subseteq \delta(s,x)$ for all $(s,x)\in S\ \times X$.

4.2   Distinguishing Automaton of Two FSMs

Given two FSMs $A=(S,X,Y,h,s_0)$ and $B=(T,X,Y,g,t_0)$, we construct an automaton $((S \times T)\ \ \{Fail\},X,f,s_0 t_0)$, such that for all $(s,t)\in S \times T$ and all $x \in X$ it holds that:

- $f(st,x) = \begin{cases} \Sigma(st,x), \text{ if } h^2(s,x)=g^2(t,x) \\ \Sigma(st,x)\ \ \{Fail\}, \text{ if } h^2(s,x)\ g^2(t,x) \end{cases}$

where $\Sigma(st,x)=\{s\ t\ \ |\ \ \exists y \in h^2(s,x) \cap g^2(t,x)\ \ [(s\ ,y)\in h(s,x)\ \ \&\ (t\ ,y)\in g(t,x)]\}$;

- $f(Fail,x)=\{Fail\}$.

Let Q denote the set of all reachable states of the automaton. Then the sub-automaton (Q,X,f,q_0), where $q_0 = s_0 t_0$, is called the *distinguishing automaton*, denoted $A\ \ B$. In this definition, *Fail* is a designated state which indicates that the two machines cannot agree on outputs.

$A\ M$	P1	P2	P4	Q1	Q2	Q3	Q4	R3	R4	Fail
x	R3 R4	R3 R4 Fail	R4	Q3 Q4	Q3 Q4 Fail	Q1 Q2 Fail	Q4	P1 P2 Fail	P4	Fail
y	P1 P4 Fail	Fail	P1 Fail	P1 P4 Fail	P4	P4	P1 Fail	Q4	Q1 Fail	Fail

Figure 3: The distinguishing automaton $A\ M$.

Example. The distinguishing automaton $A\ M$ for the machines A (Figure 1) and M (Figure 2) with the initial states P and 1, respectively, is shown in Figure 3. The initial state of $A\ M$ is $P1$.

For deterministic machines A and B, the distinguishing automaton $A\ B$ is often used to verify whether machines A and B are equivalent. In fact, the automaton $A\ B$ has the state *Fail* if and only if the machines A and B are distinguishable. The language accepted by the state *Fail* is the set of all sequences distinguishing A and B.

Proposition 4.1. Given a deterministic FSM A and an FSM M over the same input and output alphabets, let $A\ M$ be the distinguishing automaton for A and M. For any machine $B \in Sub(M)$, the distinguishing automaton $A\ B$ is a sub-automaton of $A\ M$.

Corollary 4.2. Given a distinguishing automaton $A\ B$, $B \in Sub(M)$, let $\eta_* \beta$ be a path of $A\ B$. Then $\eta_* \beta$ is also a path of the distinguishing automaton $A\ M$.

For any state $q=st$ of the distinguishing automaton $A\ M$, the states s and t of machines A and M are S-projection and T-projection of q, respectively. Two paths $\sigma_* \alpha$ and $\rho_* \gamma$ at arbitrary states of the distinguishing automaton $A\ M$ are said to be *compatible*, if for any two segments qxq of $\sigma_* \alpha$ and pxp of $\rho_* \gamma$ such that states q and p have the same T-projection, either one of states q and p is *Fail*, or the T-projections of states q and p also coincide. Path $\sigma_* \alpha$ at an arbitrary state of the distinguishing automaton $A\ M$ is called *deterministic* if it is compatible with itself.

Proposition 4.3. For any implementation $B \in Sub(M)$, each path of the distinguishing automaton $A\ B$ is deterministic and any two paths of $A\ B$ are compatible.

4.3   Detecting Nonconforming Machines

In this section, we introduce a notion of a sequence set that "traps" a nonconforming implementation $B \in Sub(M)$ just as the set $VX^{m-|n(V)|+1}$ does for machines in the set $Sub(Ch_m(X,Y)$ (Proposition 3.1). Due to Corollary 4.2, pairs of states where input sequences simultaneously take a specification machine A and an implementation $B \in Sub(M)$, are states of the distinguishing automaton $A \otimes M$; thus, we need to study properties of the states first.

Let $A \otimes M=(Q,X,H,q_0)$ be the distinguishing automaton of FSMs A and M, and $q \otimes Fail$ be a state of $A \otimes M$. The state q of $A \otimes M$ is 1-*forbidden* if there exists a symbol $x \in X$ such that $H(q,x)=\{Fail\}$. Suppose we have determined all $(k-1)$-forbidden states of the automaton $A \otimes M$, $k>1$. The state q of $A \otimes M$ is *k-forbidden* if it is $(k-1)$-forbidden or there exists a symbol $x \in X$ such that each state of the set $H(q,x)$ is $(k-1)$-forbidden.

Since the set Q of states of the distinguishing automaton $A \otimes M$ is finite there exists an integer k such that the sets of k-forbidden and $(k+1)$-forbidden states coincide, i.e. the procedure is exhaustive. We call a state *forbidden* if there exists an integer k such that it is k-forbidden. As an example, state $P2$ of the distinguishing automaton $A \otimes M$ (Figure 3) is forbidden.

Proposition 4.4. If the initial state of the automaton $A \otimes M$ is forbidden then each implementation machine $B \in Sub(M)$ is nonconforming.

For each forbidden state q of $A \otimes M$, there exists a so-called *distinguishing set* $W(q)$ such that for any sub-automaton $A \otimes B$, $B \in Sub(M)$, with the state q, at least one sequence $\alpha \in W(q)$ takes $A \otimes B$ from the state q to the state *Fail*. A set $W(q)$ can be constructed as follows.

Let q be an 1-forbidden state of $A \otimes M=(Q,X,H,q_0\}$. Then there exists a symbol $x \in X$ such that $H(q,x)=\{Fail\}$. The set $\{x\}$ may serve a distinguishing set $W(q)$. Suppose we have determined distinguishing sets for all $(k-1)$-forbidden states of the automaton $A \otimes M$, $k \geq 1$, and state q is k-forbidden. If q is $(k-1)$-forbidden then its distinguishing set is determined. Otherwise, there exists a symbol $x \in X$ such that each state of the set $H(q,x)$ is $(k-1)$-forbidden. The union $W(q)$ of the sets $xW(q')$ over all states $q' \in H(q,x)$ can serve a distinguishing set for the state q. As an example, the forbidden state $P2$ has a distinguishing set $W(P2)=\{y\}$.

By definition, if a distinguishing automaton $A \otimes B$, $B \in Sub(M)$, has a forbidden state then it has the state *Fail*. We also introduce a notion of conflicting states in the automaton A M. Any distinguishing automaton A B with conflicting states also has the state *Fail*.

Given a specification FSM A and a mutation FSM M, two states $s_1 t$ and $s_2 t$ for different states s_1 and s_2 of A and some state t of M are called *conflicting* states. The states $s_1 t$ and $s_2 t$ are said to *conflict* on the sequence ω if the sequence ω distinguishes states s_1 and s_2 in the FSM A. For any distinguishing automaton A B with states $s_1 t$ and $s_2 t$, the sequence ω takes A B to the state *Fail* from at least one of states $s_1 t$ and $s_2 t$.

A set $L \subseteq X^*$ of sequences is said to *trap* a machine $B \in Sub(M)$ (w.r.t. the specification machine A), if the set of tail states of paths of A B created by the sequences in L has two conflicting states or a state that is *Fail* or forbidden. Given a set $L_B \subseteq X^*$ that traps B, we can easily construct a set of sequences distinguishing B and A. In fact, if there exists a path $\eta \cdot \beta$, $\beta \in L_B$, of A B, whose tail state is *Fail* then the sequence β distinguishes machines A and B. If some path $\eta \cdot \beta$, $\beta \in L_B$, of A B has a forbidden tail state q then at least one sequence of the set $\beta W(q)$, where $W(q)$ is a distinguishing set of the state q, distinguishes A and B. Finally, if A B has two paths $\eta_1 \cdot \beta_1$, $\eta_2 \cdot \beta_2$, $\beta_1, \beta_2 \in L_B$, such that their tail states conflict, then one of sequences $\beta_1 \omega_{12}$ and $\beta_2 \omega_{12}$, where ω_{12} is a sequence on which the states conflict, distinguishes A and B.

A path $\eta \cdot \beta$ of A M is called *nonconforming* if η includes two conflicting states or a state that is *Fail* or forbidden; otherwise the path is *conforming*. Given an implementation FSM B, $B \in Sub(M)$, if the distinguishing automaton A B has a nonconforming path then B is a nonconforming implementation and the set $Pref(\beta)$ of all prefixes of the sequence β traps B.

5. CONSTRUCTING TESTS BASED ON DISTINGUISHING AUTOMATA

5.1 Algorithm for Test Derivation

In this section, we develop the notion of a family of M-traversal sets by generalizing universal traversal sets to incorporate the mutation machine M. A family of M-traversal sets serves a basis for deriving tests complete w.r.t. the fault model $\langle A, \cong, Sub(M) \rangle$.

First, we give some additional notations. For any sequence $\alpha \in X^*$, we use $Path(\alpha)$ to denote the set of all deterministic paths of the distinguishing automaton A M created by α. Given a sequence α and a deterministic path

$\rho_* \gamma$ of A M, we denote $Path(\alpha)|(\rho_* \gamma)$ the set of all paths of $Path(\alpha)$ compatible with $\rho_* \gamma$. For any set $V \subseteq X^*$, let $Path(V)$ $(Path(V)|(\rho_* \gamma))$ denote the union of the sets $Path(\alpha)$ $(Path(\alpha)|(\rho_* \gamma))$ over all $\alpha \in V$.

Definition 5.1. Given the distinguishing automaton A M of FSMs A and M, a finite set V of sequences and the set $Path(V)$ of all paths of A M created by V, let $\mathcal{T}$ be a family of sets $Tr(\sigma_* \alpha)$ of paths $q\eta_* \beta$ at the tail state q of $\sigma_* \alpha$, for all $\sigma_* \alpha \in Path(V)$. The family $\mathcal{T}$ is said to be a *family of M-traversal sets* w.r.t. the set V if for each nonconforming implementation $B \in Sub(M)$, there exist paths $\sigma_* \alpha \in Path(V)$ and $q\eta_* \beta \in Tr(\sigma_* \alpha)$ such that $\sigma\eta_* \alpha\beta$ is a path of the distinguishing automaton A B and the set $Pref(\alpha\beta)$ or the set V $\{\alpha\beta\}$ traps the machine B w.r.t. A.

Any mutation machine with up to m states is a submachine of the FSM $Ch_m(X,Y)$. Thus, due to Proposition 3.1, a family of the sets $Tr(\sigma_* \alpha)$, where for each $\sigma_* \alpha \in Path(V)$, the set $Tr(\sigma_* \alpha)$ comprises all paths of length up to $m-n(V)+1$ at the tail state of $\sigma_* \alpha$, is a family of M-traversal set w.r.t. V. However, for a "non-chaotic" machine M, such M-traversal sets may be redundant as it was demonstrated in Section 3.2.

Based on the notion of a family of M-traversal sets, we propose a method for deriving a test suite complete w.r.t. the fault model $<A, \cong, Sub(M)>$. We further assume that the set $Sub(M)$ has at least one nonconforming implementation; otherwise no implementation needs any test at all.

Algorithm 5.1.

Input: The distinguishing automaton A M, the set of all forbidden states of A M with their distinguishing sets, the set of all pairs of conflicting states of A M, for each pair of conflicting states, a sequence on which the states conflict, a finite set V X^* and a family of M-traversal sets $Tr(\sigma_* \alpha)$, $\sigma_* \alpha \in Path(V)$, w.r.t. the set V.

Output: A complete test suite w.r.t. the fault model $<A, \cong, Sub(M)>$.

Method. Step 1. For each $\alpha \in V$, construct the set E_α X^* of sequences as follows. Consider each deterministic path $\sigma\eta_* \alpha\beta$, where $\sigma_* \alpha \in Path(V)$ and $q\eta_* \beta \in Tr(\sigma_* \alpha)$.

- If the path $\sigma\eta_* \alpha\beta$ is nonconforming, determine its shortest nonconforming prefix $\eta_* \beta$; let p denote the tail state of $\eta_* \beta$. If $p = Fail$ then include β into the set E_α. If p is forbidden then include into the set E_α each sequence of the set β $W(p)$, where $W(p)$ is a distinguishing set of the state p. If p is neither forbidden state nor *Fail*, determine a prefix $\eta_* \beta$ of $\eta_* \beta$, whose tail state conflicts with the state p and include into the set E_α the sequences $\beta\,\omega$ and $\beta\,\omega$, where ω is the sequence, on which the states conflict.

- If the path $\sigma\eta \cdot \alpha\beta$ is conforming, then for each pair of compatible conforming paths $\eta \cdot \beta$, $\eta \cdot \beta \in Path(V)|(\sigma\eta \cdot \alpha\beta)\,\{\sigma\eta \cdot \alpha\beta\}$, whose tail states conflict, include into the set E_α the sequences $\beta\,\omega$ and $\beta\,\omega$, where ω is a sequence on which the states conflict.

Step 2. Find the set TS as union of sets E_α over $\alpha \in V$.

$\square$

The following theorem immediately results from Algorithm 5.1 and the definition of a family of M-traversal sets.

Theorem 5.1. Given the distinguishing automaton $A\,M$ of a specification machine A and a mutation machine M, and a finite set $V\,X^*$ of sequences, let a family of sets $Tr(\sigma \cdot \alpha)$, $\sigma \cdot \alpha \in Path(V)$, of paths of $A\,M$ be a family of M-traversal sets w.r.t. V. Then the set TS derived by Algorithm 5.1 is a test suite complete w.r.t. the fault model $<A, \cong, Sub(M)>$.

$\square$

Total length of a test suite delivered by Algorithm 5.1 essentially depends on total length of a given family of M-traversal sets. If the distinguishing automaton $A\,M$ has no forbidden states and the sets of X-projections of M-traversal sets of a given family are proper subsets of the set $X^{m-|n(V)|+1}$ then Algorithm 5.1 provides tests shorter than W-method. In the next section, given a distinguishing automaton $A\,M$, we present an algorithm for deriving a family of M-traversal sets.

5.2   Constructing M-traversal sets

Given a distinguishing automaton $A\,M$ and a finite set $V\,X^*$, as mentioned in Section 5.1, the family of all paths of length up to $m-|n(V)|+1$ at tail states of the paths $\sigma \cdot \alpha \in Path(V)$ is a family of M-traversal sets w.r.t. the set V. Moreover, for certain mutation machines this length can be reduced as follows.

For any sequence $\alpha \in X^*$ and any path $\rho \cdot \gamma$ at the initial state of the distinguishing automaton $A\,M$, let $P(\alpha)$ (or $P(\alpha, \rho \cdot \gamma)$) denote the set of tail states of all conforming paths of the set $Path(\alpha)$ (or $Path(\alpha)|(\rho \cdot \gamma)$, respectively). As an example, for the distinguishing automaton $A\,M$ (Figure 3), the sequence $xyxx$ and the path $P1xR3yQ4yP1xR3$, we have $P(xyxx, P1xR3yQ4yP1xR3)=\{R3\}$.

Proposition 5.2. Given the distinguishing automaton $A\,M$ and a set V of sequences such that the sets $P(\alpha)$, $\alpha \in V$, are pairwise disjoint. A family of the sets $Tr(\sigma_* \alpha)$, $\sigma_* \alpha \in Path(V)$, where $Tr(\sigma_* \alpha)$ is the set of all paths of length up to $m-|V|+1$ at the tail state of the path $\sigma_* \alpha$, is a family of M-traversal set w.r.t. V.

Given an arbitrary set $V \subseteq X^*$, let $\hat{V}$ denote a maximal subset of V such that the subsets $P(\alpha)$, $\alpha \in \hat{V}$, are pairwise disjoint. Then, due to Proposition 5.2, the family of all paths of length up to $m-|\hat{V}|+1$ at tail states of paths $\sigma_* \alpha \in Path(\hat{V})$ is a family of M-traversal sets w.r.t. the set $\hat{V}$ and, since $\hat{V}$ is a subset of V, is a family of M-traversal sets w.r.t. the set V, as well. Obviously, cardinality of the set $\hat{V}$ is not less than the value of $n(V)$, and thus, total length of a family of M-traversal sets of Proposition 5.2 is not more than that stated in Proposition 3.1.

Due to these considerations, to construct a family of M-traversal sets w.r.t. V, we may determine the maximal subset $\hat{V}$ with the above property, and construct a family of M-traversal sets w.r.t. $\hat{V}$. Moreover, considering each particular path we may further reduce the family. Given the distinguishing automaton of machines A and M, and a set $V \subseteq X^*$ such that the subsets $P(\alpha)$, $\alpha \in V$, are pairwise disjoint, we now present an algorithm for deriving a family of M-traversal sets w.r.t. V.

Algorithm 5.2.

Input: The distinguishing automaton $A \cap M$ and a finite set $V \subseteq X^*$ with the empty sequence ε, such that the sets $P(\alpha)$, $\alpha \in V$, are pairwise disjoint.

Output: A family of M-traversal sets $Tr(\sigma_* \alpha)$, $\sigma_* \alpha \in Path(V)$, w.r.t. V.

Method. Step 1. For each deterministic nonconforming path $\sigma_* \alpha \in Path(V)$ with the tail state q, assign $Tr(\sigma_* \alpha) = \{q_* \varepsilon\}$.

Step 2. For each deterministic conforming path $\sigma_* \alpha \in Path(V)$ construct the set $Tr(\sigma_* \alpha)$ as follows. If there exists a proper prefix of the path $\sigma_* \alpha$ that is in the set $Path(V)$ and has the tail state q, where q is the tail state of $\sigma_* \alpha$, then assign $Tr(\sigma_* \alpha) = \varnothing$. Otherwise, consider each deterministic path $q \eta_* \beta$ of length $m-|V|+1$ at the state q of $A \cap M$.

- If the path $\sigma\eta_* \alpha\beta$ is conforming, the states of the sequence η are distinct and for each $\gamma \in V$, the set $P(\gamma, \sigma\eta_* \alpha\beta)$ is not empty and is not a subset of states of η, then include into the set $Tr(\sigma_* \alpha)$ each prefix of $q\eta_* \beta$.

- If the path $\sigma\eta_* \alpha\beta$ is nonconforming then determine the shortest prefix $q\eta'_*\beta'$ of $q\eta_*\beta$ such that the path $\sigma\eta'_*\alpha\beta'$ is still nonconforming. If the states of the sequence η' are distinct and for each $\gamma \in V$, the set $P(\gamma, \sigma\eta * \alpha\beta)$ is not empty and is not a subset of states of η, then include the path $q\eta * \beta$ into the set $Tr(\sigma * \alpha)$.

$\square$

Theorem 5.3. Given the distinguishing automaton $A \cap M$ and a finite set $V \subseteq X^*$ such that $\varepsilon \in V$ and the sets $P(\alpha)$, $\alpha \in V$, are pairwise disjoint, a family of subsets $Tr(\sigma * \alpha)$ of paths of $A \cap M$ derived by Algorithm 5.2 is a family of M-traversal sets w.r.t. V.

Proof. Consider a nonconforming implementation $B \in Sub(M)$. Due to Proposition 4.1, the distinguishing automaton $A \cap B$ is a sub-automaton of $A \cap M$ with the state *Fail*. Thus, each path at the initial state of $A \cap B$ created by a sequence of the set V is a deterministic path of the set $Path(V)$. Let $P_{A \cap B}(V)$ denote the set of tail states of all paths, where sequences of V take $A \cap B$ from the initial state.

If for some sequence $\alpha \in V$ the path $\sigma * \alpha$ at the initial state of $A \cap B$ is nonconforming then, due to Step 1, there exist paths $\sigma * \alpha \in Path(V)$ and $q * \varepsilon \in Tr(\sigma * \alpha)$ such that $\sigma * \alpha\varepsilon$ is a path of the distinguishing automaton $A \cap B$ and the set $Pref(\alpha\varepsilon)$ traps the machine B w.r.t. the specification machine A.

Suppose now each path $\sigma * \alpha$, $\alpha \in V$, at the initial state of $A \cap B$ is conforming, i.e. *Fail* $\notin P_{A \cap B}(V)$. Let the path $q\chi * \delta$, $q \in P_{A \cap B}(V)$, be the shortest one of all paths at states of the set $P_{A \cap B}(V)$ such that their tail state is *Fail*; and $\sigma * \alpha$ be the shortest path at the initial state of $A \cap B$ such that $\alpha \in V$ and the tail state of $\sigma * \alpha$ is q. Obviously, states of the sequence χ are distinct and the sequence χ does not contain the states of the set $P_{A \cap B}(V)$. This means that for each $\gamma \in V$, the set $P(\gamma, \sigma\chi * \alpha\delta)$ is not empty and is not a subset of states of χ; and the same holds for each proper prefix of $q\chi * \delta$. Since $\sigma * \alpha$ is the shortest path of $A \cap B$ with the tail state q, such that $\alpha \in V$, each proper prefix of the path $\sigma * \alpha$ that is in the set $Path(V)$ has the tail state different from q; thus, the set $Tr(\sigma * \alpha)$ is not empty.

Suppose the path $q\chi * \delta$ has prefixes of length of up to $m - |V| + 1$ that constitute nonconforming paths when concatenated to $\sigma * \alpha$. Then let $q\eta * \beta$ denote the shortest such prefix, i.e. the path $\sigma\eta * \alpha\beta$ is nonconforming. Due to Step 2, $q\eta * \beta \in Tr(\sigma * \alpha)$. Thus, there exist paths $\sigma * \alpha \in Path(V)$ and $q\eta * \beta \in Tr(\sigma * \alpha)$ such that $\sigma\eta * \alpha\beta$ is a path of the distinguishing automaton $A \cap B$ and the set $Pref(\alpha\beta)$ traps the machine B.

Suppose that for any prefix $q\eta * \beta$ of length of up to $m - |V| + 1$ of the path $q\chi * \delta$, the path $\sigma\eta * \alpha\beta$ is conforming. This means that length of $q\chi * \delta$ exceeds $m - |V| + 1$. In this case, let $q\eta * \beta$, $\eta = q_1 q_2 \ldots q_k$, denote a prefix of the path $q\chi * \delta$ of length $k = m - |V| + 1$. Due to Step 2, each prefix of $q\eta * \beta$ is in the set $Tr(\sigma * \alpha)$. Since the sets $P(\alpha)$, $\alpha \in V$, are pairwise disjoint, it holds that $|P_{A \cap B}(V)| = |V|$; thus, the cardinality of the set $P_{A \cap B}(V) \cup \{q_1, \ldots, q_k\}$ is

$m+1$. The states of the set $P_{A\ B}(V)$ $\{q_1,\ldots,\ q_k\}$ are distinct; therefore, there are two conflicting states in the set. This means, that there exists a prefix $q\,\eta\cdot\beta\in Tr(\sigma\cdot\alpha)$ of $q\,\eta\cdot\beta$ such that the set $V\ \{\alpha\beta\}$ traps B w.r.t. A. Thus the family of sets derived by Algorithm 5.2 is a family of M-traversal sets w.r.t. V.

❑

Example. Consider the distinguishing automaton $A\ M$ in Figure 3 and the set $V=\{\varepsilon,x,xy\}$ of sequences that is a state cover of A. The path $P1xR4yQ1\in Path(V)$ is nonconforming, thus, $P(xy)=\{Q4\}$. By direct inspection, one can assure that the sets $P(\varepsilon)=\{P1\}$, $P(x)=\{R3,\ R4\}$, $P(xy)=\{Q4\}$ are pairwise disjoint, i.e. $\hat{V}=V$ and length of a path of any M-traversal set derived by Algorithm 5.2 does not exceed $m-|V|+1=2$. We illustrate Algorithm 5.2 constructing the set $Tr(P1xR3yQ4)$ of paths at the tail state $Q4$ of the path $P1xR3yQ4$. Consider all paths of length 2 at the state $Q4$, i.e. the paths (1) $Q4xQ4xQ4$, (2) $Q4xQ4yP1$, (3) $Q4xQ4yFail$, (4) $Q4yP1xR3$, (5) $Q4yP1xR4$, (6) $Q4yP1yP1$, (7) $Q4yP1yP4$, (8) $Q4yP1yFail$, (9) $Q4yFailxFail$, (10) $Q4yFailyFail$.

(1), (6): the paths $P1xR3yQ4xQ4xQ4$ and $P1xR3yQ4yP1yP1$ are conforming and the states of the sequences $Q4Q4$ and $P1P1$, respectively, are not distinct. In this case, we do nothing.

(2), (4): the paths $P1xR3yQ4xQ4yP1$ and $P1xR3yQ4yP1xR3$ are conforming and the sets $P(xy,P1xR3yQ4xQ4yP1)=0Q4\}$ and $P(\varepsilon,P1xR3yQ4yP1xR3)=\{P1\}$ are subsets of states of the sequences $Q4P1$ and $P1R3$, respectively. We do nothing.

(3), (5), (7), (8): the paths $P1xR3yQ4xQ4yFail$, $P1xR3yQ4yP1xR4$, $P1xR3yQ4yP1yP4$, $P1xR3yQ4yP1yFail$ are nonconforming while all their proper prefixes are conforming. The sets $P(xy,P1xR3yQ4xQ4yFail)=\{Q4\}$, $P(\varepsilon,P1xR3yQ4yP1xR4)=\{P1\}$, $P(\varepsilon,P1xR3yQ4yP1yP4)=\{P1\}$, $P(\varepsilon,P1xR3yQ4yP1yFail)=\{P1\}$ are subsets of states of the sequences $Q4Fail$, $P1R4$, $P1P4$, $P1Fail$, respectively. We do nothing.

(9), (10): the paths $P1xR3yQ4yFailxFail$ and $P1xR3yQ4yFailyFail$ are nonconforming and $Q4yFail$ is the shortest prefix of $Q4yFailxFail$ and $Q4yFailyFail$ such that the path $P1xR3yQ4yFail$ is also nonconforming. The states of the sequence $Q4Fail$ are distinct and for any sequence $y\in V$ the set $P(\gamma,P1xR3yQ4yFail)$ is not empty and is not a subset of states of $Q4Fail$. We include the path $Q4yFail$ into the set $Tr(P1xR3yQ4)$.

Thus $Tr(P1xR3yQ4)=\{Q4yFail\}$. Algorithm 5.2 results in the following family of M-traversal sets:

$Tr(P1\cdot\varepsilon)=\{\ P1yP4xR4,\ P1yP4yFail,\ P1yFail\ \}$;

$Tr(P1xR3)=\{\ R3xP2,\ R3xFail\ \}$;

$Tr(P1xR4)=\ \ $;

$Tr(P1xR3yQ4) = \{ Q4yFail \};$
$Tr(P1xR4yQ1) = \{ Q1{*}\varepsilon \};$
$Tr(P1xR4yFail) = \{ Fail{*}\varepsilon \}.$

We now call Algorithm 5.1 to derive a test suite complete w.r.t. the fault model $<A,\cong,Sub(M)>$. The states P and Q, as well as the states P and R, are distinguished by the sequence y. The sequence yy may serve as a distinguishing sequence for Q and R. At Step 1 of the algorithm, we obtain $E_\varepsilon=\{yxy,\ yy\}$, $E_x=\{xxy\}$, $E_{xy}=\{xyy,\ y\}$. Due to Theorem 5.3, the union $E=\{xxy,\ xyy,\ yxy,\ yy\}$ of the sets E_ε, E_x and E_{xy} is a test suite complete w.r.t. $<A,\cong,Sub(M)>$.

The Algorithm 5.2 usually delivers a shorter family of M-traversal sets than that created by all sequences of length up to $m-|V|+1$. However, the resulting family can further be reduced if one undertakes a more complicated analysis of the obtained paths. For example, consider the traversal set $Tr(P1{*}\varepsilon)=\{P1yP4xR4,\ P1yP4yFail,\ P1yFail\}$. In fact, for each implementation $B\in Sub(M)$ such that the distinguishing automaton $A\ \ M$ has the path $P1yP4$, the sequence $xy\in V$ either creates a nonconforming path in $A\ \ M$, or takes $A\ \ M$ from the initial state to the state $Q4$ conflicting with the tail state of $P1yP4$. Thus, replacing the set $Tr(P1{*}\varepsilon)$ with the set $Tr(P1{*}\varepsilon)=\{P1yP4,\ P1yFail\}$, we obtain a shorter family that is also a family of M-traversal sets. Applying Algorithm 5.1, we derive a set $E=\{xxy,\ xyy,\ yy\}$ that is a test suite complete w.r.t. $<A,\cong,Sub(M)>$. This example indicates that more research is needed to optimise the family of M-traversal sets and eventually a complete test suite.

6.   CONCLUSION

In this paper, we presented a test derivation approach to deal with situations when the specification FSM is a reduced completely specified FSM, while any implementation machine is a deterministic submachine of some nondeterministic FSM, called a mutation FSM. The mutation machine allows the user to model faults in a very compact way. Compared to the previous work based on a fault function, we allow for an implementation to have more states than its specification. Similar to other existing methods, the key step in our method is construction of traversal sets. We extended the notion of universal traversal set to a so-called family of M-traversal sets and based on this new notion, we proposed a method for test generation that can deliver shorter (at least never longer) tests than the other existing methods. The method is also more flexible than the traditional FSM-based methods, which embody a universal fault model defined only by the state number. Our future work is related to generalization of the proposed approach to nondeterministic partially defined specification and weakly initialized mutation FSMs.

Acknowledgments. This work was in part supported by the NSERC grant OGP0194381.

REFERENCES

[ACY95] R. Alur, C. Courcoubetis, and M. Yannakakis, *Distinguishing tests for nondeterministic and probabilistic machines*, Proceedings of the 27th ACM Symposium on Theory of Computing, 1995, pp. 363-372.

[BrJu92] J. A. Brzozowski, H. Jurgensen, *A Model for sequential machine testing and diagnosis*, Journal of Electronic Testing: Theory and Applications, 2, 1992, pp. 219-234.

[Chow78] T. S. Chow, *Test software design modeled by finite state machines*, IEEE Transactions, SE-4, No. 3, 1978, pp. 178-187.

[FBKA91] S. Fujiwara, G. v. Bochmann, F. Khendek, M. Amalou, A. Ghedamsi, *Test selection based on finite state models*, IEEE Trans. SE-17, No. 6, 1991, pp. 591-603.

[Gill62] A. Gill, *Introduction to the theory of finite-state machines*, McGraw-Hill, 1962, 207 p.

[GrPe88] I. Grunskij, A. Petrenko, *Design of checking experiments with automata describing protocols*, in Automatic Control and Computer Science, Allerton Press Inc., N.Y., 1988, No.4, pp. 7-14.

[John74] D. S. Johnson, *Approximation algorithms for combinatorial problems*, Journal of Computer and System Sciences, 9, 1974, pp. 256-278.

[Kozl81] V. A. Kozlovsky, *On the recognition of an automaton relative to a locally generated class*, Soviet Math. Dokl., 1981, vol.23, No.3, pp. 625-628.

[NaTs81] S. Naito and M. Tsunoyama, *Fault detection for sequential machines by transition tours*, Proceedings of Fault Tolerant Comp. Syst., 1981, pp. 238-243.

[Petr91] A. Petrenko, *Checking experiments with protocol machines*, IFIP Transactions, Protocol Testing Systems IV (the Proceedings if IFIP TC6 4th International Workshop on Protocol Test Systems, 1991), Ed. by Jan Kroon, Rudolf J. Heijink and Ed Brinksma, 1992, North-Holland, pp. 83-94.

[PeYe92] A. Petrenko, N. Yevtushenko, *Test suite generation for a given type of implementation errors*, Proceedings of the IFIP XII International Conference on Protocol Specification, Testing and Verification, 1992, pp. 229-243.

[PYD94] A. Petrenko, N. Yevtushenko, and R. Dssouli, *Testing strategies for communicating FSMs*, Proceedings of the 7th IWTCS, 1994, pp. 193-208.

[PYB96] A. Petrenko, N. Yevtushenko, G. v. Bochmann, and R. Dssouli. *Testing in context: framework and test derivation*, Computer communications, Vol. 19, pp. 1236-1249, 1996.

[PoMc64] J. F. Poage, E. J. McCluskey, *Derivation of optimum test sequences for sequential machines*, Proceedings of the 5th Annu. Symp. On Switching Theory and Logical Design, 1964.

[PoRe97] I. Pomeranz, S. M. Reddy, *Test generation for multiple state-table faults in finite-state machines*, IEEE Transactions on Computers, Vol. 46, No 7, pp. 782-794, 1997.

[Vasi73] M. P. Vasilevsky, *Failure diagnosis of automata*, Cybernetics, Plenum Publishing Corporation, NY, No. 4, 1973, pp. 653-665.

[VCl89] S. T. Vuong, W. W. L. Chan, M. R. Ito, *The UIO-method for protocol test sequence generation*, Proceedings of the 2nd IFIP International Workshop on Protocol Test Systems, 1989, pp. 161-175.

[YaLe95] M. Yannakakis, D. Lee, *Testing finite state machines: fault detection*, Journal of Computer and System Sciences, 1995, 50, pp. 209-227.

[YePe90] N. Yevtushenko, A. Petrenko, *A method of constructing a test experiment for an arbitrary deterministic automaton*, Automatic Control and Computer Science, Allerton Press Inc., N.Y., 1990, Vol. 24, No. 5, pp.65-68.

BIOGRAPHY

Irina Koufareva received the Dipl. degree in computer sciences in 1994 from Tomsk State University, Russia. At the moment, she is an assistant professor at that university. Her current research interests include the automata theory and formal methods in conformance testing.

Alexandre Petrenko received the Dipl. degree in electrical and computer engineering from Riga Polytechnic Institute and the Ph.D. degree in computer science from the Institute of Electronics and Computer Science, Riga, USSR. In 1996, he has joined CRIM, Centre de Recherche Informatique de Montréal, Canada. He is also an adjunct professor of the Université de Montréal, where he was a visiting professor / researcher from 1992 to 1996. From 1982 to 1992, he was the head of a research department of the Institute of Electronics and Computer Science in Riga. From 1979 to 1982, he was with the Networking Task Force of the International Institute for Applied System Analysis (IIASA), Vienna, Austria. His current research interests include high-speed networks, communication software engineering, formal methods, conformance testing, and testability.

Nina Yevtushenko received the Dipl. degree in radio-physics in 1971 and Ph.D. degree in computer sciences in 1983, both from Tomsk State University. She is currently a professor at that University. Her research interests include the automata and FSM theory and testing problems

VI
TEST OPTIMIZATION

15

TEST SUITE MINIMIZATION FOR EMBEDDED NONDETERMINISTIC FINITE STATE MACHINES

Nina Yevtushenko
Tomsk State University
36 Lenin av., 634050 Tomsk, Russia
yevtushenko@elefot.tsu.ru

Ana Cavalli
National Institute of Telecommunications
9, rue Charles Fourier, 91011, Evry Cedex, France
Ana.Cavalli@int-evry.fr

Ricardo Anido*
University of Campinas
Cx Postal 6167 UNICAMP, 13083-970 Campinas SP, Brazil
ranido@dcc.unicamp.br

Abstract This paper presents a method for minimizing test suites for embedded, nondeterministic Finite State Machines. The method preserves the fault coverage of the original test suite, and can be used in conjunction with any technique for generating test suites. The minimization is achieved by detecting and deleting redundant test cases in the test suite. The proposed method is an extension of the work presented in [Yevtushenko et al., 1998].

Keywords: Embedded testing, Automatic test generation, finite state machines, non-determinism.

*Work supported by FAPESP and Pronex/Finep/SAI N: 76.97.1022.00

1.   INTRODUCTION

One important aspect of automatic test generation that has been investigated recently is the derivation of tests for a component embedded within a complex system. This problem is known as gray-box testing, embedded testing ([ISO, 1991]) or testing in context ([Petrenko et al., 1996b]), and corresponds to a situation quite common in practice, where the system to be tested (the component) is part of a larger system, and can only be tested in the context of another part assumed to be fault-free (the context). A number of approaches have been developed for testing in context ([Petrenko et al., 1996b, Petrenko et al., 1996a, Lee et al., 1996, Lima and Cavalli, 1997, Lima and Cavalli, 1998, Petrenko et al., 1997, Yevtushenko et al., 1998]), some of which are heuristic and do not guarantee complete fault coverage ([Lee et al., 1996]). Other approaches deliver complete test suites with respect to various fault domains (i.e., w.r.t. sets of possible implementations of a component under test). Examples of the latter approach are methods which start with a test suite either derived by classical procedures such as W-, Wp, UIO, SC methods ([Vasilevsky, 1973, Chow, 1978, Fujiwara et al., 1991, Vuong et al., 1989, Yannakakis and Lee, 1995, Lee and Yannakakis, 1996]) or given by human experts; the derived test suite is afterwards reduced without loss of its completeness.

In this paper we propose a method for minimizing a test suite for communicating nondeterministic Finite State Machines (FSMs) while maintaining its fault-coverage. We consider a specification system composed of two communicating FSMs. One of these machines, called component, is the machine that needs testing. The other machine, called context, describes the behavior of the part of the system which is assumed to be correctly implemented. Component, context or the overall system may be nondeterministic.

The method proposed extends the work presented in ([Yevtushenko et al., 1998]), for deterministic FSMs, to nondeterministic FSMs. For each test sequence of a given test suite, we determine a regular set of internal traces that can be induced within an arbitrary implementation system. The regular sets thus obtained completely characterize the fault detection power of the test suite. We then present a procedure for determining a minimal proper subset of the test suite which accepts all nondeterministic behaviors accepted by the original test suite while maintaining its fault-detection power. The rest of the paper is structured as follows. Section 2. presents some basic notions. Section 3. analyzes the conditions under which a test case can be removed from a test suite without loss of its completeness. Section 4. describes how the problem of minimizing a given test suite can be reduced to determining a minimal column coverage of a Boolean matrix, and in Section 5. we present some conclusions.

2.   PRELIMINARIES

This section presents some basic notions on composition of FSMs and conformance testing.

2.1   FINITE STATE MACHINES

A Finite State Machine (FSM, often simply called a machine throughout this paper) is an initialized (possibly, nondeterministic) machine denoted by a 5-tuple $A = (S, X, Y, h, s_0)$, where S is a finite nonempty set of states with $s_0 \in S$ as the initial state, X is a finite nonempty set of inputs, Y is a finite nonempty set of outputs, and h is a behavior function, $h : S \times X \to P(S \times Y)$ where $P(S \times Y)$ is a set of all nonempty subsets of $S \times Y$. FSM A is *deterministic* if $\mid h(s, x) \mid = 1$ for all $(s, x) \in S \times X$. The machine is *observable* if for all $(s, x) \in S \times X$ and $y \in Y$ it holds that $\mid (s, y) \mid (s, y) \in h(s, x) \mid \leq 1$. In the usual way, the function h is extended to a function on the set $S \times X^*$ with results in the set $P(S \times Y*)$ where X^* is the set of all finite sequences in X containing the empty sequence ϵ. The function h has two projections h^s and h^y, usually called the next state function and the output function. FSM A is said to be *connected* if each state of A is reachable from the initial state, i.e. for each state s of A there exists an input sequence α such that $s \in h^s(s_0, \alpha)$.

Given sequences $\alpha = x_1 \ldots x_k \in X^*$ and $\beta = y_1 \ldots y_k \in Y^*$, the sequence $x_1 y_1 \ldots x_k y_k$ is called a trace at state $s \in S$ of FSM $A = (S, X, Y, h, s0)$ if $\beta \in h^2(s, \alpha)$. We call the sequence $x_1 y_1 \ldots x_k y_k$ a trace of A if $\beta \in h^2(s, \alpha)$. Given two states s of FSM $A = (S, X, Y, h, s_0)$ and t of FSM $B = (T, X, Y, g, t_0)$, state s is said to be equivalent to state t, written $s \leq t$, if $h^2(s, \alpha) = g^2(t, \alpha)$ for each input sequence $\alpha \in X^*$; otherwise, state s is *distinguishable* from state t and the sequence α is said to *distinguish* state s from state t. Machines A and B are *equivalent* if their initial states are equivalent, written $A \equiv B$; otherwise, they are *distinguishable*, written $A \neq B$. An input sequence is said to *distinguish* two machines if it distinguishes their initial states. Machine A is said to be *reduced* if every pair of its states is distinguishable. It is well known that each machine A is equivalent to some reduced observable FSM that is called a reduced observable form of machine A ([Starke, 1972, Luo et al., 1994]). In other words, given FSMs A and B over the same input alphabet, A and B are equivalent if they exhibit the same behavior under all input sequences; otherwise, they are distinguishable. Protocol conformance testing is often formalized as a problem of testing whether an implementation FSM is equivalent to a given reference FSM.

2.2   COMPOSITION OF FSM'S

We consider a system composed by two FSMs, as shown in Figure 1. We will use the expression *system under test* to designate a system composed by some implementation *Imp* of the component *Comp* and the context *C*. For the sake of simplicity, we consider pairwise disjoint sets X, Y, U and V. The system under test has always a single message in transit, i.e. the environment submits the next input only when the system has produced an output to the previous input. We refer to symbols of the alphabet X as external inputs, to symbols of the alphabet Y as external outputs, and to symbols of the alphabets U and V as internal actions. As it is shown in a number of publications ([Yevtushenko et al., 1998]) such a composition is general enough to discuss problems of testing in context. When there are no livelocks in the composition, we can derive the composite FSM of FSMs C and $Comp$ using various algorithms ([Petrenko et al., 1996b, Lima and Cavalli, 1997]).

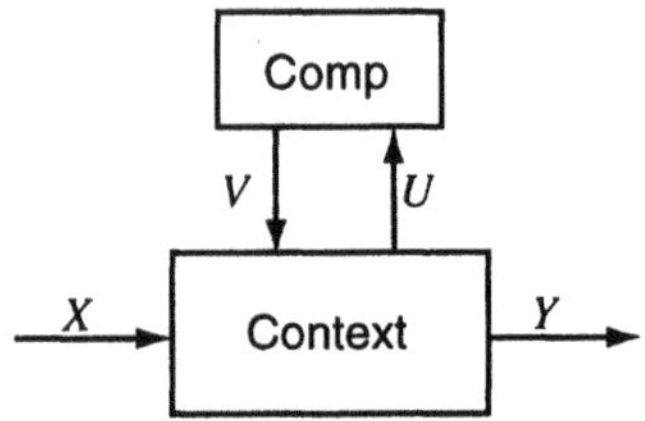

Figure 1   System model

Example. We use as an example of a coffee shop system which is inspired from the one proposed in [Petrenko et al., 1998]. The coffee shop is composed by a waiter and a coffee machine. We have modified the global behavior of the coffee shop to introduce nondeterminism. The nondeterminism is caused by the context, represented in our example by a forgetful waiter, which may forget a request for coffee, as described below. In order to simplify the description of the components we consider that internal actions are denoted by elements of disjoint sets U and V. The behavior of the coffee shop is as follows: if a customer introduces money (M), the shop answers with thanks (T) and is ready to receive the following inputs: M, which it refuses saying No (N); Espresso Please (E_p), producing two nondeterministic outputs: either Yes (Y), remaining in the same state (forgets the request), or Espresso Served (E_s), returning to the initial state. At the initial state a customer can also request a coffee (E_p), but the answer in this case is sorry (S). The coffee-machine has two inputs and two outputs. When the waiter presses Button (B) at the initial state 0 the coffee-machine responses turning on a Lamp (L). If Button (B) is pressed at the state 1 the coffee-machine produces Espresso (E). An input Coin (C) causes a looping transition and an output Espresso (E) at each state.

The FSMs representing the waiter and coffee-machine are presented in Figure 2.

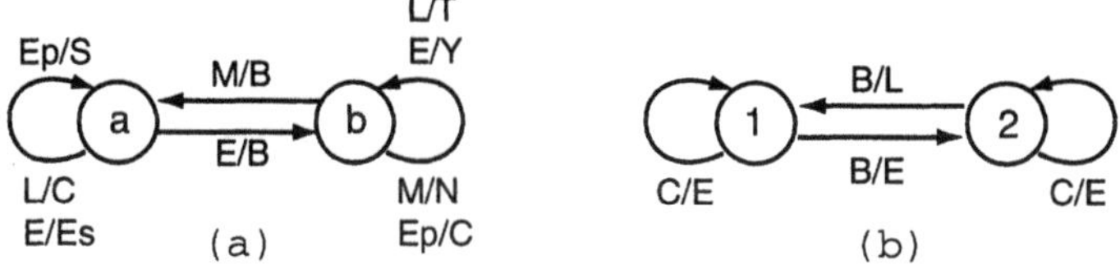

Figure 2   *(a)* Context FSM *Waiter* and *(b)* Component FSM *CoffeeMachine*

The combined system of waiter and coffee machine has no livelocks and its composite FSM RS is shown in Figure 3.

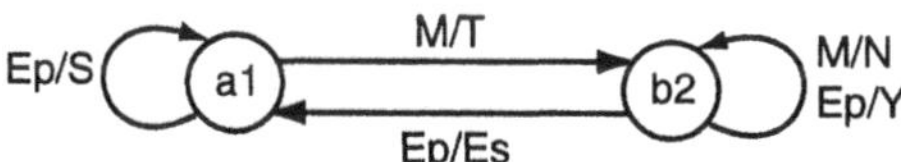

Figure 3   The FSM RS modeling the behavior of the coffee shop

2.3   FAULT MODEL FOR TESTING IN CONTEXT

We consider a reference system that is a composition of two FSMs, and assume that at most one of the two machines, named the component machine, denoted $Comp$, can be faulty. The other machine, named the context, denoted C, is fault-free. In other words, any implementation system is a composition of the context and some (possibly faulty) implementation component machine Imp combined, as shown in Figure 1. We also assume the behavior of the reference system as well as the behavior of any (possibly faulty) implementation system is described by a reduced observable FSM with at most n states. The set of all possible implementation systems is called a fault domain, and is denoted $\Re_{n,C}$.

To test a system with a nondeterministic behavior each test sequence usually is submitted to the system under test several times, under diverse conditions, until a test engineer verifies the system has produced each possible output to the test sequence. Under this so-called "all-weather conditions" assumption one can use the same notion of a complete test suite for deterministic and nondeterministic machines.

A *test suite* for the reference FSM RS is a finite set of finite input sequences of RS. A test suite for the FSM RS is *complete* w.r.t. the fault domain $\Re_{n,C}$ if for any FSM $B \in \Re_{n,C}$ such that B and the reference machine are distinguishable the test suite has a sequence distinguishing the machines; otherwise, the test suite is not complete w.r.t. the fault domain $\Re_{n,C}$.

One can use a straightforward approach to derive a complete test suite w.r.t. the fault domain $\Re_{n,C}$: explicitly enumerate all FSMs of $\Re_{n,C}$ and for each FSM $B \in \Re_{n,C}$ that is distinguishable from the reference machine determine a sequence distinguishing the machines. The set of all distinguishing sequences is a complete test suite w.r.t. $\Re_{n,C}$. When it is impossible to explicitly enumerate all FSMs of the set $\Re_{n,C}$ we can expand $\Re_{n,C}$ to the fault domain $\Re_n$ comprising any FSM that has a reduced observable form with n states, and use existing methods for complete test derivation w.r.t. the fault domain $\Re_n$. Below we sketch a method proposed in [Luo et al., 1994]; in fact, this method is similar to the well known W or Wp-method for deterministic FSMs when the reference FSM is reduced and completely specified.

Let RS be a reduced connected FSM with n states. A finite subset V, $|V| \le n$, including the empty sequence, is called a *state cover* of RS if for each state $s \in S$ there exists an input sequence α such that $s \in h^s(s, \alpha)$. We further denote VX the set obtained by concatenating each sequence of the set V with each sequence of the set X. The set VX is usually called a transition cover of the FSM RS since each transition of the FSM RS is traversed with an appropriate sequence of the set VX. Here we notice that differently from deterministic FSMs the state cover V of a nondeterministic FSM RS can have less sequences than number n of states of RS ([Luo et al., 1994]).

Given a state $s \in S$ of the FSM $RS = (S, X, Y, h, s_0)$, a finite set W_s of finite input sequences is said to be a state identifier of s if for each state $p \in S$, $p \ne s$, there exists an input sequence α distinguishing state s from state p, i.e. $h^y(s, \alpha) \ne h^y(p, \alpha)$. An input sequence is said to be a *distinguishing sequence* for the FSM RS if it is a state identifier for every state of RS. Similar to the W-method when a state identifier is fixed for each state of the FSM RS, a procedure for derivation of a complete test suite w.r.t. the fault domain $\Re_n$ comprises two phases.

In the first phase, part T_1 of a test suite is derived. T_1 verifies that an implementation FSM has at least n states and that each state identifier correctly identifies the corresponding state. This part of the test suite is obtained by concatenating each sequence α of the state cover set V with each sequence of the state identifier for each last state of α, that is for each state where α takes the reference FSM, starting from the initial state. If the implementation FSM has the reference output response to each input sequence of the set T_1 then every state of the reference FSM has a corresponding state in the implementation FSM with the same state identifier.

Part T_2 of the test suite derived in the second phase. T_2 checks whether each transition in the implementation FSM is correctly implemented. The set T_2 is obtained by concatenating each sequence $\alpha \in VX$ with each sequence of the state identifier for each state where α takes the reference FSM from the initial state. Merging T_1 and T_2 we obtain a complete test suite $TS = T_1 \cup T_2$

w.r.t. the fault domain $\Re_n$. Each sequence that is a prefix of another sequence can be deleted from TS, without loss of completeness of TS w.r.t. the fault domain $\Re_n$. If we then apply each sequence of TS to an implementation under test, several times, to observe each possible response, and if the set of responses coincides with that of the reference machine then we conclude the implementation has a reference behavior.

Example. The reference machine in Figure 3 is nondeterministic, reduced and observable. We assume the waiter is fault-free and any system with a (possibly faulty) coffee-machine has an observable reduced form with at most two states. An input sequence M is a distinguishing sequence of the reference FSM. We select a state cover set $V = \{\epsilon, \mathsf{M}\}$. Then $T_1 = \{\mathsf{E_p}, \mathsf{M}, \mathsf{ME_p}, \mathsf{MM}\}$ while $T_2 = \{\mathsf{E_pM}, \mathsf{MM}, \mathsf{ME_pM}, \mathsf{MMM}\}$, and $TS = \{\mathsf{E_pM}, \mathsf{ME_pM}, \mathsf{MMM}\}$.

A test suite TS complete w.r.t. the fault domain $\Re_n$ is also complete w.r.t. the fault domain $\Re_{n,C}$. However, it usually is redundant, because some of the test cases included are (unnecessarily) testing only the context, which is assumed fault-free. Furthermore, some test cases may be testing parts of the implementation already tested by other test cases.

Given a test suite TS, complete w.r.t. the fault domain $\Re_n$, a proper subset T of TS is complete w.r.t. $\Re_{n,C}$ if, for every possible implementation FSM $B \in \Re_{n,C}$, when B has the expected set of output responses to every sequence of T, it has also the expected set of output responses to every sequence of TS. As shown in [Yevtushenko et al., 1998], the detecting power of each test case can be characterized through a set of internal traces that are detectable by the test case. We can delete from TS every test case traversing only transitions of the context, since the set of internal traces that are detectable by this test case is empty, i.e. the test case checks nothing in the component. For example, the suffix MM of the test case MMM tests nothing in the component ([Lima and Cavalli, 1997, Yevtushenko et al., 1998]). Therefore, the test suite $\{\mathsf{E_pM}, \mathsf{ME_pM}, \mathsf{M}\}$ is also complete w.r.t. the fault domain $\Re_{n,C}$. If the test case $\alpha \in TS$ traverses transitions in which the component is involved a more detailed analysis of the test case must be performed. In the next sections we show that $TS \backslash \{\alpha\}$ is complete w.r.t. $\Re_{n,C}$ if the following two properties hold: *(i)* there exists a test case $\alpha \in TS \backslash \{\alpha\}$ that induces an unexpected external input for each internal trace that can induce an unexpected external output to α; *(ii)* for each reference output sequence β to α there exists a test case $\alpha \in TS \backslash \{\alpha\}$ which induces an internal trace such that the system under test produces the output response β when α is submitted.

3.   MAINTAINING THE FAULT COVERAGE

In this section, we analyze the conditions to reduce the set TS by deleting some of its sequences without loss of its completeness w.r.t. the fault domain

$\Re_{n,C}$. Given an external input sequence α, an internal trace $\gamma = x_1 y_1 \ldots x_k y_k$ over alphabets U and V is said to be *detectable* by α if, when α is applied to a system under test where the component Imp has the trace $x_1 y_1 \ldots x_k y_k$, the system produces an unexpected output response to α *YCL98*. Otherwise, a trace is said to be *undetectable* by α. Given a set T of external input sequences, an internal trace γ over alphabets U and V is said to be detectable by T if there exists $\alpha \in T$ such that the internal trace γ is detectable by α.

In [Yevtushenko et al., 1998] we show that any prolongation of a trace detectable by α is also detectable by α, and that the set of all internal traces detectable by α, in the case of systems composed by deterministic context and component FSMs, is a regular set ([Hopkroft and Ulman, 1979]). The same results are valid in the case of nondeterministic context and/or component.

3.1   DETERMINING DETECTABLE INTERNAL TRACES

Procedure 1 below describes a step by step derivation of the regular set of detectable internal traces for given context. The key construction of the procedure is the acceptor $LC(\alpha)$ of all possible traces that can be induced by an external input sequence when the context is combined with any implementation of the component machine Comp. Traces detectable by α are designated in the $(U \cup V)$-projection of the acceptor $LC(\alpha)$ by a dead state *fail*. Below we briefly sketch how the acceptor $LC(\alpha)$ can be constructed.

Given an external input sequence $\alpha = x_1 \ldots x_k$, the acceptor $LC(\alpha)$ can be derived step by step by use of the deterministic Label Transition System (LTS) LC representing all traces of the context C. States of the acceptor $LC(x_1 \ldots x_k)$ are states of LC. Given an input x_1, we construct the acceptor $LC(x_1)$ starting from the initial state of the LC. There is a transition labeled with x_1 from the initial state to state p if x_1 takes LC from the initial state to state p. For two intermediate states p and r, there is a transition labeled with internal action $a \in U \cup V$ if there is a transition labeled with a from state p to state r in LC. There is a transition labeled with $y \in Y$ in the $LC(x_1)$ from intermediate state p to a final state if there exists an outgoing transition labeled y from state p. The acceptor $LC(x_2)$ is constructed at each final state of the acceptor $LC(x_1)$ and so on. The final states of the acceptor $LC(x_k)$ are declared the final states of the acceptor $LC(x_\alpha)$.

Procedure 1 *Derivation of a regular set of detectable internal traces.*

Input: The composite FSM RS of a reference system, the deterministic LTS LC representing all traces of the context C and an external input sequence α.
Output: The regular set $D(\alpha)$ of internal traces detectable by α.

Step 1. Construct the acceptor $LC(\alpha)$ using the deterministic context LTS LC.

Step 2. For each path of the acceptor $LC(\alpha)$ from the initial state to a final state such that the projection of the sequence labeled the path is not a trace of the reference FSM, replace the final state with a final state *fail*.

Step 3. If for some transition labeled with an external input $x \in X$ or with an internal action $v \in V$, all the subsequent paths have a final state *fail* then replace the final state of the transition with final state *fail*.

Step 4. Construct the $(U \cup V)$-projection of the obtained acceptor by a subset construction, replacing with a designated fail-state without outgoing transitions each subset that has the fail-state. Construct the regular set $D(\alpha)$ as the set of all sequences of the $(U \cup V)$-projection labeled paths from the initial to the *fail* state.

By construction of the set $D(\alpha)$, the following statements hold.

Proposition 1 *Given a reference system RS and an implementation system IS with a component Imp, the system IS produces an unexpected response to the input sequence α if and only if the set of traces of the machine Imp intersects the set $D(\alpha)$ derived by Procedure 1 ([Yevtushenko et al., 1998]).*

Corollary 2 *Given a reference FSM RS, an implementation system IS, a complete test suite TS w.r.t. the fault domain $\Re_{n,C}$ and a sequence $\alpha \in TS$, let the system IS have an unexpected response to the sequence α. If for each trace $u_1v_1 \ldots u_kv_k \in D(\alpha)$ there exists a sequence $\gamma \in TS\backslash\{\alpha\}$ such that the set $D(\alpha)$ comprises a prefix of the trace $u_1v_1 \ldots u_kv_k$, then the system IS has an unexpected response to some sequence of the set $TS\backslash\{\alpha\}$.*

In fact, let all the conditions of Corollary 2 hold. If the IS has an unexpected response to the sequence α then the set of traces of the component Imp has a trace $u_1v_1 \ldots u_kv_k$ of the set $D(\alpha)$ (Proposition 1), along with its prefix $u_1v_1 \ldots u_lv_l, l \leq k$, being in the set $D(\gamma)$ for some sequence $\gamma \in TS\backslash\{\alpha\}$ i.e. IS has an unexpected response to the sequence γ.

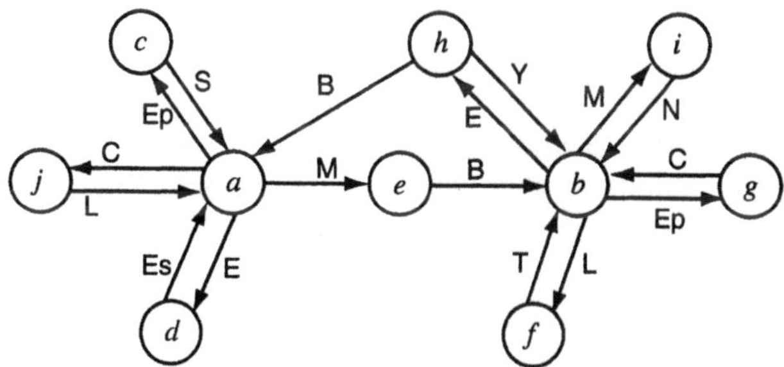

Figure 4   The context LTS LC

Example. Consider an input sequence $M \in TS$. The deterministic context LTS LC is shown in Figure 4. The acceptor $LC(M)$ and its $(U \cup V)$-projection with the designated fail-state are shown in Figures 5 and 6. If the component implementation Imp has a trace BE then the implementation system IS has an unexpected output response Y to input M. On the other hand, if the component implementation Imp has no trace BE then the implementation system IS always produces the expected output response T to the input M. Thus, the implementation system IS always produces the expected output response T to the input M if and only if the component implementation Imp answers with output E to the input B. Thus, $D(M) = \{BE\}$. Similarly, we obtain $D(E_pM) = \{BE\}$ and $D(ME_pM) = \{BE, BLCL, BLCEB(LC)^*EBE\}$. By direct inspection, one can assure that if the implementation system IS has only expected output responses to the input sequence ME_p M then the implementation component machine Imp has no traces of the set $D(ME_pM)$, i.e. IS has the expected output responses to the input sequences E_pM and M.

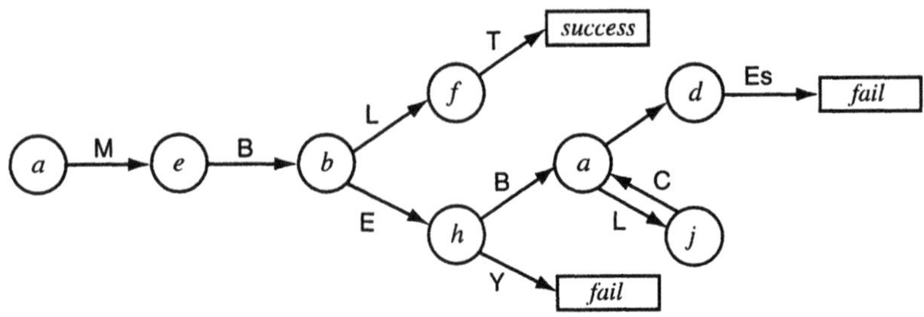

Figure 5   Acceptor LC(M)

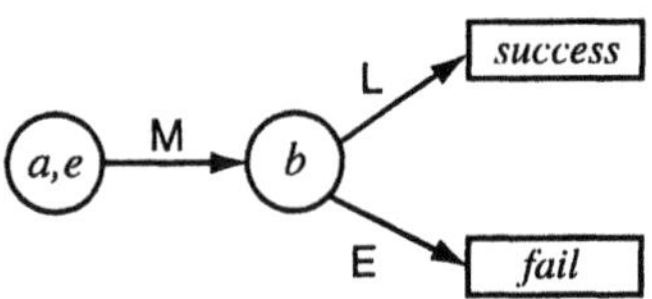

Figure 6 $(U \cup V)$–projection of acceptor LC(M)

Corollary 2 specifies a necessary and sufficient condition for minimizing a test suite while maintaining its fault detection power when the context and the component are deterministic. Although this condition is insufficient in the case of nondeterministic systems, since the minimized test must also accept all possible behaviors accepted by the original test suite. That property is analyzed in the next section.

3.2   PRESERVING ALL POSSIBLE BEHAVIORS

Given an external input sequence $\alpha = x_1 \ldots x_k$, let $LC(\alpha)$ be an acceptor of all internal traces that can be induced in the context when α is submitted to an implementation system. Let also $\beta = y_1 \ldots y_k$ be an output sequence that can be produced by the reference system in response to α. We denote $C(\alpha, \beta)$ the regular set of $(U \cup V)$-projections of all sequences that label paths from the initial to the final state in $LC(\alpha)$ and have the $(X \cup Y)$-projection $x_1 y_1 \ldots x_k y_k$. If a system under test has produced an output response β to the sequence α, it means that the set of traces of the implementation component machine intersects the set $C(\alpha, \beta)$.

> **Procedure 2** *Derivation of the regular set of all internal traces which are possible in an implementation system with a given reference trace.*
>
> **Input:** The composite FSM RS of a reference system, the LTS LC representing all traces of the context C, an external input sequence α and the output response β of the reference system RS to α.
>
> **Output:** The regular set $C(alpha, \beta)$ of all internal traces which are possible to occur in an implementation system with the reference trace (α, β).
>
> **Step 1.** Construct the acceptor $LC(\alpha)$ using the deterministic context LTS LC.
>
> **Step 2.** For each path of the acceptor $LC(\alpha)$ from the initial state to a final state such that the projection of the sequence labeled the path is a trace (α, β), replace the final state with a designated state *success*.
>
> **Step 3.** Construct the $(U \cup V)$-projection of the obtained acceptor by a subset construction replacing with a designated state *success* each subset that has the state *success*. Derive the regular set $C(\alpha, \beta)$ as the set of all sequences of the $(U \cup V)$-projection labeled paths from the initial to the state *success*.

Proposition 3 *Given two external input sequences α and β, the sets of output responses $h^y(s_0, \alpha)$ and $h^y(s_0, \delta)$ of the reference system RS to the sequences α and δ, let IS be an implementation system such that the set of output responses of IS to δ coincides with $h^y(s_0, \delta)$. If for every $\beta \in h^y(s_0, \alpha)$ there exists a sequence $h^y(s_0, \delta)$ such that each sequence in the set $C(\delta, \lambda)$ has a prefix in the set $C(\alpha, \beta)$ then the set of output responses of IS to α coincides with $h^y(s_0, \alpha)$.*

As a corollary to Proposition 3, another condition for a proper subset T of the complete test suite TS to provide a complete set of reference outputs to each test case in TS can be established.

Corollary 4 *Given the reference FSM RS, an implementation system IS, a complete test suite TS w.r.t. the fault domain $\Re_{n,C}$ and a sequence $\alpha \in TS$, let the system IS have only expected output responses to the sequence α. Let also for each reference output response β to the sequence α, there exist a sequence $\delta \in TS \setminus \{\alpha\}$ and $\lambda \in h^y(s_0, \delta)$ such that each trace in the set $C(\delta, \gamma)$ has a prefix in the set $C(\alpha, \beta)$. If for each sequence $\delta \in TS \setminus \{\alpha\}$, the set of output responses of IS to δ coincides with that of the reference system then the set of output responses of the system IS to the sequence α also coincides with that of the reference system.*

Example. The set $C(\mathsf{ME_p M}, \mathsf{TYN}) = \{\mathsf{BLCE}\}$ while $C(\mathsf{ME_p M}, \mathsf{TE_s T}) = \{\mathsf{BLCEB(LC)^*EBL}\}$. In other words, an implementation system IS has all expected output responses to the input sequence $\mathsf{ME_p M}$ if and only if the implementation component machine has the trace BLCE as well as at least one trace of the regular set $\mathsf{BLCEB(LC)^*EBL}$.

4.   MINIMIZATION OF A TEST SUITE

By combining the results of Corollaries 2 and 4 we obtain a sufficient condition for a proper subset of a complete test suite to be also complete w.r.t. the fault domain $\Re_{n,C}$.

Proposition 5 *Given a reference FSM RS, a complete test suite TS w.r.t. the fault domain $\Re_{n,C}$, let $\alpha \in TS$. The set $TS \backslash \{\alpha\}$ is also complete w.r.t. $\Re_{n,C}$ if the following conditions hold.*

a) For every trace $u_1 v_1 \ldots u_k v_k \in D(\alpha)$, there exists $\delta \in TS \backslash \{\alpha\}$ such that the set $D(\delta)$ comprises a prefix of the trace $u_1 v_1 \ldots u_k v_k$.

b) For every reference output response $\beta \in h^y(s_0, \alpha)$ there exist $\delta \in TS \backslash \{\alpha\}$ and $\lambda \in h^y(s_0, \delta)$ such that for every trace $u_1 v_1 \ldots u_k v_k \in C(\delta, \lambda)$, the set $C(\alpha, \beta)$ comprises a prefix of the trace $u_1 v_1 \ldots u_k v_k$.

Example. In our working example, only a single input sequence $\mathsf{ME_pM}$ of the test suite induces two reference output responses. Thus, the set $\{\mathsf{ME_pM}\}$ is a test suite complete w.r.t. the fault domain $\Re_{n,C}$.

Proposition 5 shows that a given test suite can be minimized by examining the regular sets detectable by the test cases and the sets of internal traces providing reference output sequences. The problem of comparing arbitrary regular expressions is out of the scope of this paper. When the regular sets C and D derived by Procedures 1 and 2 are finite; in that case, similarly to [Yevtushenko et al., 1998], we can reduce the problem of test minimization to the problem of determining a minimal column coverage in a Boolean matrix, as described below.

Given a complete test suite TS, let $D(TS)$ be the set of internal traces detectable by the set TS derived by Procedure 1. For each $\alpha \in TS$ that induces at least two reference output responses and each $\beta \in h^y(s_0, \alpha)$, we derive the set $C(\alpha, \beta)$ of internal traces which provide the external trace (α, β) (Procedure 2) and denote $C(TS)$ the collection of such sets. We construct a Boolean matrix B as follows. Rows of the matrix B correspond to sequences of the prefixes of all sequences in the test suite TS; columns of B correspond to the items in $D(TS)$ and in $C(TS)$. That is, the columns of B are all internal traces detectable by TS, and all subsets $C(\alpha, \beta)$ of internal traces induced by each trace (α, β) of the reference FSM RS for each $\alpha \in TS$ providing at least two reference output sequences.

Element b corresponding to a test sequence $\alpha \in TS$ and an internal trace $\gamma \in D(TS)$ has value '1' if and only if there exists a prefix of γ detectable by α. Element b corresponding to a test sequence $\alpha \in TS$ and a subset $C(\delta, \lambda), \delta \in TS$, has value '1' if and only if there exists a response β of the reference FSM RS to α such that each sequence in the set $C(\alpha, \beta)$ has a prefix in the set $C(\delta, \lambda)$. The set of rows of the matrix corresponding to its minimal column coverage is a test suite complete w.r.t. the fault domain $\Re_{n,C}$.

5.   CONCLUSION

In this paper, the approach proposed in [Yevtushenko et al., 1998] is extended to a system of communicating nondeterministic FSMs. The system under test is composed of two FSMs, a context that is assumed to be fault-free and a component that needs testing. We assume that the behavior of both the reference and the implementation system are described by observable FSMs. In order to minimize a given test suite, complete under the above assumptions, we construct two regular sets. One of them is the set of all internal traces detectable by each test case, i.e. a set of internal traces which cause an unexpected output response of an implementation system. Another regular set is the set of all internal traces that cause a given reference output to the test case. If a system under test produces every reference output response to a test case one guarantees that the component machine has at least one trace of such set for each reference output response. After the regular sets have been derived the problem of determining a minimal subset of a given test suite that is also complete w.r.t. the chosen fault domain can be solved as a problem of finding a minimal column coverage of a Boolean matrix. We can apply the approach for test minimization w.r.t. the reduction relation. In this case, similar to a system of communicating deterministic finite state machines the Procedure 2 becomes unnecessary. The proposed method can also be used to reduce a test suite given by a human expert while preserving its fault detection power under an assumption that the context is fault-free.

References

[Chow, 1978] Chow, T. S. (1978). Test software design modeled by finite state machines. *IEEE Transactions on Software Engineering*, 4(3):178–187.

[Fujiwara et al., 1991] Fujiwara, S., v. Bochmann, G., Khendek, F., Amalou, M., and Ghendamsi, A. (1991). Test selection based on finite state models. *IEEE Transactions on Software Engineering*, 17(6):591–603.

[Hopkroft and Ulman, 1979] Hopkroft, J. E. and Ulman, J. D. (1979). *Introduction to automata theory, languages and computation*. Addison-Welsey, NY.

[ISO, 1991] ISO (1991). *Information technology, Open systems interaction, Conformance testing methodology and framework*. International Standard IS-9646.

[Lee et al., 1996] Lee, D., Sabnani, K. K., Kristol, D. M., and Paul, S. (1996). Conformance testing of protocols specified as communicating finite state machines – a guided random walk based approach. *IEEE Transations on Communications*, 44(5):631–640.

[Lee and Yannakakis, 1996] Lee, D. and Yannakakis, M. (1996). Principles and methods of testing finite state machines, a survey. *IEEE Transactions*, 84(8):1090–1123.

[Lima and Cavalli, 1997] Lima, L. P. and Cavalli, A. R. (1997). A pragmatic approach to generating test sequences for embedded systems. In *10th IWTCS*, pages 125–140.

[Lima and Cavalli, 1998] Lima, L. P. and Cavalli, A. R. (1998). Application of embedded testing methods to service validation. In *submitted to 2nd IEEE Intern. Conf. On Formal Engineering methods*.

[Luo et al., 1994] Luo, G., Petrenko, A., , and v. Bochmann, G. (1994). Selecting test sequences for partially-specified nondeterministic finite state machines. In *7th IWTCS*, pages 95–110.

[Petrenko et al., 1997] Petrenko, A., , and Yevtushenko, N. (1997). Testing faults in embedded components. In *10th IWTCS*, pages 125–140.

[Petrenko et al., 1998] Petrenko, A., , and Yevtushenko, N. (1998). Solving asynchronous equations. In *Joint International Conference FORTE/PSTV98*, pages 231–247.

[Petrenko et al., 1996a] Petrenko, A., Yevtushenko, N., , and v. Bochmann, G. (1996a). Fault models for testing in context. In *1st Joint International Conference FORTE/PSTV96*, pages 125–140.

[Petrenko et al., 1996b] Petrenko, A., Yevtushenko, N., v. Bochmann, G., and Dssouli, R. (1996b). Testing in context: framework and test derivation. *Computer communications*, 19:1236–1249.

[Starke, 1972] Starke, P. H. (1972). *Abstract automata*. American Elsevier Publishing Company, Inc. New York.

[Vasilevsky, 1973] Vasilevsky, M. P. (1973). Failure diagnosis of automata. *Cybernetics*, (4):653–665.

[Vuong et al., 1989] Vuong, S. T., Chan, W. W. L., and Ito, M. R. (1989). The uio-method for protocol test sequence generation. In *IFIP TC6 Second International Workshop on Protocol Test Systems*, pages 161–175.

[Yannakakis and Lee, 1995] Yannakakis, M. and Lee, D. (1995). Testing finite state machines: fault detection. *Journal of Computer and System Sciences*, (50):209–227.

[Yevtushenko et al., 1998] Yevtushenko, N., Cavalli, A. R., and Lima, L. P. (1998). Test minimization for testing in context. In *11th IWTCS*, pages 127–145.

16

AUTOMATED TEST CASE SELECTION BASED ON SUBPURPOSES

Tibor Csöndes
Conformance Center, Ericsson Ltd.
Laborc u. 1., Budapest, 1037, Hungary
Tibor.Csondes@eth.ericsson.se

Balázs Kotnyek
Eötvös Loránd University
Budapest, Hungary
kotnyekb@cs.elte.hu

Abstract In this paper we present an approach to test selection from ATS. The method can help test laboratories to decide which Test Cases should be executed. The method uses TTCN, it is flexible in terms of the different preferences of the laboratories, it is efficient and mostly automated. We introduce the new concept of *subpurpose*, which is a detectable part of the Test Purpose and convey some parts of the parallelism between the tests. We define three different models and two types of optimization problem, and give their mathematical formulations. In the second half of the paper we present empirical results obtained by applying the method to a quite complex real-life protocol.

Keywords: Protocol testing, test selection, ATS, TTCN

1. INTRODUCTION

Conformance Testing is usually put into practice by test laboratories. Their task is to check whether the Implementation Under Test (IUT) is conform to the standards or not. In order to establish their verdict they have to use a standardized test suite the Abstract Test Suite (ATS). The ATS contains a considerable number of Test Cases (TC), which focus on specific parts of protocol behaviour, and can be executed independently from each other. The time test laboratories can spend on testing of an IUT is usually strongly limited. That is the reason why they have to make a selection, to decide which TCs will be executed and which ones will be omitted. Of course, they try to select a set of tests that covers as widely the protocol behaviour as possible, in other words, the coverage of which is as big as possible. A good selection has important economic consequences. Test laboratories can complete testing faster or, in other point of view, they can achieve bigger coverage in a given time. Therefore a method that can help them in efficient selection has considerable practical importance. Such a method should have the following properties:

- It is efficient as it makes the test process faster by losing only a relatively small amount of coverage.

- It is worth doing because it needs less time than it can save.

- It is applicable to different types of real-life protocols.

- It is flexible as special preferences of the test laboratories can be incorporated as well as their knowledge of the specific protocol.

- It is reproducible, the selection can be repeated if it is needed.

- It selects TCs from the ATS, so the testers can execute them straightforwardly.

So far, to our knowledge, there has not been reported such a method. The approach that is the closest to these aims is the metric based test selection [1, 2]. This method selects from a pre-generated set of "execution sequences" based on their "distances". Although it is surely possible to convert the ATS into a set of execution sequences, to define an adequate distance function, and then to select a set of sequences to execution with the metric based method, it needs considerable work and time. Furthermore, the execution sequences do not correspond exactly to Test Cases, so the testers cannot execute them straightforwardly.

Our proposed approach is based on the formal specification of the ATS, the TTCN format [3]. In general, this is the only formal specification that is available. Most of the test laboratories know this notation and use it for testing, so they do not need to learn a new notation and create the model of the ATS in that format. As the testers are able to use TTCN it is possible to carry on the whole process of selection and testing automatically and continuously having input the limits and demands at the beginning. Our method decides which TCs are the best to execute within a time limit or which ones should be executed to reach a given lower bound for the coverage in the shortest possible time.

In this paper we present our method and show how it can be applied to a real-life protocol, the ISDN DSS1 Layer 3 Protocol [4]. Although the results may not be extrapolated to all other protocols, we believe it can demonstrate the efficiency of the method.

The rest of the paper is organized as follows. In next section the basic theory of the method is presented as well as the possible models and their mathematical formulations. In Section 3 we present how the theory can be turned into practice, and Section 4 gives a series of experiments. Finally, we conclude by discussing on observations and further research work.

2.   THEORETICAL BASIS

The aim of Conformance Testing is to check whether the Implementation Under Test (IUT) fulfils the requirements it should fulfil. The set of requirements a Test Case (TC) can check is described in its Test Purpose (TP). The TPs of different Test Cases usually overlap, there are requirements more tests can check. If we have to select Test Cases to execute, we want to leave out tests the requirements of which will be checked by the selected ones. That is the reason why we would like to detect the connections between the tests on the basis of their purposes. For this we should determine the requirements related to a TC.

There are protocols the conformance requirements of which are given in the ATS standard accompanied by the Test Cases which are able to check them. We have shown an example of such protocols (ISDN DSS1 Layer 2 protocol [5]) in a former paper [6].

Most of the protocols do not have this pleasant property, the only thing given is the Test Purpose of every Test Case. The TP is written in informal language, so it is practically impossible to detect the requirements and the connection between different TCs using only their TPs. We should find the building blocks of the purposes, and we should do it automatically in order to make the selection faster and more reproducible avoiding the different interpretations of the prose TP. These building blocks have to characterize the requirements and convey some part of the parallelism between the tests. We will call such a block **subpurpose.**

Subpurpose can be the requirement itself if we are able to detect them. Otherwise we have to choose them depending on the type of the ATS. For example the Protocol Data Units (PDU) used in the TC or the parameters of the PDUs can play the role of subpurposes. Subpurposes are properly chosen if they help us to make efficient selection. One of the main aims of this paper is to present an example where the PDUs are the subpurposes.

2.1   THE MODEL

Let us define the **Subpurpose - Test Case Incidence Matrix** and let us denote it A. Subpurposes $(S_1, S_2, \ldots, S_k)$ are placed in the rows, Test Cases $(T_1, T_2, \ldots, T_n)$ are placed in the columns. The entry of the matrix standing in the (i, j) cell is 1 if S_i is a subpurpose of T_j. Otherwise the entry is 0. If we find some 1 entries in a given row (a_i denotes the ith row), then the TCs related to the 1s are connected in terms of their subpurposes. We can assume that executing one of these tests we get some information about the possible result of the execution of the others.

We append a **cost** to every TC. The cost of the jth test, $c(T_j)$ measures the amount of resources (usually the time) needed to execute it. We may be able to determine it exactly or, more frequently, we may use estimation. The cost of a set of TCs can be defined as the sum of the individual TCs in the set.

We define the **coverage** function of the subpurposes. The coverage of a subpurpose ($cov(S_i)$) may be obtained from theoretical consideration or we can simply mark the priority of the subpurpose with it. The test laboratories can express their possible individual preferences or knowledge of the specific protocols and potential faults by choosing an appropriate coverage function.

Let $\mathcal{T}$ be a set of the Test Cases, $\mathcal{T} \subset \{T_1, T_2, \ldots, T_n\}$. Let us denote the coverage we get if we execute $\mathcal{T}$ with $cov(\mathcal{T})$. This coverage is composed of the values $cov(S_i, \mathcal{T})$, which equal the partial coverage we achieve of $cov(S_i)$ ($i = 1, 2 \ldots, k$) executing only $\mathcal{T}$.

$$cov(\mathcal{T}) = \sum_{i=1}^{k} cov(S_i, \mathcal{T})$$

If we execute all the tests related to the ith subpurpose $cov(S_i, \mathcal{T})$ equals $cov(S_i)$, but executing only parts of the tests we may achieve smaller coverage. Let b_i be the number of TCs related to S_i (equals the number of the 1 entries in the ith row of the matrix). The increasing function $f_i : \{0, 1, \ldots, b_i\} \rightarrow [0, cov(S_i)]$ measures the coverage that can be obtained executing m TCs of ones related to S_i: $f_i(m) = cov(S_i, \mathcal{T})$. Naturally, $f_i(0) = 0$ and $f_i(b_i) = cov(S_i)$. We define three models that differ in choosing functions f_i.

1. **"Linear" model:** The coverage is in direct proportion to the number of executed tests i.e.

$$f_i(m) = \frac{m}{b_i} cov(S_i) \qquad m = 0, \ldots, b_i$$

2. **"All or nothing" model:** We consider the subpurpose being checked only when we have executed all the related tests.

$$f_i(0) = f_i(1) = \ldots = f_i(b_i - 1) = 0 \quad \text{and} \quad f_i(b_i) = cov(S_i)$$

3 **"One is enough" model:** If only one test is executed among the ones that correspond to S_i, we will get the whole coverage.

$$f_i(0) = 0 \quad \text{and} \quad f_i(1) = \ldots = f_i(b_i) = cov(S_i)$$

2.2   OPTIMIZATION

Let x be the characteristic vector of $\mathcal{T}$, so $x \in \{0, 1\}^n$, and $x_j = 1$ if $T_j \in \mathcal{T}$ and $x_i = 0$ if $T_i \notin \mathcal{T}$. The cost of $\mathcal{T}$

$$c(\mathcal{T}) = \sum (c(T_j) \mid T_j \in \mathcal{T}) = \sum_{j=1}^{n} c(T_j)x_j = cx$$

where $c = (c(T_1), \ldots, c(T_n))$ is the cost vector. The number of tests that are related to S_i and are in $\mathcal{T}$ (above we used the notation m for this number) is $a_i x = \sum_{j=1}^{n} a_{ij}x_j$ where a_i is the row of the matrix corresponding to S_i so

$$cov(\mathcal{T}) = \sum_{i=1}^{k} cov(S_i, \mathcal{T}) = \sum_{i=1}^{k} f_i(a_i x)$$

We will use the notation $v = (cov(S_1), \ldots, cov(S_k))$ for the coverage vector.

We introduce two possible optimization problems. In the first one our aim is to select a test set from the test suite with **minimal cost**, supposing a constraint (K) bounding the coverage from below.

$$\begin{aligned} \min \quad & c(\mathcal{T}) \\ \text{subject to} \quad & cov(\mathcal{T}) \geq K \\ & \mathcal{T} \subset \{T_1, T_2, \ldots, T_n\} \end{aligned}$$

Using the characteristic vector and functions f_i this problem can be written as

$$\begin{aligned} \min \quad & cx \\ \text{subject to} \quad & \sum_{i=1}^{k} f_i(a_i x) \geq K \\ & x \in \{0, 1\}^n \end{aligned} \tag{1}$$

Let us see now how this formula looks like in the case of the three mentioned models.

"Linear" model for the minimal cost problem.

$$\sum_{i=1}^{k} f_i(a_i x) = \sum_{i=1}^{k} \frac{a_i x}{b_i} cov(S_i) = \left(\sum_{i=1}^{k} \frac{a_i cov(S_i)}{b_i} \right) x$$

Thus (1) turns into a binary minimization with a single linear constraint:

$$\min \quad cx$$
$$\text{subject to} \quad \left(\sum_{i=1}^{k} \frac{a_i cov(S_i)}{b_i}\right) x \geq K \qquad (2)$$
$$x \in \{0,1\}^n$$

"All or nothing" model for the minimal cost problem.   Let us introduce a new variable vector $z = (z_1, \ldots, z_k)$ defined in the following manner:

$$z_i = \begin{cases} 1 & \text{if } a_i x = b_i \\ 0 & \text{if } a_i x < b_i \end{cases}$$

In other words, $z = \max\{Ax - b + e, 0\}$, where $e = (1, 1, \ldots, 1)$ and the maximization is made componentwise. Using this vector problem (1) can be written in the following manner

$$\begin{array}{rrcl} \min & cx & & \\ \text{subject to} & vz & \geq & K \\ & z & = & \max\{Ax - b + e, 0\} \\ & x & \in & \{0,1\}^n \end{array}$$

It is easy to see that this is equivalent to

$$\begin{array}{rrcll} \min & cx & & & \\ \text{subject to} & z_i & \leq & \frac{a_i}{b_i}x & \qquad i = 1, \ldots, k \\ & z & \geq & Ax - b + e & \\ & vz & \geq & K & \qquad (3) \\ & x & \in & \{0,1\}^n & \\ & z & \in & \{0,1\}^k & \end{array}$$

This is a binary minimization with linear constraints. The number of the variables is $n + k$, the number of the constraints is $2k + 1$.

"One is enough" model for the minimal cost problem.   This model can be handled similarly to the previous one. Let now z be the following vector:

$$z_i = \begin{cases} 0 & \text{if } a_i x = 0 \\ 1 & \text{if } a_i x \geq 1 \end{cases}$$

namely $z = \min\{Ax, e\}$. In this case (3) can be transformed into the following problem:

$$\begin{array}{rrcll} \min & cx & & \\ \text{subject to} & z_i & \geq & \frac{a_i}{b_i}x & i = 1, \ldots, k \\ & z & \leq & Ax & \\ & vz & \geq & K & \\ & x & \in & \{0,1\}^n & \\ & z & \in & \{0,1\}^k & \end{array} \qquad (4)$$

Above we assumed homogenous models, but we are not restricted to them. We defined the functions f_i separately for each subpurpose, so one can imagine an optimization problem where different subpurposes follow different models. These "mixed" models do not cause new problems, that is why we omit them.

Our second optimization problem is to find a **maximal coverage** test set supposing an upper bound (L) for the cost.

$$\begin{array}{rl} \max & cov(\mathcal{T}) \\ \text{subject to} & c(\mathcal{T}) \leq L \\ & \mathcal{T} \subset \{T_1, T_2, \ldots, T_n\} \end{array}$$

This optimization problem, just like the previous one, can be formulated as a binary minimization problem with functions f_i:

$$\begin{array}{rl} \max & \sum_{i=1}^{k} f_i(a_i x) \\ \text{subject to} & cx \leq L \\ & x \in \{0,1\}^n \end{array} \qquad (5)$$

Without further details let us see the formulae for the three models.

"Linear" model for the maximal coverage problem.

$$\begin{array}{rl} \max & \left(\sum_{i=1}^{k} \frac{a_i cov(S_i)}{b_i} \right) x \\ \text{subject to} & cx \leq L \\ & x \in \{0,1\}^n \end{array} \qquad (6)$$

"All or nothing" model for the maximal coverage problem.

$$
\begin{aligned}
\max \quad & vz \\
\text{subject to} \quad z_i &\leq \tfrac{a_i}{b_i}x & i = 1,\ldots,k \\
z &\geq Ax - b + e \\
cx &\leq L \\
x &\in \{0,1\}^n \\
z &\in \{0,1\}^k
\end{aligned}
\tag{7}
$$

"One is enough" model for the maximal coverage problem.

$$
\begin{aligned}
\max \quad & vz \\
\text{subject to} \quad z_i &\geq \tfrac{a_i}{b_i}x & i = 1,\ldots,k \\
z &\leq Ax \\
cx &\leq L \\
x &\in \{0,1\}^n \\
z &\in \{0,1\}^k
\end{aligned}
\tag{8}
$$

We have transformed each type of the optimal selection problem into linear integer (binary) optimization problems. These problems, in theory at least, can be solved by adequate mathematical methods. Although integer programming is NP-complete, we have not met a problem we could not solve so far. In the following sections we are going to present a protocol the ATS of which contains 668 Test Cases, and we will show that our method is able to find relatively small test set to execute. Though much bigger ATSs rarely exist, we are ready to use heuristic methods in order to find suboptimal solutions in case we are not able to find optimal ones.

3.   REALIZATION

In the previous section we introduced subpurposes. In this section we would like to show how the theory based on them can be turned into practice. We have to decide first what segment of the protocol is going to play the role of subpurposes.

Requirements describe well-defined parts of protocol behaviour. They can be classified according to their internal properties, which consists of the messages and their parameters, the time periods, the states, and the protocol mechanism [7]. To test the messages (Protocol Data Units, PDUs) and their parameters is the most fundamental control. That is the reason why we choose the PDUs to be the subpurposes. Of course, the whole purpose is not covered entirely by the PDUs but they form a good approximation, and unlike requirements the PDUs related to a TC are easy to detect. The connection of purposes, requirements and PDUs is illustrated in Figure 1.

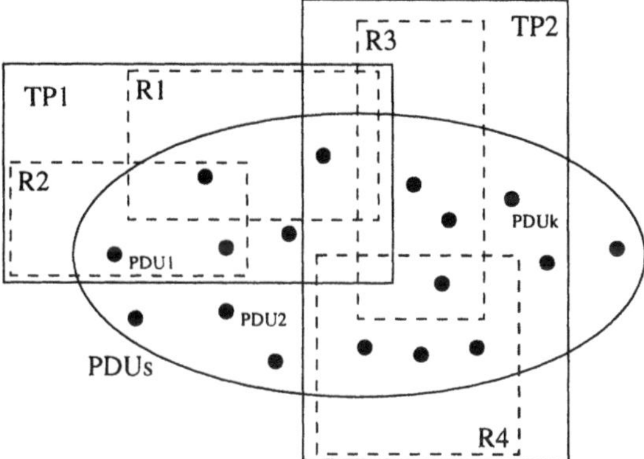

Figure 1   The connection of Test Purposes (TP), requirements (R) and Protocol Data Units (PDU)

Although PDUs are the most natural candidates for subpurposes, sometimes it is worth stepping one layer up or down if it results in more efficient selection. That is, choosing the PDU types or the parameters of the PDUs respectively to be subpurposes. Which the most appropriate subpurpose is depends on the structure of the ATS.

The fact that we choose PDUs to be subpurposes has an other important advantage, namely the Subpurpose-Test Case Incidence Matrix can be filled up automatically using the TTCN-mp format of the ATS. We implemented a program that is able to find the PDUs used in the TCs and to create the incidence matrix independently of the size of the ATS. This program can also detect PDUs that are used in the Test Steps or Defaults appearing in the Test Case. So the PDUs related to a TC are sent either in its main body or in a referred Test Step or Default. The program deletes the all zero rows if there are any, and merge the rows which are equal, because in this latter case the PDUs related to these rows can be supposed to compose one subpurpose together.

Determining the cost of the tests and the coverage of the subpurposes is up to the test laboratories' individual preferences, but we feel that choosing them to be the same for each TC and PDU respectively is the most natural solution. As for the coverage this is because at first sight there is no theoretical basis to distinguish between the PDUs. The costs can be chosen to be the same because the most time consuming task during the conformance assessment process is preparation and parametrization, the time of which is in direct proportion to the number of the executed tests.

We introduced three different models regarding the coverage: the "linear", the "all or nothing", and the "one is enough" models. Supposing that the PDUs play the role of the subpurposes the "linear" model means that the more PDUs

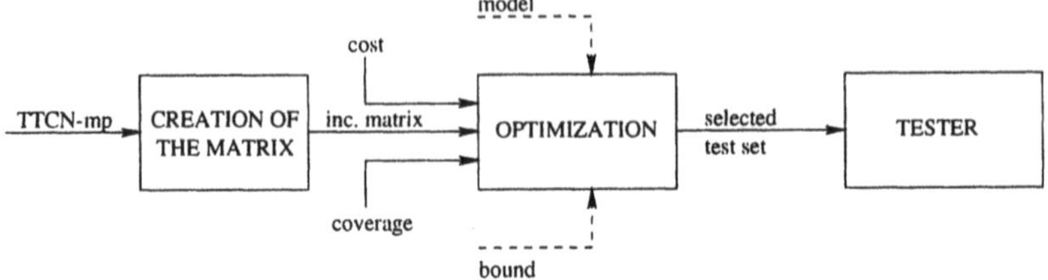

Figure 2   The process of selection

were sent in the executed TCs, the bigger coverage we obtain. In the "all or nothing" model we regard a PDU to be checked if we execute all the TCs related to it. In the case of the "one is enough" model a Test Case is already assumed to be checked if we execute only one TC that contains the PDU. This third model, we think, is not adequate.

Figure 2 illustrates the process of selection. The main steps are (1) creating the Subpurpose-Test Case Incidence matrix using the TTCN-mp format of the ATS; (2) determining the cost and coverage functions, deciding which model will be used (minimal cost/maximal coverage, "linear","all or nothing" or "one is enough"), and inputting the cost or coverage bound; (3) loading the selected TCs into the tester. For a protocol the first step should be done only once. If the incidence matrix is available, we can try and carry out the optimization for different cost, coverage functions, or different models and bounds in order to find the best test set that meet our aims.

4.   EXPERIMENT RESULTS

Having described the method let us look at the experiments now. We applied the method to the ATS of ISDN DSS1 Layer 3 protocol [4]. We decided to use this protocol for it is well-known, widely used, and relatively complex.

First we applied the above mentioned program to find the PDUs, the Test Cases, the Test Steps (TS) and the Defaults (DEF) defined in the ATS. The standard contains 229 PDUs, 668 TCs, 46 TSs and 5 DEFs. The program also determines which PDUs are used in which TC, TS or DEF, and which TSs or DEFs are included in which TC. Using this information we could create the PDU-TC incidence matrix. The $(i, j)^{th}$ entry of this matrix is 1 either if the ith PDU is sent in the jth TC or if it is sent in a TS (DEF) that is referred in the jth TC or, even deeper, if it is sent in a TS (DEF) referred in an other TS (DEF) which is a part of the jth TC, and so on. Then we merged the equal rows and deleted the all zero rows (there were some). By means of these operations we reduced the size of the incidence matrix to 192 rows (subpurposes=PDUs or more PDUs together) and 668 columns (Test Cases).

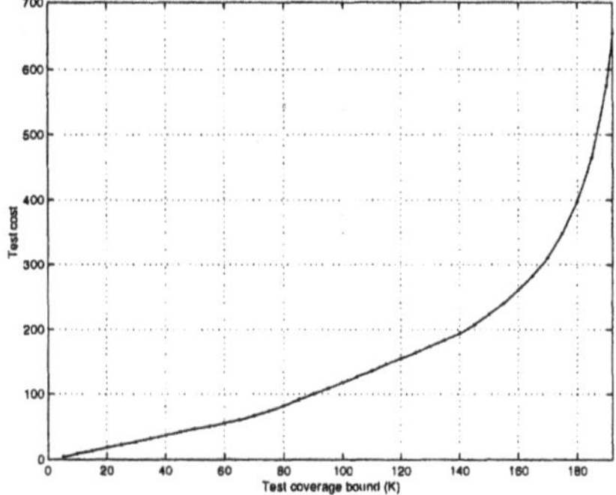

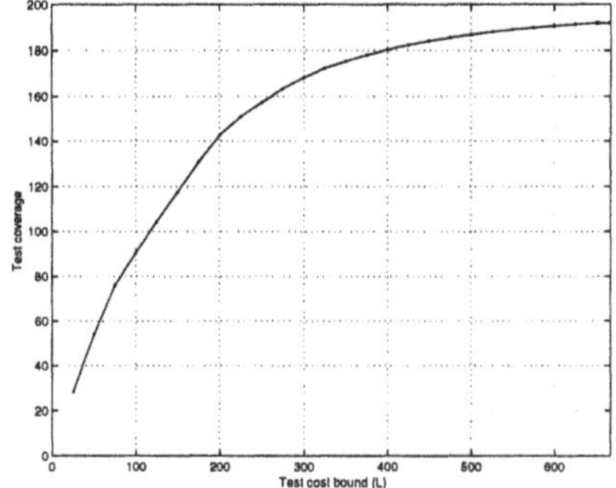

Figure 3 Minimal cost in "linear" model   *Figure 4* Maximal coverage in "linear" model

For the reasons we have mentioned in the previous section we defined the cost and the coverage to be the same for each TC and subpurpose respectively, and we chose the "linear" and the "all or nothing" models in optimization. To solve the optimization problems we used CPLEX MIP Solver [8].

4.1   RESULTS USING THE "LINEAR" MODEL

We calculated the minimal cost test set for coverage bounds $K = 5, 10, \ldots,$ $185, 190, 192$ in problem (2). The results are shown in Figure 3. As it can be seen, increasing the coverage bound the cost of the optimal test set does not increase linearly. This means that using our method we can obtain better (shorter in time) test sets to execute than we would get if we chose it at random. In fact, our method gives us the best possible test set within the constraints.

In problem (6) we calculated the maximal coverage test set for cost bounds $L = 25, 50, 75, \ldots, 625, 650, 653, 668$. Figure 4 shows the results. There are two things that are worth remarking. The first is the same phenomenon we mentioned above: we got better results than they would have been in case of random selection. The second is that we reached the whole coverage at $L = 653$, so 15 Test Cases are unnecessary. Indeed, we found 15 empty TCs, which do not contain any PDUs.

Let us see a specific example. If we let 10% of the coverage be lost (so $K = 173$ in the minimal cost problem), then it is enough to execute only 331 of the 668 tests (49.5%). While the whole coverage of these TCs is 173.11 the individual coverages of different subpurposes vary between 0 and 1. There are 155 subpurposes the coverages of which are equal to 1, which means that they are checked completely. All the subpurposes that appears in less than 11 TCs are that kind. The remaining 37 incompletely checked subpurposes are mostly those that appear in lots of TCs (the average number is 267). If a PDU appears

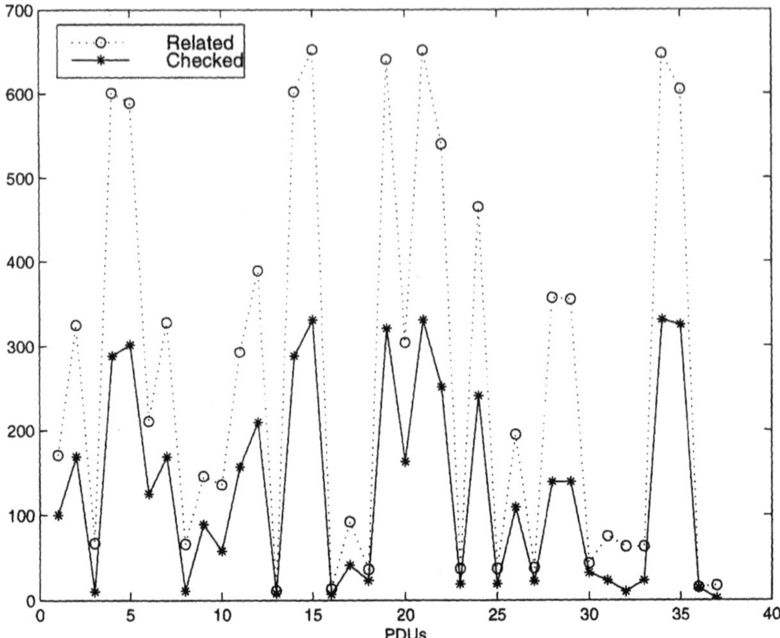

Figure 5   The number of the related TCs to the 37 incompletely checked PDUs, and the number of the checked ones among them.

in such an enormous number of TCs, then it is possibly sent in a Test Step or Default that is used in most of the Test Cases. Therefore a coverage of 50% for this PDU is a promising result. Figure 5 presents how many TCs are related to these 37 subpurpose and how many times they were checked.

Let us examine now the results from the point of view of the TCs. For example the 32nd Test Case - L3N_N00_V_032 - is not selected. Figure 6 presents this TC in TTCN.

This TC first initiates a parallel test component (PTC1_IN), takes the IUT into Null call state N00 with the preamble L3N_PR_N00, indicates a SETUP message, then checks whether the valid (SU_R10) or the invalid (SU_R12) PDU is received, checks the final state (Call Present call state N06) with L3N_CS(6,1), and in the end finishes with the postamble L3N_PO(1). All these stages can be characterized by PDUs or Test Steps. We can characterize the Test Steps by the PDUs that appear in them, therefore the subpurposes related to this TC characterize its behaviour. Of course, they do not tell us much about the structure, but they do carry an important part of the information contained by the Test Case.

Test Case Dyamic Behaviour				
Test Case Name : L3N_N00_V_032				
Group : N00/Valid/Incoming_point_to_point/				
Purpose : Ensure that the IUT in the Null call state N00, to indicate the arrival of a call to a point–to–point configuration, sends a SETUP message using the point–to–point data link and enters the Call Present call state N06.				
Configuration : CONFIG1				
Default : L3N_DEF(1)				
Comments : subclause 5.2.1				

Nr	Label	Behaviour Description	Constraints Ref	Verdict	Comments
1		CREATE(PTC1:PTC1_IN)			(1)
2		+L3N_PR_N00			preamble N00
3		CPA1!CP_M START TWAIT	S_SETUP		(2)
4		L0?SETUPr (CREF := SETUPr.mun.cr.cr_r) CANCEL TWAIT	Sr(SU_R10(CHlb_R1,CHlp_R1))	(P)	(3)
5		+L3N_CS(6,1)			state = 6?
6		L0?SETUPr (CREF := SETUPr.mun.cr.cr_r) CANCEL TWAIT	Sr(SU_R12)	(F)	(4)
7		+L3N_PO(1)			postamble N0
8		?TIMEOUT TWAIT		I	no response

Detailed Comments : (1) The slave component PTC1 is started.
(2) This coordination message indicates to PTC1 to send a SETUP message.
(3) A valid SETUP message is received.
(4) A SETUP with an invalid Channel identification information element is received. The test fails.

Figure 6   The TTCN format of the L3N_N00_V_032 Test Case.

The 225th Test Case - L3N_N06_I_002 - is selected. The purpose of this TC assumes that the IUT is in Call Present call state N06, and ensures it with the preamble L3N_PR_N06. This preamble is basically contains the 32nd TC. It can be said that the 225th TC is a continuation of the 32nd one, thus if it is executed, then the other one is (indirectly) executed as well. That means, although the 32nd TC is not selected (so it is not executed), we do check its purpose by executing the 225th TC.

The situation is very similar in the case of the 148th TC (L3N_N04_V_001). Its preamble L3N_PR_N04 and the TC itself together is the same word for word as the preamble L3N_PR_N010O. Therefore, though the 148th TC is not selected, every selected test which contains the latter preamble checks its purpose.

4.2   RESULTS USING THE "ALL OR NOTHING" MODEL

We calculated the minimal cost test set for coverage bounds $K = 5, 10, \ldots,$ $185, 190, 192$ in problem (3) and the maximal coverage test set for cost bounds $L = 25, 50, 75, \ldots, 625, 650, 653, 668$ in problem (7). Figure 7 and Figure 8 show the results.

The node limit of the branch-and-bound algorithm implemented in CPLEX was 20000. There are values of K and L for which we did not get optimal

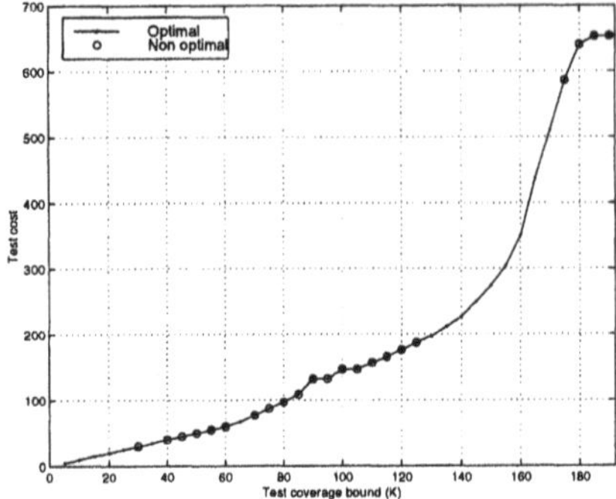

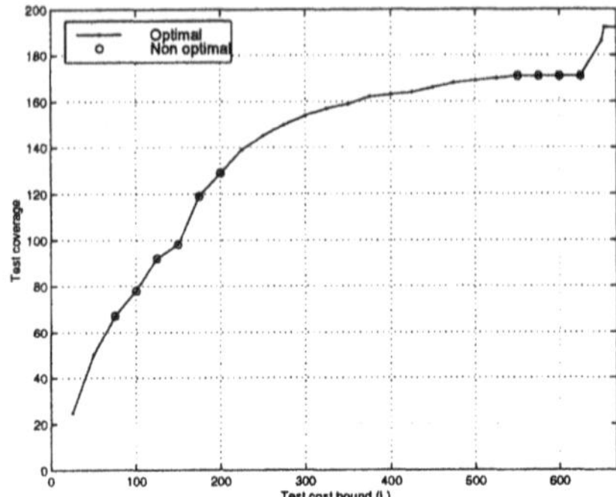

Figure 7 Minimal cost in "all or nothing" model

Figure 8 Maximal coverage in "all or nothing" model

solutions with this small node limit. But the suboptimal feasible solutions give us attractive results. The values we got were better in each cases than we would have obtained by random selection.

5.   CONCLUSIONS

We have presented the theory and practice of a new selection method which selects Test Cases from the Abstract Test Suite. We have performed experiments to explore its efficiency. The results show that the method can reduce the time of testing. We have applied the method to a fairly complex, widely used, real-life protocol, the ISDN DSS1 Layer 3 protocol. We find our method to be empirically promising, however, further experiments can be conducted for other protocols.

Further research may be devoted to the possible other subpurpose candidates, like the parameters. We plan to refine the method in order to make it capable to choose the best candidate.

References

[1] S.T. Vuong, J.A. Curgus. On test coverage metrics for communication protocols. In *4th Int. Workshop on Protocol Testing System*, 1991.

[2] J.A. Curgus, S.T. Vuong and J. Zhu. Sensitivity analysis of the metric based test selection. In *IFIP 10th Int. Workshop on Testing of Communicating Systems*, Cheju Island, Korea, 1997.

[3] ISO/IEC 9646-3, Tree and Tabular Combined Notation (TTCN), 1998.

[4] ETSI draft prETS 300 403-7 (1998): "Integrated Services Digital Network (ISDN); Digital Subscriber Signalling System No. one (DSS1) protocol; Signalling network layer for circuit-mode basic call control; Part 7: Abstract

Test Suite (ATS) and partial Protocol Implementation eXtra Information for Testing (PIXIT) proforma specification for the network".

[5] ETSI Draft prTBR4 Integrated Services Digital Network (ISDN); Attachment requirements for terminal equipment to connect to an ISDN using ISDN primary rate acces.

[6] T. Csöndes, S. Dibuz and B. Kotnyek. Test Suite Reduction in Conformance Testing, *Acta Cybernetica*, to appear in 1999.

[7] K. Tarnay. *Protocol Specification and Testing*, Akadémiai Kiadó, Budapest, 1991.

[8] CPLEX, Version 3.0, 1989-1994, CPLEX Optimization Inc.

[9] ISO/IEC DIS 13245/ ITU-T Z.500: Formal methods in conformance testing, 1997

[10] G.L. Nemhauser and L.A. Wolsey. *Integer and Combinatorial Optimization*, Wiley, New York, 1988.

17

CONFORMANCE TESTING OF MULTI-PROTOCOL IUTs

Yongbum Park[1]
Protocol Engineering Center, Electronics and Telecommunications Research Institute
161 Kajong-Dong, Yusong-Gu, Taejon, 305-350, Korea
E-mail: ybpark@pec.etri.re.kr

Myungchul Kim
Information and Communications University
58-4 Hwaam-Dong, Yusong-Gu, Taejon, 305-348, Korea
E-mail: mckim@icu.ac.kr

Sungwon Kang
Korea Telecom Research & Development Group
463-1 Junmin-Dong, Yusong-Gu, Taejon, 305-390, Korea
E-mail: kangsw@sava.kotel.co.kr

Abstract   To declare conformance of multi-protocol Implementation Under Test (IUT), every layer of the multi-protocol IUT should be tested. According to ISO9646, single-layer test method is applied to testing the highest layer of multi-protocol IUT and single-layer embedded test method is used for the layers other than the highest layer because the interfaces between layers are not exposed. So far the conventional researches have focused on testing layer by layer all the protocols in a multi-protocol IUT.

This paper proposes a new method for testing a multi-protocol IUT. The proposed test method assumes that a multi-protocol IUT is under test and that the interfaces between the layers can not be controlled or observed by the tester. We apply the proposed test method to TCP/IP and compare the application results with those of the existing test methods in terms of various criteria such as the number of test cases[2], the number of test events and test

---

[1] He is also a Ph.D. student of Information and Communications University.
[2] Refer to ISO9646 for the definitions of test case, test event, and behavior line.

coverage. It turns out that the proposed test method significantly reduces the number of test cases as well as the number of test events while providing the same test coverage. In addition, the proposed test method shows the capability to locate the layer that is the source of failure in testing multi-protocol IUTs.

Keywords: Conformance testing, multi-protocol, embedded test, test in context

1.   INTRODUCTION

Conformance testing methodology and framework has been published as an International Standard by ISO/IEC JTC1 [ISO9646]. This standard defines the concept of conformance testing, test methods, test specifications, test realization, and the requirements for the third party testing and also provides the Tree and Tabular Combined Notation (TTCN) for test scenario description. It is mainly for the case where a single-layer protocol is under test. It proposes a successive use of single-layer embedded test method for testing a multi-protocol IUT, i.e. an IUT that is a stack of protocols or profile. This method is referred to as incremental testing of a multi-protocol IUT.

According to [ISO 9646], testing is performed for a single specific layer and its lower layers play the role of a service provider. It is generally assumed that the lower layers are correct and all the causes of failures reside in the target layer. As a matter of fact, sometimes it is very difficult to determine which layer is the source of failure when it occurs during testing.

The single-layer embedded test method is applied to every layer of a multi-protocol IUT except the highest layer. In the method, test is specified for each single-layer and it need include specification of protocol activity above the one being tested but need not specify control or observation at layer boundaries within the multi-protocol IUT. Thus for a multi-protocol IUT consisting of protocol N_1 through upto protocol N_m, the test cases for protocol $N_i, 1 <= i < m$, shall include the specifications of the PDUs of protocol N_{i+1} through upto to N_m as well as protocol N_i. The existing approaches to the test coverage and the test sequence generation for the embedded test method ([Petr96], [Petr97], [Zhu98], [Yevt98]) so far focused on testing one by one single-layer protocols within a multi-protocol IUT.

In the embedded test method (also called testing in context), System Under Test (SUT) is considered as a composition of two Finite State Machines (FSMs) where one FSM to be tested is called component and the other is called context. Context is an FSM for protocols that are not being tested and interacts with component through internal interfaces that are not directly controllable or observable by the tester. In addition context is assumed error-free. [Petr96] and [Petr97] modeled testing in context as a communicating FSM and proposed a method to generate test sequences. Since in the embedded test method the interfaces between protocol layers are not directly controllable or observable, the test coverage of the method is limited. [Zhu98] proposed an approach and developed a tool to evaluate test coverage of the embedded test method. [Yevt98] proposed an approach to minimizing the test suite for testing in context. All these researches assume error-free context and focus on testing one by one single layer protocols of a multi-protocol IUT. Incremental testing starts from the lower layer and proceeds toward the higher layer. If errors were found at the lower layer, testing upper layers could be meaningless because the errors affect the behavior of the upper layer. In this case, the existing test method may assign an inconclusive verdict to the test case for the upper layer. This is the problem that is dealt with in this paper.

In this paper we propose a new test method for testing a multi-protocol IUT. The method yields one single test suite which has the same effect as applying both of the test suites; one for the single-layer test method and the other for the single-layer embedded test method. Comparing with the existing test methods, the proposed one has the following advantages:

* The upper layer and the embedded layers can be tested simultaneously with a single test suite.

* Producing a smaller size of test suite and shorter lengths of test cases, it reduces the load of test suite description and test execution.

* The proposed method provides a means of testing multi-protocol IUTs with inaccessible internal interfaces.

The rest of the paper is organized as follows: Section 2 proposes a new test method for testing multi-protocol IUTs. Section 3 presents application of the proposed test method to TCP/IP and provides some results of comparing this method with the existing ones. In Section 4, we conclude this paper with some discussion on the future work.

2.   CONFORMANCE TEST METHOD FOR MULTI-PROTOCOL IUTS

According to the conformance test framework of [ISO9646], testing of multi-protocol IUTs should be carried out by a successive use of the single-layer embedded test method to every layer of multi-protocol IUT. In this case, when a specific protocol is tested, it is assumed that every protocol except the target protocol is assumed to be error-free. However, it is possible to determine conformance of a multi-protocol IUT by a single testing assessment of applying a single test suite to the multi-protocol IUT if we can locate which layer is the source of failure in testing of multi-protocol IUT. In general, two adjacent layers interact with each other and upper-layer protocol PDU is included in the user data field of its lower-layer PDU. With the knowledge of interaction and relationship of PDUs, we are able to observe and control both layers at once by observing and controlling only the lower layer. This is the main idea of this paper.

B-ISDN ATM Adaptation Layer (AAL) [ITU94] is an example of a multi-protocol. In AAL, Service Specific Connection Oriented Protocol (SSCOP) and Common Part Convergence Sublayer (CP-AAL) are below Service Specific Coordination Function (SSCF), which is the highest sublayer of AAL. The interfaces of SSCOP are not open and can be indirectly accessed only through SSCF and CP-AAL. Another example of a multi-protocol is Multiprotocol Label Switching (MPLS) of IETF, which is an integration of layers 2 and 3 to improve performance of routers [IETF97]. In MPLS implementations of commercial routers, the interface between layr 2 and layer 3 is not open.

For our test method for multi-protocol IUTs, we assume the following:
* protocol for a layer or a sublayer is represented as an FSM
* the highest and the lowest interfaces of a multi-protocol IUT are open

In spite of the second assumption, the behavior of multi-protocol can be inferred indirectly by controlling and observing the lowest interface. With this insight, we propose a new test method to test several protocols at the same time with a single test suite.

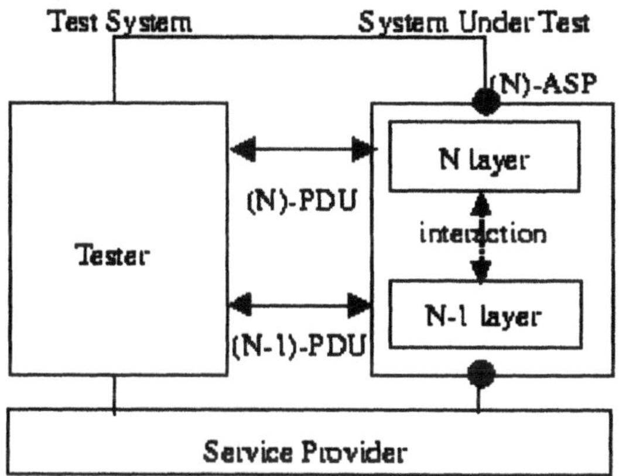

Figure 1. Configuration of a multi-protocol IUT

Using Figure 1 we explain our method. In Figure 1 the upper interface of N layer and the lower interface of N-1 layer are open but the interface between N layer and N-1 layer is not directly controllable or observable by the tester.

According to the existing test method in the standard [ISO9646], N layer and N-1 layer protocols of the System Under Tester (SUT) in Figure 1 should be tested separately. The single-layer test method is applied to N layer protocol and the single-layer embedded test method is applied to N-1 layer protocol. While the test event for N layer protocol testing is described with (N)-ASP and (N)-PDU, the test event for N-1 layer protocol testing is described with (N)-ASP and (N-1)-PDU.

Test suite from the multi-protocol test method uses test events used in the single-layer test method and the single-layer embedded test method. In the test configuration of Figure 1, the test event for multi-protocol test method is described with (N)-ASP, (N)-PDU and (N-1)-PDU. Note that (N)-PDU is included in the user data field of (N-1)-PDU.

In the existing test method of [ISO9646], the upper and the lower interfaces of IUT are controlled and observed by tester. These two interfaces are called Points of Control and Observation (PCOs). In the multi-protocol test method proposed in this paper, one input or output event at a lower PCO, say PCO3 in Figure 2, includes the behaviors of the two protocol layers. Let us explain the Figure 2, which shows the PCO structure of the multi-protocol test method for TCP/IP.

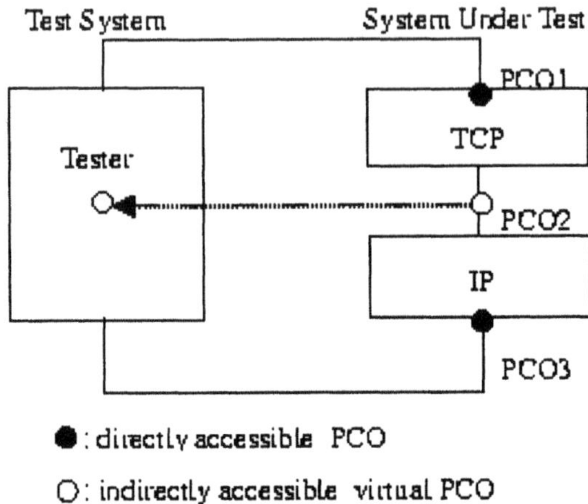

Figure 2. PCO structure example in the multi-protocol test method

In Figure 2, PCO1 and PCO3 are directly accessible by the tester whereas PCO2 is a virtual PCO not directly accessible by the tester. The test events at PCO1 are presented with TCP Abstract Service Primitives (ASPs) and the test events at PCO3 are presented with IP PDUs. Since IP PDU includes TCP PDU in its data (payload) field, it is possible to control and observe TCP PDU as well as IP PDU at PCO3. As this example demonstrates, the multi-protocol test method is able to test both TCP and IP protocols together at the same time.

3.   APPLICATION OF THE MULTI-PROTOCOL TEST METHOD TO THE TCP/IP

This section presents an application of the proposed test method to TCP/IP and compares the results of this application with those of the existing test methods in terms of criteria such as the number of test cases, the number of test events and the test coverage.

3.1   FSM for the simplified TCP/IP

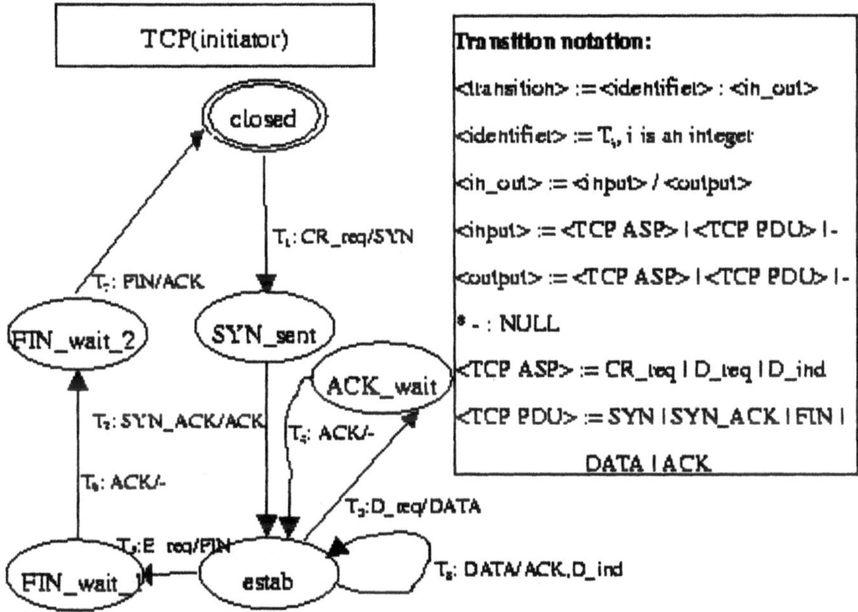

Figure 3. FSM for a simplified TCP

Figure 3 is a simplified FSM of TCP [Stev94]. TCP supports connection management, timer management, reliable data transfer through error control, and flow control functions. In the FSM of Figure 3, connection management and data transfer functions are represented but the states relating to timers in connection management phase and error control in data transfer phase are omitted. In this paper only the initiator role of TCP is used to explain the multi-protocol test method. Test cases for the responder role of TCP can be obtained with the same approach.

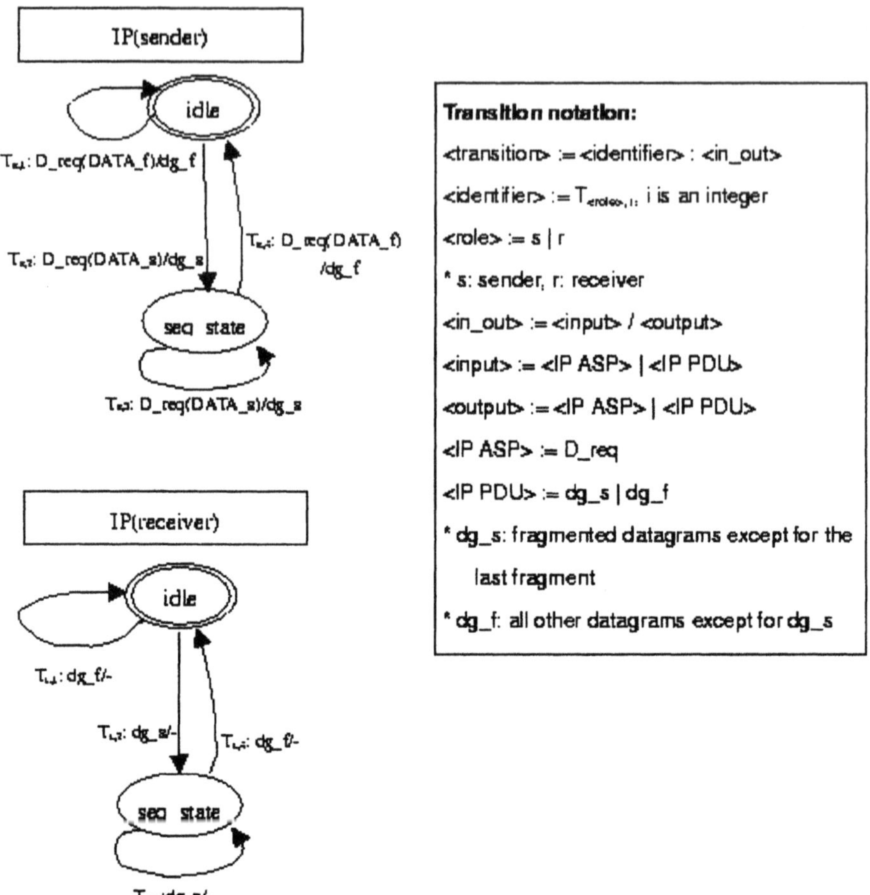

Figure 4. FSM for IP with fragmentation

IP can be represented with a single state FSM because it has no connection management function and supports only a simple datagram delivery service. However, because of fragmentation, it is useful to represent IP as an FSM with two states. When the IP transmits datagrams on demand of TCP and the size of requested data is greater than the size of IP datagram, IP does fragmentation to deliver the data with smaller datagrams. In this case the protocol behaviors for the last fragment and for the other fragments are different. Thus we use two states to distinguish these different behaviors. In this way, we come up with the FSM for IP as in Figure 4. In Figure 4 IP datagram is classified into dg_f and dg_s. dg_s stands for a datagram for a fragment except the last one and dg_f all the other datagrams except dg_s.

TCP and IP above it interact with each other and the interaction is dependent upon the size of TCP PDU. For example, TCP PDUs such as SYN, SYN_ACK, FIN and ACK in Figure 3 are delivered with dg_f PDU in Figure 4. DATA PDU of the TCP in Figure 3 is delivered with either dg_s or dg_f PDU depending on the size of the DATA PDU of TCP. Therefore, the transition T_3 in TCP interacts with either the transition $T_{s,1}$ or the transitions $T_{s,2}$, $T_{s,3}$ and $T_{s,4}$ in IP depending on the size of DATA.

3.2   Test cases for TCP in the multi-protocol test method

This section presents examples of test cases obtained by our multi-protocol test method. Naming convention for test cases is as follows:

<test case name> := <test method>_tc_<transition>[_<seq_no>]

<test method> := m | s | e

* m: multi-protocol test method

* s: single-layer test method

* e: single-layer embedded test method

<transition> := T_i | $T_{\langle role\rangle, i}$, i is an integer

<role> := S | R

* S: sender, R: receiver

<seq_no> := integer

The interaction between T_3 in Figure 3 and the transitions in IP FSM of Figure 4 results in two varieties depending on the size of DATA. Thus, the number of test cases for T_3 in the multi-protocol test method becomes also two as m_tc_T_3_1 and m_tc_T_3_2. In this paper we use a simplified TTCN to describe test cases. Let us put line numbers in the left-most column.

m_tc_T_3_1: test case for T_3 - 1

1 +m_tc_T_2 /* preamble to put IUT to state estab */

2   PCO1 ! D_req(DATA) /* small DATA, no fragmentation */

3   PCO3 ? dg_f(DATA) /* PCO2 observation included */

In the above test case, the first behavior line is a test step to put IUT in "estab" state in order to prepare for execution of T_3. The second behavior line sends a small data not requiring fragmentation in IP and is input to the transition T_3. By feeding in a small data, not only the transition T_3 in the TCP FSM but also the transition $T_{s,1}$ in the IP FSM of Figure 4 is executed by the interaction between TCP and IP. These transition executions are presented in the third behavior line. Here receiving of dg_f is observed at PCO3 and thus the tester is able to observe PCO2 indirectly through observing the data (payload) field included in dg_f. As the result, this single test case covers two different transitions in two different protocols respectively (namely, T_3 of TCP and $T_{s,1}$ of IP). This is summarized in Table 1.

Table 1. The relationship between transitions and behavior lines in the test case
m_tc_T₃_1

Behavior	transition of TCP	transition of IP
PCO1 ! D_req(DATA)		
PCO3 ? dg_f(DATA)	T_3	$T_{s,1}$

m_tc_T₃_2: test case for T₃ - 2

```
1 +m_tc_T₂
2    PCO1 ! D_req(DATA) /* large DATA, fragmentation */
3       PCO3 ? dg_s(DATA_s) /* PCO2 observation included */
4          PCO3 ? dg_s(DATA_s) /* PCO2 observation included */
5             PCO3 ? dg_f(DATA_f) /* PCO2 observation included */
```

In the test case m_tc_T₃_2, the second behavior line sends a large data requiring fragmentation in IP. By interacting with this behavior line, transitions such as $T_{s,2}$, $T_{s,3}$ and $T_{s,4}$ are executed and these interactions can be observed indirectly through the observation at PCO3 of TCP's DATA PDUs included in IP datagrams. As shown in this test case, multiple observations at the lower layer protocol may be required for single stimulus at the upper layer protocol in test case of multi-protocol test method. This test case covers transition T_3 of TCP and the transitions $T_{s,2}$, $T_{s,3}$, and $T_{s,4}$ of IP as shown in Table 2.

Table 2. The relationship between transitions and behavior lines in the test case
m_tc_T₃_2

Behavior	transition of TCP	transition of IP
PCO1 ! D_req(DATA)		
PCO3 ? dg_f(DATA_s)		$T_{s,2}$
PCO3 ? dg_f(DATA_s)		$T_{s,3}$
PCO3 ? dg_f(DATA_f)	T_3	$T_{s,4}$

Two test cases shown above are for testing behavior of the initiator role of the TCP. Testing behavior of the responder role of the TCP can be explained with the test case for transition T_8 in Figure 3. The following test case is one for receiving large TCP data requiring fragmentation in IP.

m_tc_T_8_2: test case for T_8
```
1 +m_tc_T₂
2    PCO3 ! dg_s(DATA_s)
3       PCO3 ! dg_s(DATA_s)
4          PCO3 ! dg_f(DATA_f) /* large DATA, fragmentation, PCO2
control included */
5             PCO3 ? dg_f(ACK)
```

In the test case m_tc_T_8_2, stimuli for transition T_8 of TCP FSM corresponds to the three behavior lines (lines 2, 3 and 4) describing the behavior of IP protocol. And the stimuli of these three behavior lines result in observing ACK PDU of the TCP. This test case provides an example demonstrating multiple stimuli at the lower layer protocol result in a single observation at the upper layer protocol. This test case covers the transition T_8 of TCP and the transitions $T_{r,2}$, $T_{r,3}$, $T_{r,4}$ and $T_{s,1}$ of IP.

Appendix 1 shows all the test cases for TCP FSM in Figure 3 obtained as the result of applying our multi-protocol test method.

3.3   Test case by the single-layer test method for the transition in TCP FSM

This section presents an example test case by the single-layer test method. In the test configuration of Figure 2, the highest layer protocol, TCP, can be tested by the single-layer test method because its upper and lower interfaces can be accessed by the tester [ISO9646]. In general, in the single-layer test method there are two PCOs that are located at the upper and lower interfaces respectively. These PCOs are PCO1 and PCO2 in the configuration of Figure 2 for testing the TCP. Following is the test case for transition T_3 in Figure 3.

s_tc_T_3: test case for T_3
```
1 +s_tc_T₂
2    PCO1 ! D_req(DATA)
3       PCO2 ? DATA
```

In the single-layer test method we are able to control and observe only the behavior of the TCP but IP. Therefore, IP PDU such as dg_f and dg_s used in test case m_tc_T_3_2 in the section 3.2 is not used in test case s_tc_T_3. In test case s_tc_T_3, the second behavior line may invoke the fragmentation in sender IP depending on the size of the DATA. In turn receiver IP delivers DATA to TCP after reassembling the fragmented data. Because the single-layer test method on TCP does not handle the behavior of IP, test case s_tc_T_3 does not include transitions $T_{s,1}$ and transitions $T_{s,2}$, $T_{s,3}$, $T_{s,4}$ of IP FSM. That result is different from the multi-protocol test method shown in the previous section.

Appendix 2 shows all the test cases for all transitions of TCP FSM in Figure 3 as a result of applying the single-layer test method presented above.

3.4   Test case by the single-layer embedded method for the transition in IP FSM

This section presents an example test case for IP FSM by the single-layer embedded test method. PCOs used in the embedded test method are PCO1 and PCO3 in the configuration in Figure 2.

e_tc_$T_{s,2}$: test case for transition $T_{s,2}$

```
1 +e_tc_T_s,1
2    PCO3 ! dg_f(SYN_ACK)
3       PCO3 ? dg_f(ACK)
4          PCO1 ! D_req(DATA) /* large DATA, fragmentation */
5             PCO3 ? dg_s(DATA_s)
```

Test case above is to test IP. The fourth behavior line is a stimulus for transition $T_{s,2}$ of the IP. To send a large DATA PDU of TCP requiring fragmentation in the IP, TCP connection is to be established. The behavior related with TCP connection establishment is described in the second and third behavior lines.

Appendix 3 shows all the test cases for all transitions of IP FSM in Figure 4 as a result of applying the single-layer embedded test method.

3.5   Comparison the multi-protocol test method with the existing methods

As shown in the previous sections, the test cases for the multi-protocol test method include the test cases for the single-layer test method and for the single-layer embedded test method. And the relationship of those test cases is as follows:

$$m\_tc\_T_1 = s\_tc\_T_1 + e\_tc\_T_{s,1}$$
$$m\_tc\_T_2 = s\_tc\_T_2 + e\_tc\_T_{s,1} + e\_tc\_T_{r,1}$$
$$m\_tc\_T_3\_1 = s\_tc\_T_3 + e\_tc\_T_{s,1}$$
$$m\_tc\_T_3\_2 = s\_tc\_T_3 + e\_tc\_T_{s,2} + e\_tc\_T_{s,3} + e\_tc\_T_{s,4}$$
$$m\_tc\_T_4 = s\_tc\_T_4 + e\_tc\_T_{s,1}$$
$$m\_tc\_T_5 = s\_tc\_T_5 + e\_tc\_T_{s,1}$$
$$m\_tc\_T_6 = s\_tc\_T_6 + e\_tc\_T_{r,1}$$

$$m\_tc\_T_7 = s\_tc\_T_7 + e\_tc\_T_{r,1} + e\_tc\_T_{s,1}$$
$$m\_tc\_T_8\_1 = s\_tc\_T_8 + e\_tc\_T_{r,1} + e\_tc\_T_{s,1}$$
$$m\_tc\_T_8\_2 = s\_tc\_T_8 + e\_tc\_T_{r,2} + e\_tc\_T_{r,3} + e\_tc\_T_{r,4} + e\_tc\_T_{s,1}$$

For example, the test case $m\_tc\_T_3\_2$ by the multi-protocol test method includes the test case $s\_tc\_T_3$ for testing TCP by the single-layer test method and the test cases $e\_tc\_T_{s,2}$, $e\_tc\_T_{s,4}$ and $e\_tc\_T_{s,4}$ for testing IP by the single-layer embedded test method. Table 3 summarizes this relationship and other comparison results in terms of the number of test cases, the number of test events and test coverage.

Table 3. Comparison of test cases for each test method

test method / measure	proposed multi-protocol test method	existing test methods		
		single-layer test method for TCP	single-layer embedded test method for IP	Combination of two existing test methods
no. of test cases	10	8	8	16
no. of behavior lines	32	22	23	45
no. of events	64	48	47	95
test coverage	all transitions			all transitions

As shown in the Table 3, the multi-protocol test method significantly reduces the number of test cases compared with the existing method which combines single-layer test method and single-layer embedded test method while providing the same test coverage. The size of the test case is decreased and the load for test suite description is reduced because the number of behavior lines is reduced by 29%. And the test execution time is reduced because the number of test events in all test cases is reduced by 33%.

This overhead of the existing test method is caused by duplicated execution of the transitions of the lower layer protocol. The reason is that the transitions of the lower layer protocol executed in the single-layer test method for testing the upper layer protocol is also executed in the single-layer embedded test method for testing the lower layer protocol. However, in the multi-protocol test method, we are able to test two-layer protocols at once and thus avoid duplicated execution of transitions. As the result, the number of behavior lines and the number of test events are reduced.

Another advantage of the multi-protocol test method is the ability to locate the cause of failure in a multi-protocol IUT. In the existing test method it is assumed that every layer below the target protocol is correct and all the cause of failure is from the target protocol. However, real causes of the failure may be from the lower layers other than the target protocol. In this case, the multi-protocol test method provides a capability to locate the exact source of the failure.

The multi-protocol test method has a limitation in testing thebehavior of the lower layer protocol on errors. This is a problem caused from that the tester can not directly access the internal interfaces but control and observe those interfaces indirectly via the upper layer protocols. This problem also exists in the embedded test method.

4.   CONCLUSION AND FUTURE WORK

In this paper we proposed a new test method for testing the multi-protocol IUT. This test method is able to test multiple protocols with a single test suite and the test suite of the proposed test method is described with combination of test events that are used in the single-layer test method and the single-layer embedded test method. Application of the proposed test method to the simplified TCP/IP is presented as an example and some results of comparing this test method with the existing test methods are shown. By applying the multi-protocol test method, the number of test events, the number of test cases and the number of behavior lines are significantly reduced comparing to the existing test method by the single-layer test method and the single-layer embedded test method. Also the proposed test method is able to locate the exact source of failure in a specific layer in testing multi-protocol IUT.

In this paper we showed an example of application of the proposed test method to the multi-protocol IUT with two protocols in stack. It is future work to generalize the proposed test method to be applicable to the multi-protocol IUT consisted of more than two protocols in stack. Another further work could be applying the proposed test method to the real protocols such as ATM AAL and MPLS.

REFERENCES

[ISO9646] ISO 9646, "Information Technology - OSI - Conformance Testing Methodology and Framework," 1992.

[ITU94] ITU-T Recommendation Q.2110, "B-ISDN ATM Adaptation Layer - Service Specific Connection-Oriented Protocol(SSCOP)," 1994.

[IETF97] IETF draft-ietf-mpls-framework-02.txt, "A Framework for Multiprotocol Label Switching," 1997.

[Stev94] W. Richard Stevens, TCP/IP Illustrated, Addison-Wesley, 1994.

[Petr96] Alexandre Petrenko, Nina Yevtushenko, Gregor v. Bochmann and Rachida Dssouli, "Testing in context: framework and test derivation," Computer Communications 19, pp. 1236-1249, 1996.

[Petr97] Alexandre Petrenko and Nina Yevtushenko, "Fault detection in embedded components," IWTCS'97, pp. 272-287, Cheju Island, Korea, 1997.

[Zhu98] Jinsong Zhu, Son T. Voung and Samuel T. Chanson, "Evaluation of test coverage for embedded system testing," IWTCS'98, pp. 111-126, Tomsk, Russia, 1998.

[Yevt98] Nina Yevtushenko and Ana Cavali, "Test suite minimization for testing in context," IWTCS'98, pp. 127-145, Tomsk, Russia, 1998.

APPENDIX 1. TEST CASES FOR TCP BY THE MULTI-PROTOCOL TEST METHOD

m_tc_T1: test case for T_1
PCO1 ! CR_req
   PCO3 ? dg_f(SYN) /* PCO2 observation included */
- covered transitions:T_1, $T_{s,1}$

m_tc_T_2: test case for T_2
+m_tc_T_1
   PCO3 ! dg_f(SYN_ACK) /* PCO2 control included */
      PCO3 ? dg_f(ACK) /* PCO2 observation included */
- covered transitions: T_2, $T_{r,1}$, $T_{s,1}$

m_tc_T3_1: test case for T3_1
+m_tc_T_2
   PCO1 ! D_req(DATA) /* small DATA, no fragmentation */
      PCO3 ? dg_f(DATA) /* PCO2 observation included */
- covered transitions: T_3, $T_{s,1}$

m_tc_T3_2: test case for T3_2
+m_tc_T_2
   PCO1 ! D_req(DATA) /* large DATA, fragmentation */
      PCO3 ? dg_s(DATA_s) /* PCO2 observation included */
      PCO3 ? dg_s(DATA_s) /* PCO2 observation included */
      PCO3 ? dg_f(DATA_f) /* PCO2 observation included */
- covered transitions: T_3, $T_{s,2}$, $T_{s,3}$, $T_{s,4}$

m_tc_T4: test case for T4
+m_tc_T3
   PCO1 ! D_req(ACK)
      PCO3 ? dg_f(ACK) /* PCO2 observation included */
- covered transitions: T_4, $T_{s,1}$

m_tc_T_5: test case for T_5

+m_tc_T_2
 PCO1 ! E_req
  PCO3 ? dg_f(FIN) /* PCO2
observation included */
- covered transitions: T_5, $T_{s,1}$

m_tc_T_6: test case for T_6
+m_tc_T_5
 PCO3 ! dg_f(ACK) /* PCO2
control included */
- covered transitions: T_6, $T_{r,1}$

m_tc_T_7: test case for T_7
+m_tc_T_6
 PCO3 ! dg_f(FIN) /* PCO2
control included */
  PCO3 ? dg_f(ACK) /* PCO2
observation included */
- covered transitions: T_7, $T_{r,1}$, $T_{s,1}$

m_tc_T_8_1: test case for T_8_1
+m_tc_T_2
 PCO3 ! dg_f(DATA) /* small
DATA, no fragmentation, PCO2
control included */
  PCO3 ? dg_f(ACK)
- covered transitions: T_8, $T_{r,1}$, $T_{s,1}$

m_tc_T_8_2: test case for T_8_2
+m_tc_T2
 PCO3 ! dg_s(DATA_s)
  PCO3 ! dg_s(DATA_s)
   PCO3 ! dg_f(DATA_f) /*
large DATA, fragmentation, PCO2
control included */
  PCO3 ? dg_f(ACK)
- covered transitions: T_8, $T_{r,2}$, $T_{r,3}$, $T_{r,4}$, $T_{s,1}$

APPENDIX 2. TEST CASES FOR TCP BY THE SINGLE-LAYER TEST METHOD

s_tc_T_1: test case for T_1
PCO1 ! CR_req
 PCO2 ? SYN
- covered transitions: T_1

s_tc_T_2: test case for T_2
+s_tc_T_1
 PCO2 ! SYN_ACK
  PCO2 ? ACK
- covered transitions: T_2

s_tc_T_3: test case for T_3
+s_tc_T_2
 PCO1 ! D_req(DATA)
  PCO2 ? DATA
- covered transitions: T_3

s_tc_T_4: test case for T_4
+s_tc_T_3
 PCO1 ! D_req(ACK)
  PCO2 ? ACK
- covered transitions: T_4

s_tc_T_5: test case for T_5
+s_tc_T_2
 PCO1 ! E_req
  PCO2 ? FIN
- covered transitions: T_5

s_tc_T6: test case for T_6
+s_tc_T_5
 PCO2 ! ACK
- covered transitions: T_6

s_tc_T_7: test case for T_7
+m_tc_T_6
 PCO2 ! FIN
  PCO2 ? ACK
- covered transitions: T_7

s_tc_T_8: test case for T_8
+s_tc_T_2
   PCO2 ! DATA

        PCO2 ? ACK
        - covered transitions: T_8

APPENDIX 3. TEST CASES FOR IP BY THE SINGLE-LAYER EMBEDDED TEST METHOD

e_tc_$T_{s,1}$: test case for $T_{s,1}$
PCO1 ! CR_req
   PCO3 ? dg_f(SYN)

e_tc_$T_{s,2}$: test case for $T_{s,2}$
+e_tc_$T_{s,1}$
   PCO3 ! dg_f(SYN_ACK)
      PCO3 ? dg_f(ACK)
         PCO1 ! D_req(DATA) /*
large DATA, fragmentation */
         PCO3 ? dg_s(DATA_s)

e_tc_$T_{s,3}$: test case for $T_{s,3}$
+e_tc_$T_{s,2}$
   PCO3 ? dg_s(DATA_s)

e_tc_$T_{s,4}$: test case for $T_{s,4}$
+e_tc_$T_{s,3}$
   PCO3 ? dg_f(DATA_f)

e_tc_$T_{r,1}$: test case for $T_{r,1}$
PCO3 ! dg_f(SYN)
   PCO1 ? CR_ind

e_tc_$T_{r,2}$: test case for $T_{r,2}$
+e_tc_$T_{r,1}$
   PCO1 ! CR_resp
      PCO3 ? dg_f(SYN_ACK)
         PCO3 ! dg_f(ACK)
           PCO3 ! dg_s(DATA_s)

e_tc_$T_{r,3}$: test case for $T_{r,3}$
+e_tc_$T_{r,2}$
   PCO3 ! dg_s(DATA_s)

e_tc_$T_{r,4}$: test case for $T_{r,4}$
+e_tc_$T_{r,3}$
   PCO3 ! dg_f(DATA_f)
   PCO1 ? DATA_ind

BIOGRAPHY

Mr. Yongbum Park received B.S. in electronic engineering from Kyungbook National University in Korea in 1986 and received M.S. degree in computer science from Korea Advanced Institute of Science and Technology in 1989. Since 1989, he has been a senior research staff of Electronics and Telecommunications Research Institute. In 1995-1996, he was a guest researcher at National Institute of Standards and Technology of U.S.A. His research interests include communication protocol testing, Internet, wireless communication and mobile computing.

Dr. Myungchul Kim received B.A. in Electronics Engineering from Ajou University in 1982, M.S. in Computer Science from the Korea Advanced Institute of Science and Technology in 1984, and Ph. D in Computer Science from the University of British Columbia, Vancouver, Canada, in 1993. Currently he is with the faculty of the Information and Communications University, Taejon, Korea. Before joining the university, he was a managing director in Korea Telecom Research and Development Group for 1984 - 1997 where he was in charge of research and development of protocol and QoS testing on ATM/B-ISDN, IN, PCS and Internet. He has also served as a member of Program Committees for IFIP International Workshop on Testing of Communicating Systems, IEEE International Conference on Distributed Computing Systems, and IFIP International Conference on Formal Description Technique / Protocol Specification, Testing and Verification, the chairman of Profile Test Specifications - Special Interest Group of Asia-Oceania Workshop (for 1994 - 1997), and co-chair of the 10th IWTCS'97. His research interests include Internet, protocol engineering, telecommunications, and mobile computing.

Dr. Sungwon Kang received B.A. from Seoul National University in Korea in 1982 and received M.S. and Ph.D. in computer science from the University of Iowa in U.S.A. in 1989 and 1992, respectively. Since 1993, he has been a senior researcher at Korean Telecom. In 1995-1996, he was a guest researcher at National Institute of Standards and Technology of U.S.A. In 1997, he was the co-chair of the 10th International Workshop on Testing of Communication Systems. Currently he is the head of the Protocol Engineering Team at Korean Telecom R&D Group. His research interests include communication protocol testing, program optimization and programming languages.

VII

NEW AREAS FOR TESTING

18

THE CHALLENGE OF QOS VERIFICATION

Jan de Meer
GMD-FOKUS Berlin, D10589, Germany
jdm@fokus.gmd.de

Son Vuong
University of British Columbia, Vancouver, B.C. Canada V6T 1Z4
vuong@cs.ubc.ca

Abstract   Quality of Service Engineering deals with the management and control of distributed applications coping with multiple media. Controlling concludes the discriminating of qualities obtained and qualities specified. Thus, this is a kind of verification. Since qualities describe how things have to be done, the verification of qualities depend on the running environment. Conformance between a specification and its realization is necessary but not sufficient. In the OSI framework conformance verification methodologies - by testing - have almost maturely been developed. Quality verification methodologies - by controlling - are lacking. Thus, we have to investigate into the requirements of the new multimedia and high speed communication technology. These technologies are the means by which new systems and services will be built. The actual verification of the Quality of Service (QoS) depends from the embedding system and thus can only be checked during system operation time. In order to understand and to identify verification mechanisms and architectural observation or control points, a generic QoS model is envisaged. The new model is based on the basic concept of continuity respectively of processes in order to model continuous behaviour, i.e. streams adequately.

Keywords: Quality of Service, Multimedia Services, Conformance, QoS Verification and Testing, Continuity

1.   INTRODUCTION

The liberalisation of the telecommunication market and the wide-spread availability of communication technology has emerged as an enormous market for new various multimedia services to be realized and verified imposing adequate techniques and methods. The new services are expected to be easy deployable and customizable in order to match with users' flexibility. Furthermore, these services involve more and more continuous interaction using streams of data preferably in one direction and to be distributed to more than one customer. So, the advent of new technology will likely change the current paradigm of conformance verification from a discrete point-to-point view to a continuous multi-point view.

A new notion seems to be required to adopt the increase in dynamics and flexibility of communication. The new notion in mind is continuous behaviour and its verification, related to QoS testing.

The paradigm change might best be described by looking to the current trials of introducing flexibility to communication protocols. It begun with parameterization of the point-to-point network and transport protocols with so-called service classes. By such a service class, protocol service users are allowed to make - prior to the use - a selection on optional protocol functions, i.e. retransmission of lost protocol units, fixed throughput classes, etc. This is comparable to the invention of differentiated services to internet service users. Differentiated services provide service classes ranging from best effort to some type of guaranteed services.

In conventional transmission services time-consuming protocol functions like retransmission or time-outs have been possible, because transmission was not constrained by strong time requirements. Contrarily, for the new multimedia transportation services, timeliness and continuity became indispensable qualities. Hence, any appropriate QoS verification methodology has to go far beyond pure functional checks. Advanced transportation environments may include adaptation mechanisms to deal with QoS guarantees. Dynamic features like those, are based on permanent observation and control. Consequently, QoS verification, respectively testing integrates existing conformance testing techniques with those of continuous observation and controlling. Observation copes with sensors being introduced at appropriate component or object interfaces being fed-back to controllers. Controlling copes with service adaptation using specific traffic models or resource usage policies.

In the area of QoS, standardization bodies (i.e. ISO/IEC, ITU-T, ETSI) have dealt with three documents, the *Basic Framework of QoS* [QoS95], *Methods and Mechanisms* [QoS95] and *QoS in the ODP RM* [QoS98]. The basic framework document is restricted to the OSI basic reference model and thus, to a layered communication philosophy. Consequently it characterizes connection-oriented and connection-less transportation qualities. The identified QoS functions are assigned to layered entities. End-to-end control of stream-based distributed objects is devoted to the QoS ODP RM related document under way. The document *Methods and Mechanisms* provides with a precise description of QoS negotiation mechanisms. In all provided documents the verification respectively testing of QoS is not really addressed. Initial work is provided by the ODP RM by which conformance reference points in systems architectures are identifiable.

2.   THE CHALLENGE OF QOS VERIFICATION

The presented work is aiming at the notion *quality of service verification.* By the invention of massive and continuous service application the concepts of verification respectively testing are to be re-considered and readjusted.

The phenomenon of mass service application changes the paradigm of communication from discrete message-based type to continuous stream-based type of interaction, from point-to-point distribution to multi-casting, from mono-media to multi-media. Conformance testing respectively verification techniques by which the quality of applied services shall be checked must take into account the effects of continuous processing of a stream and the mass of units transported by the stream.

By considering the above mentioned interconnection protocols, prior to deployment the protocols are to be *verified* by various conformance testing methodologies. Confidence about the protocol functionality is gained by the set of accepted test runs. Conformance Statements about the mass transfer of transported data units in a multicasting environment however, is not captured by the conformance testing approaches. In a massive and continuously communicating environment, quality constraints will replace discrete event assertions. Additionally, multicasting communication topologies is a second constraint which makes the adaptation of available verification and testing techniques to mass and continuity requirements indispensable.

The term *verification* is rather overloaded by many research activities in the telecommunication domain. Verification is mainly realized by testing, monitoring and system management techniques. Even so, it does not exclude verification techniques provided by formal models. In this work however, we would like to adopt the view on verification of QoS from the concept of permanent observation of systems in operation. In case of deviations from contracted quality during service application, a modification action has to be taken, which either tunes the service to an allowed degradation quality or provides with sufficient resources to maintain a given level of quality.

Consequently, a fundamental step needs to be taken from protocol conformance testing towards QoS verification. The observation and controlling techniques and concepts for individual message-based systems are discussed in comparison with the features of systems and services serving for mass communication. Testing and QoS verification concepts become built-in features of the envisaged QoS-aware systems. The effects of timeliness, which one of the major factor to determine the service quality, is analysed by this work. Finally, a synthesis is offered which integrates some of the identified factors of influence into a unified model of QoS. From the integrated model, language concepts can be derived to describe mass effects, their dynamic qualities and techniques for their permanent observation and control. The integrated model will serve for the analysis and simulation of the end-to-end control mechanism of the service quality with respect to perception, mass distribution, timely delivery, safety and security. It provides means to check the effects of advanced quality control and adaptation mechanisms.

3.   CONFORMANCE CONTROL FUNCTIONS

Since protocol conformance testing is a *prior-to-deployment* verification technique the expected behaviour in terms of events and responses compared to a specification is tested. Systems which relay on stream-based communication between autonomous components are interfered by many factors like the variation of background traffic. Hence, the quality of operation is to be controlled continuously during application of a service. Any observation and controlling activity with respect to QoS conformance must thus be performed on-line.

In figure 1 one can see that conformance testing of event-based systems may involve streams, but there are important difference to continuously operating systems. The streams in event-based systems do not continuously flow, they are bound to a very limited number of allowed units to be transported. The openness of a stream, i.e. the determination of the beginning of the stream and his end is - in event-based systems- restricted to a countable and fixed number of discrete events allowed to happen, respectively transportation units allowed to flow. In the flow control protocol component the units which can flow are determined by the size of a so-called window. The window is opened or closed by acknowledging or unacknowledging messages received. The controlling and decision-taking mechanism is implemented at the protocol server side and consists of a flow control buffer *FCB* and a component which decides on the information gained from the connected protocol client.

For the purpose of conformance testing, in figure 1, the role of the protocol client is played by the conformance tester. Notice, here we refrain from explaining the variations of the conformance testing architecture which provide with lower and upper tester roles, depending on the interface of the implementation observed. Instead, we restrict ourselves to the interactions between the protocol server and client which are observable at the so-called *lower interface*, where protocol operations are made observable. The tester connected to is thus also called *lower tester*. In order to check the flow control mechanism the tester can send and receive streams of a length up to the pre-defined size of the window. To the units transferred a sequence number is added to determine losses,

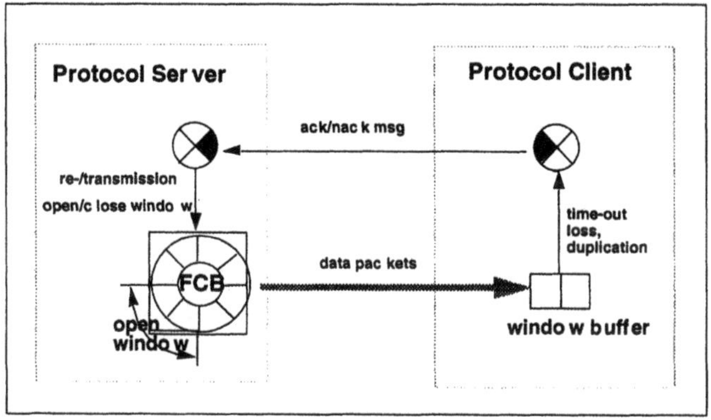

Figure 1   Discrete Event-based Interaction

duplications or misorders. Misorders and duplications can easily be resolved at the receiver's side by applying reordering the sequence or just pruning the duplicated units. Units are lost, if they do not arrive in fixed limit of time, a retransmission must be requested.

In all these cases conformance testing prior to deployment is still possible because the events are limited in time, even the event of a data unit loss virtually occurs by the elapsing of a timer, independent of the message eventually will be received after.

3.1   QOS CONTROL FUNCTIONS

In the ODP reference model it is said that events just happen which means that their beginning and ending are observable. Streams however, cannot be simply cut-off by time-outs or limited to a pre-defined length. The receiver must continuously be able to consume the stream of units over time. Furthermore, system inherent interference quantities will influence the quality of continuously ongoing behaviour. Continuous behaviour is represented by processes behaving over time and thus represented by the symbol $q(t)$. Since streams have also a direction and transport a mass of units, any assessment operation of streams must deal with all three properties, timeliness, direction and mass of transportation.

Streams are dynamic processes which behaviour varies over time and hence cannot be tested as simple events. Instead they must be observed and assessed continuously during operation. Consequently, conformance testing approaches for QoS must be developed towards a process-oriented quality assessment strategy, which means the extension of current conformance testing methodologies from single event observation to continuous process observation.

The operational quality of a system or a system component depends on many factors related to its real-time environment, like the current traffic in the network, or the actual utilization of resources. By verification prior to deployment the correct parameterization of the set of possible interactions and the causality relations between the interactions are tested. Verification after deployment, i.e. during operation, must deal with certain interference quantities like traffic, load conditions, buffer over or underflows etc. One can say pure conformance testing provides with structural and relational inter-operability while the quality of service verification provides with inter-operability under varying embedding conditions.

As it is illustrated in figure 2, the QoS control mechanism, or in other words the protocol of a stream-based system can in case of figure 2 be influenced from a single service client. In other cases this influence will be issued by the load which represents the a group of clients sharing e.g. the stream server. The stream server comprises a multimedia store and may serve the requests of many clients. In order to avoid overloading with requests, the stream server has a built-in admission control mechanism. The admission controller balances the capacity of the server by the number of out-streams requested for several clients. For this purpose the admission controller is able to vary the deadlines of the out-streaming units. In order to decide which deadlines of which stream are to be changed the admission controller needs further information from his clients. For example the urgency of a stream to be served. Thus, the deadlines of less urgent streams can be changed to a lower frequency. Or, clients may have problems in presenting the in-streaming units, thus wishing a degradation of a stream unit rate. This might be signalled to the server's admission controller, which must decide very rapidly in order to avoid losses of stream units.

A feedback link from the client's in-stream buffer to the server's admission controller might thus be realized to signal decision-making information. The continuous nature of stream-based systems induces the cooperation between the clients and their servers.

At each site, at the server's and at the client's site, locally functioning control mechanisms are installed. At the server's site the accuracy of the stream units' deadlines must be under permanent control. The accuracy of the server's deadlines influences the quality of the inter and intra-stream synchronisation at the clients' sites for example. There, the in-streaming units must be buffered prior to presentation to accommodate the differences in speed between the incoming streams and the presentation capabilities of the multimedia devices. Points of observation are installed to observe the possible skipping of stream units or their pausing. These observations are translated into a frame rate changing signal to the server, which must either speed-up or slow-down the individual streams.

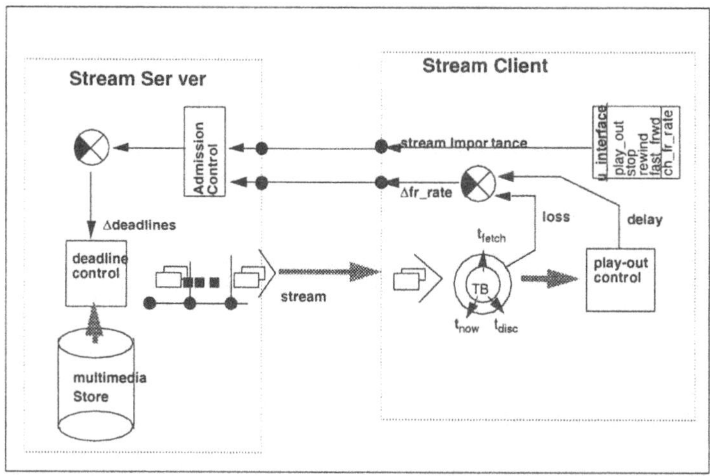

Figure 2   Continuous Stream-based Interaction

In such a self-controlling environment, the QoS specification is provided by a special specification, called the QoS contract. It contains all the values, thresholds, limits and bounds of QoS parameters agreed among a server and of every one of his clients. In case of figure 2 the QoS parameters comprise the deadlines set for the out-stream, the accuracy of the presentation device of each client or the minimum QoS of presentation to be achieved, etc. From a conformance point of view these dynamic processes must be tested. Since, the architecture provides with a closed loop the testing components must be inserted into the loop such that controlling or measurement signals can be generated. This will be the case when replacing the so-called discriminators of figure 2 by testing components which are able to take over the discriminators' roles.

In opposition to the event-based testing methodologies applied to configurations like those of figure 1, where testing components replace peer-entities of a considered communication protocol, in stream-based systems, all the communicating peer-entities will not be replaced. The system under test will not be changed, instead it must provide with interfaces at special points of interest internally to the system under test, e.g. the measurement points at the clients, or the steering and admission control points at the single server. QoS-aware system components are tightly coupled. Thus testing is not done by simply playing the role of peer communicating components but by a general purpose external signal generator, that intrudes signals in order to change global behaviour of communicating components. In practical terms the discriminator must provide with a second interface at which the externally generating testing signals will be accepted.

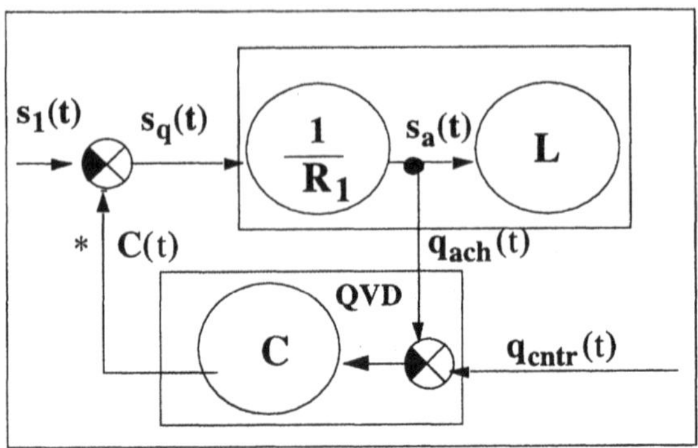

Figure 3   Integrated QoS Model

4.   CONTINUOUS VERIFICATION MODEL

To achieve time-independent quality of presentation and performance of competing processes, it is necessary to scale, to customize and to coordinate the access to and usage of global available resources. Service users and service providers together with the underlaying transportation service, need all the negotiation of their QoS requirements during connection set-up. And once the connection is established a facility for controlling the contracted QoS is required. Therefore conformance verification techniques and an architecture, that supports actively the control of QoS are needed in order to guarantee the negotiated level of quality during service application to all participating players. In the model, the streams flow - so to say - through the resource object R with a time delay but get interfered by factors, which may influence the stream quality. At the client site the achieved quality is compared with the contracted one and the comparison is fed into a controlling object C. This process informs a control mechanism at the server site and advises him which changes are to be taken to maintain the contracted quality. In figure 3 the resultant basic architecture is modelled. The streaming processes are indicated by lower case letters. E.g. s_1 and s_2 stand for two concurrent streams from the server to several clients via some transportation network. q_{ach}, q_{cntr} stand for the achieved and contracted qualities at the client's site. The resource characteristics of the streamed-through objects in the model are indicated by upper case letters. R and L stand for some resource providing objects; C is the characteristics of the controlling object being able to influence in-streams by providing a certain amount, proportional to C, of less or more resources.

Assume, at an observation point at the client's site, the achieved quality of the stream $q_{ach}(t)$ is gained. Then the observed quality must permanently be discriminated with the quality contracted $q_{cntr}(t)$, i.e. $q_{ach}(t) - q_{cntr}(t)$.

Since the consumption of resources will influence the quality of a stream, the discrimination process must act proportionally to the change of discriminated quality, i.e. $C * q(t)$. A stream passing a component, i.e. a network, at least consumes time which is considered to be a resource in terms of the model. Consuming time however, means between source and sink there is a certain amount of stream units in transition. The amount of units in transit can be calculated by measuring (testing?) the difference between the stream $s_q(t)$ fed into the network and the stream $s_a(t)$ received from the network at the client's site at some later time. The network must thus be able to store all the units of a stream during an interval of time e.g. $[0 : T]$. If the total quantity of stream units kept by a capacitive component would be stable, no jitter of transition delay per unit would occur. Since this is unrealistic, the amount of stored units depends from the system's respectively from the component's dynamic characteristic. It is a variable which must thus be kept permanently under control. In order to keep the achievable quality in acceptable limits, it may not be allowable to vary the size of stored units too much. For example, if the stream considered is a video stream to be presented with a certain contracted quality the play-out buffer must not be emptied less than a given lower limit. So possible delays occuring during decoding can be equated. From another point of view, such a system parameter can be said to describe the state of a system respectively of a component. The state is then the difference between the controlled stream $s_q(t) = C_q(t)$ and the achieved stream $s_a(t)$ having passed the system respectively the component considered:

$$\frac{1}{R}\int_0^T (Cq(t) - s_a(t))\, dt$$

In the formula above, R represents the capability of a component to store a certain amount of stream units. The quotient of the storing capability R and the difference of the in- and outstreaming quantities for a period of time T describes the dynamically changing state of the component. The componental state is the result of two dynamic processes, i.e. the instreaming process and the out-streaming one. The quantification and hence the measurability of the state is expressed by the integral.

In a further step the QoS verification model will be improved by another state variable modelling delay and jitter. The quantity of a transmitted stream is effected by a time relocation of T time units which is proportional to the time of transporting a quantity z over a distance L at a speed v. Since, there are the dependable variables L and t, delay is modelled by the partial differential equation:

$$\frac{d}{dt}z(L,t) = v\frac{d}{dL}z(L,t)$$

In the integrated model of figure 3 delay is represented by a component with the characteristic L which represents a distance. The bigger the distance modelled the greater the delay. Notice, a stream z passes unchanged the component that obeys certain delaying characteristics. One can say the in-stream and out-stream processes are tightly coupled by so-to-say a "conveyor" of length L. Contrary, a capacitive component couples in- and out-streams only loosely, i.e. both processes may obey behaviour which is independent from each other.

5. CONCLUSIONS

New approaches of QoS verification and testing, must be applied to the emerging technology of communication including multi media and multi casting. The verification techniques for protocols based on discrete message passing must be adapted to the needs of continuous interaction relaying to streams. The major challenges are the system inherent properties of continuity and quantity. In continuous systems inter-action is not only a question of correct operation but also a question of the quality of operation. That is because quality depends mainly from resource consuming streams passing through components, respectively from loads of components. Hence, verification or testing of inter-operation is twofold. Firstly, inter-operation generally is interfered by the embedding conditions and secondly, it is a real-time matters. Prior-to-deployment checks are restricted to static and functional correctness, for which conformance testing techniques are infact appropriate. Quantitative aspects play just a minor role. Contrarily, in continuous systems, quantitative aspects, i.e. resource consumption cannot any more be neglected.

In this paper we have presented a model that is based on the concept of stream processing. A stream processing object is characterized by the properties of the resources it contains. The consumption or use of these resources is not constant over time. So, certain quality constraints to the system can not be verified without observing the consumption of these resources. Quantitative constraints expressed by limits, thresholds, bounds etc., instead of sequences of events are the basic elements of testing. Thus, testing translates into a permanent process of taking probes and comparing them with limits. In the former sections, the variability of componental states is expressed in terms of integrals and differential quotients. The integral represents the variation in accumulation, i.e. the summation of transported stream units and the quotient represents the variation in transportation, i.e. the delay. Since both characteristics are variable over time they are out of scope of discrete testing procedures and are thus matters of continuously taken measurements and evaluations. To this continuous activities we would like to apply the notion of QoS verification.

In an adequate model for QoS, architectural concepts like feed-back links, suitable points of observations and control and characteristics of resources like delaying or accumulation become new elements of consideration of QoS testing. this paradigm shift from conventional conformance testing towards QoS testing has been motivated by comparing the well-known scenario of a point-to-point message transportation protocols with the advanced scenario of continuously communicating components. Whereas in the massaging scenario flow control can be evaluated by a limited set of tests based on probes and observing reactions, in the continuous communicating scenario flow control can only be evaluated by permanent observation including quantitative measurement methods.

However, a complete QoS model has not yet been developed. We have started with an evaluation of testing methods for message passing systems with the belief that there is an evolution from the current event-based techniques towards the more sophisticated stream-based ones. For the development of new verification methods including QoS verification an extension of the anticipated QoS reference model in standardisation is envisaged. By means of this generic reference model, basically quantification of resources, which is much more than just accumulating or delaying as shown in the paper will be captured. The enhancements will deal with encoding patterns, relations between two or more streams, adaptation strategies on resource usage, negotiations among groups of applicants, etc. This requires sophisticated language and modelling concepts for which we gave a motivation.

While conformance testing usually is considered an off-line activity and which is executed prior to the service or system deployment, most work of quality of service verification must be done during the operation of an application or use of a service. An overall system is decomposed into the composite application and its supporting transportation environment. Both the transportation and service application environment comprise resources, which are under the control of the *Quality Verification Discriminator (QVD)* (see figure 3). This structure has been outlined by our early RACE2 project R2088 TOPIC [R2088]. A comparable approach has also been made available early by [FraHav94] and is known as the *Performability Manager*. The *QVD* component of the proposed model communicates with the observers that measure the resource usage, averages the values over the defined period and raises an alarm signal if the value falls below a specified threshold. In existing networks, whenever packets that flow between some application objects need more resources originally allocated to the transportation, then violate the QoS requirements, and thus will be discarded or must be retransmitted. In continuous systems those failures can be avoided by setting appropriate QoS thresholds and by implementing controlling and adaptation techniques in order to react on violation of QoS border lines.

Work is underway to set-up verification experiments with conferencing, joint editing, news-on-demand services, group communication techniques etc. Qualities will be evaluated on models and checked by measurements at implementations. The information gained will be used to improve the stream process model and to develop more advanced adaptation and decision strategies for quality verification.

References

[Blair95]   Blair, Coulson Lancaster University U.K.; Stefani, Horn, Hazard CNET France: Supporting the real-time requirements of continuous media in open distributed processing. CNIS Issue No. 27 1995.

[CTMF94]   ISO/IEC 9646, 1-9. IT- OSI - Conformance Testing Methodology and Framework 1994.

[FMCT95]   ISO/IEC P1.21.54.1,2 Revised Working Draft - Framework on Formal Methods in Conformance Testing, Guidelines on test generation methods from formal descriptions, 1995.

[FraHav94]   Leonard Franken, Boudewijn Haverkort: The Performability Manager. IEEE Network Reprint Vol.8, No.1, January 1994.

[HdM95]   Herman de Meer, University of Hamburg: Modelling and Management of Responsive Systems. The Quality of Service for multimedia Applications in High-speed Networks as an Example. Stifterverband fuer die Deutsche Wissenschaft.

[ODP95]   ISO/IEC 10918, 1-4: ODP Reference Model Part I - IV, 1995

[Lazar93]   Aurel A. Lazar, Columbia University: Quality of Service Control and Management for Broadband Networks. SICON93 Tutorial, Singapore.

[NakTez94]   Tatsuo Nakajima, Hiroshi Tezuka, Japan Advanced Institute of Science and Technology: A Continuous Media Application supporting Dynamic QoS Control on Real Time Mach, 1994 ACM.

[R2088]   RACE2 R2088 Project Toolset for Protocol and Advanced Service Verification in IBC Environments 1992 - 1994. Project Deliverables ftp://ftp.fokus.gmd.de/pub/race/topic.

[QoS98]   ISO/IEC JTC1/SC7 N1996 CD15936 IT ODP RM - Quality of Service, 1998-10-16.

[QoS95]   ISO/IEC JTC1.21.57.2/.3, January 1995. Quality of Service Basic Framework, 1995. Methods and Mechanisms, 1995.

19

FAST FUNCTIONAL TEST GENERATION USING AN SDL MODEL

Robert L. Probert, Alan W. Williams
School of Information Technology and Engineering
University of Ottawa
Ottawa Ontario K1N 6N5, Canada

{bob,awilliam}@site.uottawa.ca

Abstract   This paper reports the results of a successful study undertaken to determine the suitability of CASE tools and formal methods for systematic, rapid generation of functional test cases. In particular, the study involves the use of Message Sequence Charts (MSCs) [1], Specification and Description Language (SDL) [2] and Tree and Tabular Combined Notation (TTCN) [3] and the use of the Telelogic Tau tool set [4] which supports all three of these languages.

Keywords: functional testing, SDL, TTCN, MSC, formal test design, time to market

1.   INTRODUCTION

The objective of the study was to generate functional tests for a telephone switching system. This was an unusual use of the TTCN test language, which has primarily been used for conformance testing. Recently, ITU has opened several new questions involving extensions to the definition, capabilities, and applicability of TTCN and Concurrent TTCN, including the ability to specify functional testing and performance testing. We see our work as contributing to this work on extending and improving upon the current versions of TTCN.

A secondary, but equally important objective was to demonstrate the ability of our "fast to text/first to test" approach to dramatically compress the test cycle, and so to improve the time-to-market (TTM) of the corresponding product. The advantage of our approach is that it is no longer necessary to convince design engineers of the merits of a formal description technique. Instead, the test designers are able to use it at the beginning of the development process so that designers have high-yield scenarios in MSC form to use to validate their designs, often before any code is generated. This provides a cost-effective infrastructure for scenario-directed design. Others have proclaimed the benefits of scenario-directed design, but we may be the first to suggest a cost-effective approach acceptable to industry (no overhead on designer time). Here, we utilize the scenarios (those most likely to detest the presence of serious design errors or omissions).

A tertiary objective was to demonstrate to designers the value of using an FDT-based representation method supported by an industrial-strength (scalable) CASE tool set. Eventually, if designers see the cost-effectiveness of this more formal approach, they will adopt it. Testers already know its value.

Functional tests for a basic call, plus calls using four call processing features were derived. Two different models of phone sets were used: a basic set, and a set with an alphanumeric display screen and various lamps and feature keys. By employing the CASE tool coverage analysis features, coverage of all reachable transitions was ensured. Overall, we were able to generate high quality functional test cases in a reduced time interval.

As a result of the study, we have determined recommendations for enhancements to the languages and tools. We have also developed and applied some productivity measures that support the cost-effectiveness of this approach.

1.1   The Project in Context

In the industrial environment, it is unlikely that a fully complete and correct set of requirements will be determined before developing software. Usually, there is an initial set of requirements, and some development work is done using those requirements as a starting point. As a result of the first phase of development work, modifications are often made to the requirements. Aside from uncovering errors in the first version of the requirements, the consideration of exceptional cases may lead to new scenarios being developed. An initial prototype may lead to additional or modified requirements.

After a first version of software is released, developers repeat the whole process for the next version of software, which undoubtedly will have additional features requested by customers.   Recommendations for improving the software development process have to take into account that much of industrial software development is inserting new functionality into existing code.

As a result, we wanted replicate this iterative and incremental process in the study. *Figure 1* shows the process that we want to follow. We are taking a use case-based approach, where the use cases are captured as Message Sequence Charts (MSCs). From an initial set of MSCs, we developed an executable SDL model. This allowed us to simulate and validate the model. Simulation allows us to execute the model interactively, and observe that it behaves as expected. Validation allows us to explore the state space of the model to detect properties such as deadlocks, and unspecified message receptions. Building the SDL model also produced new scenarios that were added to our set of MSC use cases. Once we were satisfied with the model, we could proceed to generate abstract TTCN test cases based on the use cases.

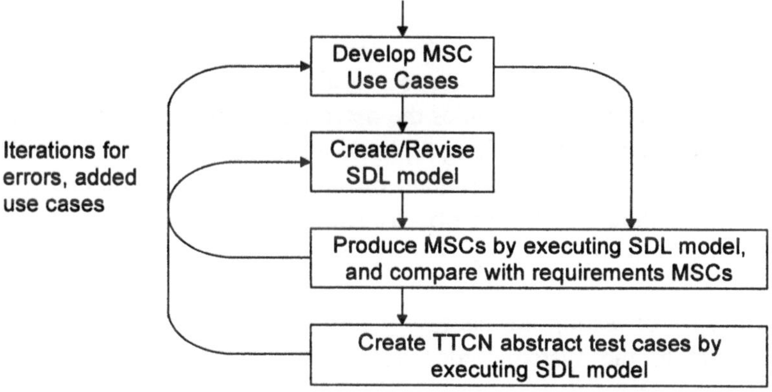

Figure 1. The Iterative, Incremental Development Process

2.   FUNCTIONAL VS. CONFORMANCE TESTING

In this study, our objective was to generate functional tests rather than conformance tests. Conformance testing typically concentrates on the exchange of protocol messages both for their sequence and content. The goal is to verify system interoperability by conformance to a standard.

The goal of functional testing is to take a user's view of the system, and check that customer requirements have been met. In particular, the actions and reactions of the system are considered from the user's viewpoint, to ensure that the system behaves as the user expects.

There is a distinction to be made between taking a "customer" viewpoint and an "end user" viewpoint in functional testing. While the direct customer of a telephone switch manufacturer is a company that offers telecommunications services, the end user is someone who makes phone calls. In our project, we took the "end user" view, and considered only the actions of the switch as could be controlled or observed from a phone. If we were taking a "customer" viewpoint, we also would have modelled and checked the behaviour of the operations and maintenance console for the switch.

These goals are reflected in the specific types of actions taken during testing, and reactions observed from the system under test. A functional test action might be to go offhook on a phone set. The corresponding conformance test action would be to construct and send a protocol message to the telephone switch as a result of going offhook. Similarly, a functional testing observation would be to check that the dial tone is heard, and various lamps and displays on the phone are updated. The conformance testing observation would be to check the exact sequence and data format of messages that are sent to the phone as a result of going offhook.

Functional testing also includes the detection of the presence and absence of persistent events. When taking a phone offhook, the dial tone should not only be started, but it should remain on until the user has dialled the first digit. Then, it should be turned off. We should be able to check that the dial tone is on at any time during the appropriate interval, and that it is off afterward.

Both types of testing are necessary. Functional testing is more indirect with respect to the system under test, in that we are using a phone to construct the protocol messages. With conformance testing, we would construct our own messages (correct, or not), and send them directly to the system under test. Conformance testing is to detect that the exchange of protocol messages is correct and robust. Functional testing is to confirm that the system behaves as according to customer or user expectations.

3. TEST MODEL ARCHITECTURE DEVELOPMENT

In this section, we explain how we decided what to include in, and exclude from, our test model in SDL.

We decided to work backwards from our goal of running TTCN test cases in an automated execution environment, and seeing what constraints we had to satisfy in both the SDL model, and in how to format our MSC use cases that would drive the test execution.

3.1 Mapping the MSC model to the test environment

Because we are taking an end user view, functional testing of a telephone switch is necessarily an indirect process. In our study, we determined that the functionality of interest was that the interaction between a phone set and a switch exhibited the correct behaviour when using a phone. Therefore, our test actions will be to take the phone on or off hook, and press keys on the set. Our test observations will consist of observing the lamps, displays, and detecting the tones produced by the phone speaker in the handset.

Another element of indirection comes from the nature of telephone calls. If we are interested in observing that an incoming call works properly on our phone set of interest, another phone must be used to stimulate the switch to produce the incoming call notification. *Figure 2* illustrates these situations.

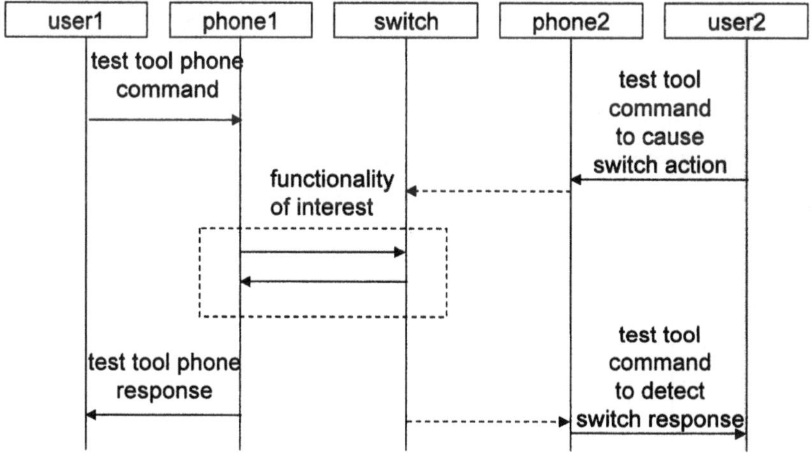

Figure 2. Mapping the MSC model to the test environment

Eventually, what is described in *Figure 2* as "test tool" commands or responses has to become TTCN send and receive commands to execute tests in the automated execution environment.

The particular environment, for which the test cases are targeted, has a TTCN compiler with an "adaption layer" that contains information about the specific protocol implemented in the system. That is, the adaption layer converts TTCN messages to specific commands that will drive the protocol interface to the system under test. *Figure 3* illustrates this architecture.

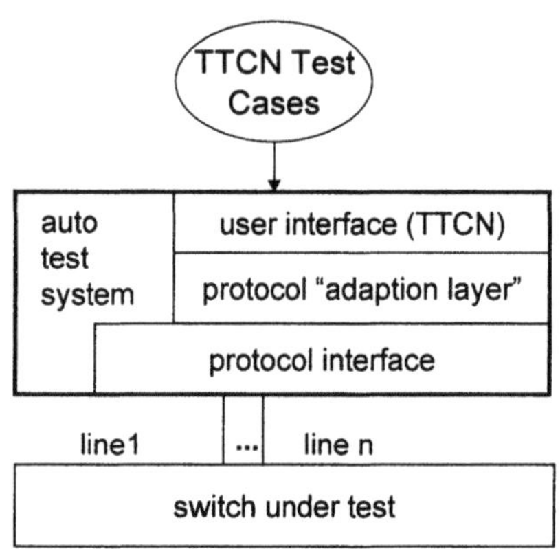

Figure 3. Test Architecture

The result, from the test case perspective, is that the adaption layer defines the legal set of TTCN messages that we are allowed to use. Therefore, our TTCN test cases must use those same message names and formats to work with the automated test execution environment. Since we are generating the TTCN tests automatically with a tool from an SDL model, this in turn means that the SDL model must also use these exact same message names and formats. *Figure 4* shows this correspondence.

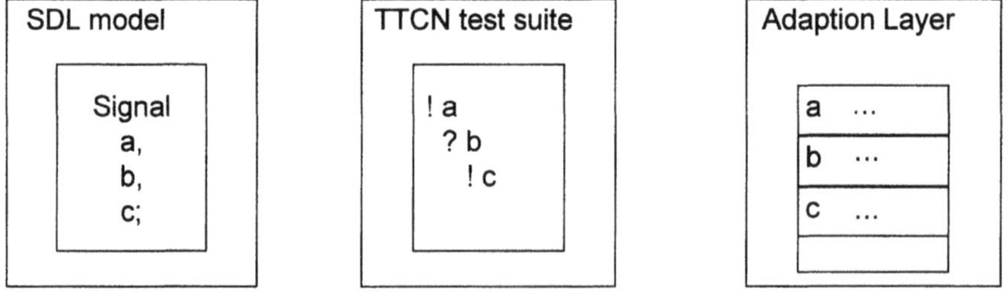

Figure 4. Determining SDL message names based on the Adaption layer

This correspondence goes one step further when MSCs are used to drive the test generation tool to produce TTCN test cases automatically. The names of the messages in the MSCs must also correspond with those in the SDL, at least at the boundary between the environment and the system. We shall discuss more about MSC creation in the next section.

4.   CREATING THE MSCS

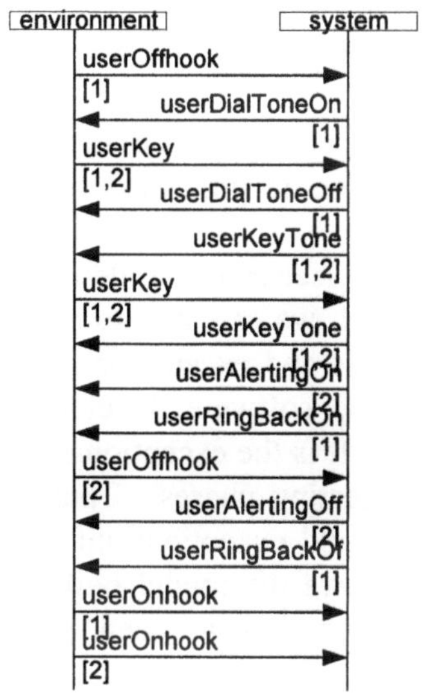

Figure 5. A system level message sequence chart

MSCs were used as our method for capturing use case scenarios for the project. MSCs are excellent for review purposes, as customers easily understand them. We used the MSCs to confirm that we had correctly captured scenarios. The first MSC we produced was for a basic normal call.

In the second stage of the project, we also captured four call processing features, to determine how difficult or easy it is to add functionality incrementally.

At this stage, we are not trying to impose architecture on the system. Therefore, the MSCs that are produced are "system level" MSCs in that they capture only the interactions between the system and the environment. For our functional testing viewpoint, this means capturing the interactions between a user and the various phones used in making phone calls. In particular, the automated execution environment would take on the role of all the phones, and therefore we are capturing the interactions between the test environment interface and the user. *Figure 5* shows the type of MSC produced for a basic normal call.

Because the system level MSC exactly captures the commands that we need to send to the automated test execution environment, it can also be used as a basis to drive the automated execution of TTCN tests. Each message in the MSC in *Figure 5* will be converted into a TTCN send or receive command by the test generation tool.

5.   CREATING THE SDL TEST MODELS

In this study, we developed several iterations and versions of SDL test models. The first version was to produce a test model for telephone calls with no additional features.

While our original MSC scenario only describes a successful call, once this is captured in the model, we can start to ask "what if...?" questions to examine possible alternatives and exceptions. This is extremely important because it is likely that the software development team will implement normal scenarios correctly. It is the exceptional cases where errors are most likely to be found during testing phases. Therefore, to produce effective, high-yield tests, we must build exceptions into our test model. The state-oriented nature of SDL allows the model builder to systematically ensure that, for example, incoming calls have been accounted for in every single state.

Once we had created a model of telephone calls with no extra features, we developed two more versions of SDL test models. One was to determine how difficult or easy it was to add additional call-processing features. In essence, this is adding to the functionality of the telephone switch. Our second additional model was to investigate adding to the functionality of the phone set. That is, we added features such as a display screen, additional function keys, and multiple lamp indicators. This still impacts the functionality of the switch, as the switch has the responsibility of controlling the operation of these additional features on the phone set.

5.1   Scope of the model

For testing purposes, the SDL model is not intended to replicate the entire system behaviour of the switch and phone sets. Instead, it should reflect the goals of requirements capture, and of testing those requirements. It is possible that the models we produced could be used as inputs to the detailed design process for further refinements, and to include behaviour that is internal to the system.

For functional testing, the model should include:

1. the functionality to be tested
2. additional behaviour extending to points of control and observation, for test tool commands
3. enough internal behaviour to execute the model

The model does not need to include internal behaviour of the system that cannot be observed by test equipment, and is not part of the protocol.

5.2   Simulated Execution of the SDL model

By using the SDL tool simulated execution facility, the user can interact with the model, and view execution graphically. The user can record the execution as an MSC, and inspect the resulting MSC for anomalies.

The execution MSC can also be compared against the equivalent requirements MSC to see if it is consistent. This allows the model builder to confirm that the model correctly implements the initial requirements.

The requirements MSCs and the execution MSCs are not identical though. They should be consistent in the exchange of messages between the system and environment. But they can differ because of extra information that results in building the SDL model. The SDL model contains states and these are automatically inserted into the execution MSCs. The SDL model implemented timers, and the setting, expiry, and cancellation of timers also appear in the execution MSCs. Also, we found that what was originally envisaged as two original messages were consolidated into a single message with parameters. Finally, the execution MSCs also contains architectural information inside the system, including internal messages sent between system components. *Figure 6* shows an example of an execution MSC that highlights differences from the original model.

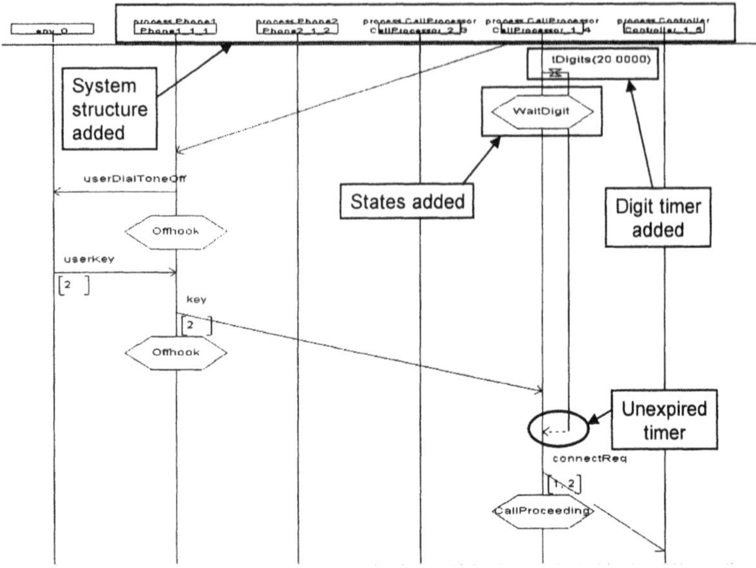

Figure 6. Differences in an execution MSC

6.   FINDING A BUG USING STATE SPACE EXPLORATION

We also checked the SDL model using a state space validation tool. The validator checks for deadlocks, unspecified receptions, and can determine the consistency of an MSC with the SDL model.

The output from the tool for any anomalies consists of a description of the problem found, and an MSC scenario that describes what had happened up to the point when the problem was discovered.

In our study, we ran the validator on our SDL model, and found that an unspecified reception turned out to be a bug in our SDL model of the call back feature (illustrated in *Figure 7*). The feature is activated after a called party did not answer a call. We forgot to have the feature activation terminate the ringing of the called phone. The (still ringing) phone was later answered, but the "answer indication" reception was unspecified at the caller.

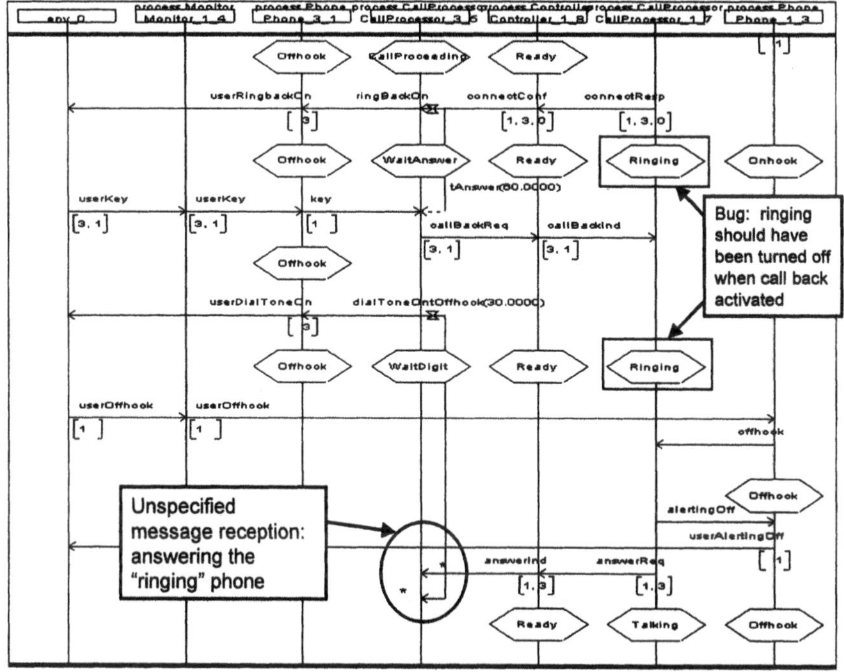

Figure 7. An MSC trace showing an unspecified reception

7.   COVERAGE-BASED TEST GENERATION

The test cases are constructed using a transition coverage based approach, based on having an SDL transition coverage tool that counted the number of times that a transition was executed during simulation.

We started by creating a test case representing a normal call. We used the coverage tool to display the coverage achieved by this scenario. In our model, a normal call achieves about 30% transition coverage.

The strategy was then to choose a "target" state and input combination, and attempt to create a TTCN test case that results in covering this particular transition. *Figure 8* shows an example of a coverage report produced during this process. After the new TTCN test case had been created, its coverage information would be merged with prior results, to get a new view of the current level of coverage. This process was repeated until all reachable transitions were covered. We found that, on average, we could produce one TTCN test case every fifteen minutes using this approach.

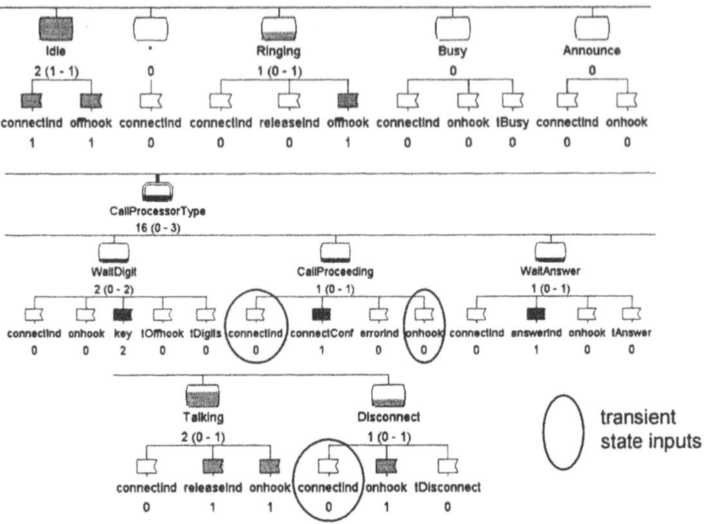

Figure 8. An SDL transition coverage report

Test Case Name	:	NormalCall
Group	:	
Purpose	:	
Configuration	:	
Default	:	OtherwiseFail
Comments	:	
Selection Ref	:	
Description	:	

Nr	Label	Behaviour Description	Constraints Ref	Verdict	Comments
1		phone1 ! userOffhook			
2		phone1 ? userDialToneOn			
3		phone1 ! userKey	userKey_2		
4		phone1 ? userDialToneOff			
5		phone1 ! userKey	userKey_2		
6		phone2 ? userAlertingOn			
7		phone1 ? userRingbackOn			
8		phone2 ! userOffhook			
9		phone1 ? userRingbackOff			
10		phone2 ? userAlertingOff			
11		phone1 ! userOnhook			
12		phone2 ! userOnhook		PASS	
13		phone2 ? userAlertingOff			
14		phone1 ? userRingbackOff			
15		phone1 ! userOnhook			
16		phone2 ! userOnhook		PASS	
17		phone1 ? userRingbackOn			
18		phone2 ? userAlertingOn			
19		phone2 ! userOffhook			
20		phone1 ? userRingbackOff			
21		phone2 ? userAlertingOff			
22		phone1 ! userOnhook			
23		phone2 ! userOnhook		PASS	
24		phone2 ? userAlertingOff			
25		phone1 ? userRingbackOff			
26		phone1 ! userOnhook			
27		phone2 ! userOnhook		PASS	

Detailed Comments	:	

Figure 9. An automatically generated TTCN test case

This approach gave us a set of tests that has the potential for high-yield during execution, because the coverage based approach gave us a series of test cases that concentrates on exceptional cases. Of the eighteen test cases we generated for call processing with no features, there are five cases where timers expire, three cases where the caller aborts before connecting, one call to a non-existent number, and eight cases where incoming calls are rejected with a busy signal.

The coverage-based approach also highlights situations where it may not be possible to create a reliable test case, due to race conditions or transient states. An example of a race condition is when the two phones hang up simultaneously. The disconnect messages may pass in transit, or one or the other might arrive first. An example of a transient state occurs just after when the user hits the final key to dial a number, and before it is determined that the called party exists and is idle. The caller could hang up, or a call could be incoming in this state, even though in real time, the call process is only in this state for a very short length of time.

Creating reliable test cases for an automated execution environment that involves either race conditions or transient states is extremely difficult, if not impossible. It requires extremely fine tolerance for timing and control of the test equipment that is used.

8.   PRODUCTIVITY RESULTS

Before presenting our productivity results, it is necessary to describe how many people were involved and how much product, tool, and methodology knowledge they had. We had one product expert working with two students. The product expert had recently taken courses on the tools and methods, and one of the students learned these while doing the project. The other had prior experience with the use of the tools and methodology, and with telephone switch testing.

During the three months of the project, the students were able to devote most of their time for two months to the project, but during one month, the students were concentrating on other academic work. *Figure 10* shows a project time line containing the time taken for various elements of the project (BC = basic call).

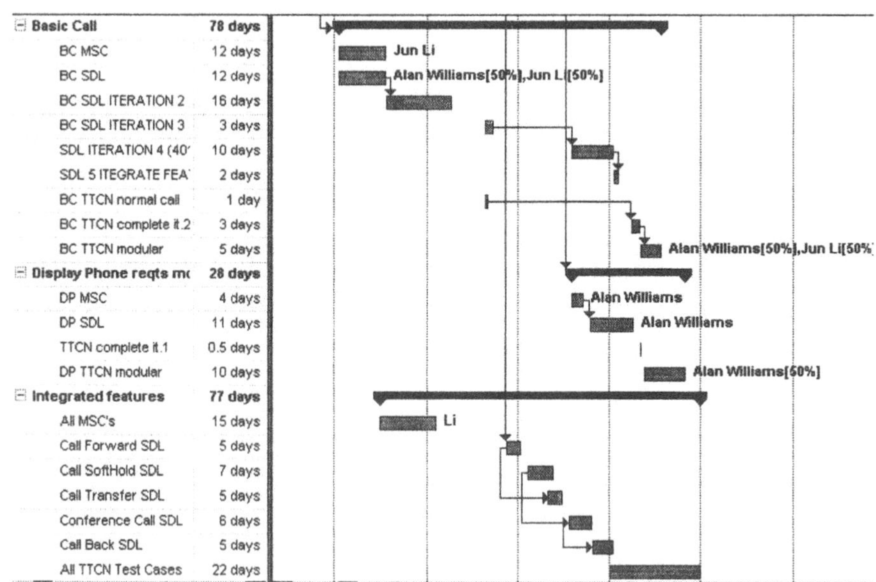

Figure 10. Productivity results

9.   ISSUES FOR FUNCTIONAL TESTING

Overall, we were very satisfied with the quality of the test cases that were produced. However, we found two areas where there are opportunities for improvement:
1. ordering of events
2. generation of valid alternatives.

9.1   Ordering of events

The sequential nature of an SDL transition imposes an ordering on two consecutive outputs within a single transition. For example, suppose that a single transition has two outputs to the **same** phone: one to turn on a lamp, and the other to update a screen. When a TTCN test case is generated automatically from this transition, the result is:

? lamp on

    ? screen update

For user oriented functional testing, the order is irrelevant, because the user is only interested that both events happen within a suitable interval. We do not want to fail the test case if these two events are detected in the opposite order. However, SDL does not provide a mechanism to specify "send two events in no particular order". In fact, forcing the ordering of these two outputs puts unnecessary constraints on the team that implements the specification.

SDL considerations aside, TTCN does allow us to explicitly provide a number of alternative event orderings. But, we found that on the phone set with many display features, there were typically four update messages for each action performed. Because the number of orderings is factorial, there are 24 possible orderings of four events. Specifying each of these every time is not a usable solution.

There needs to be a mechanism in both TTCN and SDL to specify something like the following:

! offhook

    ? screen | main lamp | line lamp | dial tone

where the interpretation is that the events separated by vertical bars could happen in any order, but without intervening messages and before subsequent actions.

9.2   Generation of valid alternatives

This point is related to the last one in that the root cause is that we want to leave some message orderings unspecified because a group of them could happen in any order.   In this case, we are discussing the process of generating tests, and how to account for such unordered events.

We are using MSCs to guide the test generator, but the MSC contains only a single ordering of possible events (although the co-region feature of MSC-96 allows multiple orderings). Using the example from the previous section, if the MSC happened to have a particular ordering of the messages to a phone's display screen, the two lamps, and the tone generator, the test generator needs to recognize that other alternative orderings are possible. The test generator typically adds a TTCN "otherwise" clause that would cause the test case to fail if any messages other than the one desired arrive at any particular time. But, if there are three other messages that could arrive, and do not represent a violation of desired behaviour, then this needs to be considered by the test generator. The TTCN test case that results should ideally add alternatives that will also lead to a pass verdict.

The tool we were using overcame part of this difficulty by using the state space search mechanism for identifying alternatives. However, instead of taking each alternative branch to its conclusion, if an alternative message arrived, the verdict of the test case was immediately declared to be "inconclusive". The result is that we would still have to generate separate test cases for each possible ordering. But, at least with this approach, valid behaviour would not result in a test case assigning a verdict of fail.

10.   CONCLUSIONS

Based on our experience with this project, we have developed a number of recommendations:

1. Our approach provides a means of considerably enhancing the productivity of test design engineers. We were able to generate a correct, value-added, TTCN test case every fifteen minutes, which is considerably faster than with a manual approach.
2. Our approach provides a systematic means of improving the quality of designs. Using the state space exploration found a bug in the functionality in a model that had previously undergone testing. More significantly, this bug could have been found long before detailed design, and coding.

3. The process should be integrated with the design process, so that both designers and testers can take advantage of having an executable model at an early stage.
4. The correct metrics framework for our project only became clear towards the end of the project. We believe that our experience will now allow us to define better metrics that are usable throughout the process.

We would like to further investigate the scalability of our approach, both in terms of the methods, and of the tools used to support them.

11.   ACKNOWLEDGEMENTS

The authors would like to thank Hasan Ural, Gregor v. Bochmann, Richard Plackowski, Jun Li, and Louise Desrochers for their help with this work, and Daisy Fung, Kelvin Steeden, Bob Armstrong, and Peter Perry for their support..

The authors gratefully acknowledge the support of Mitel Corporation, Nortel Networks, the National Sciences and Engineering Research Council (NSERC), Communications and Information Technology Ontario (CITO), and the University of Ottawa for this research.

REFERENCES

[1] International Telecommunications Union (ITU-T), Recommendation Z.120: Message Sequence Charts, revised 1996.

[2] International Telecommunications Union (ITU-T), Recommendation Z.100: Specification and Description Language, revised 1996.

[3] International Standards Organization (ISO), OSI Conformance Testing Methodology and Framework — Part 3: The Tree and Tabular Combined Notation, International Standard 9646-3, 1992.

[4] Telelogic Tau tool set, version 3.3, Telelogic AB, Malmo Sweden.

20

PERFORMANCE TESTING AT EARLY DESIGN PHASES

Péter Csurgay

Department of Telematics,
Norwegian University of Science and Technology
Peter.Csurgay@eth.ericsson.se

Mazen Malek

Conformance Center, Ericsson Ltd.
Laborc u. 1., Budapest, 1037, Hungary
Mazen.Malek@eth.ericsson.se

Abstract   Testing non-functional properties based on functional specifications at early design phases is still a challenge in teleservice and protocol engineering. Modelling and evaluating performance aspects along with the traditional functional verification oriented formal description techniques has to be integrated into one unified design process to avoid extra effort in maintaining consistency between multiple design models. The paper introduces PerfSDL, a semantic extension to SDL for modelling non-functional properties, and a performance evaluation framework based on PerfSDL and simulation. To verify the feasibility of the approach, three design alternatives in a small TCP-like packet transfer protocol above transport services of different quality were tested and evaluated by means of simulation and statistical analysis.

Keywords: Testing, performance evaluation, protocol engineering, formal description techniques, SDL, prototyping

1.   INTRODUCTION

Standardised formal techniques for specifying and functionally verifying telecommunications protocols fail to capture non-functional aspects like performance and Quality of Service (QoS). Thus, protocol engineering is abound to use separate models for the design of functional and non-functional be-

317

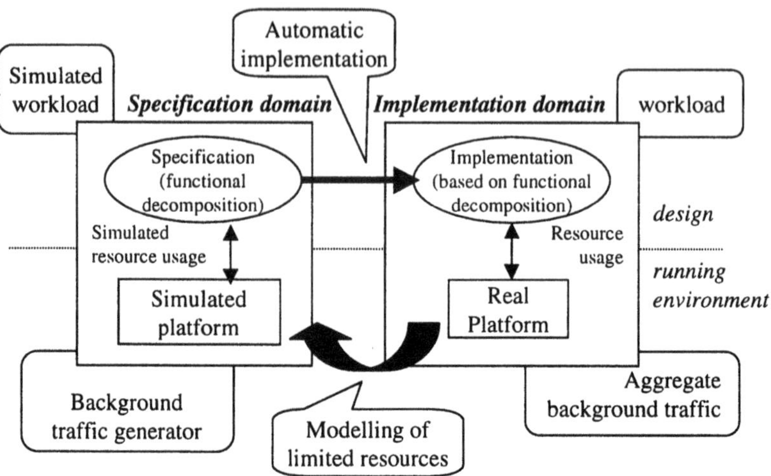

Figure 1   Integration of non-functional aspects into early design

haviour. The extra time and cost of keeping the models consistent can be saved by treating all aspects in one unified model. A feasible approach is to integrate performance issues into a well settled protocol engineering process. The ITU-T standardised formal language SDL has a growing acceptance for functional modelling and verification in practical protocol engineering [4], and ongoing research explores its applicability to performance engineering with extensions to modelling non-functional properties [10]. These efforts are summarised in Section 2.

In the background of this project lies the recognition that while originally the language was constructed for describing required and actual behaviour of reactive systems, its use spans into automated implementation of protocols based on functional specification [1]. Thus, the actual implementation of the protocol will closely reflect the functional decomposition from early design phases written in SDL. In order to be able to ask performance-related questions about the future implementation of a protocol in an earlier design phase, information about that implementation must be provided for the design model. Such information covers the scheduling algorithm of concurrent processes, limited resources of the target environment (like buffer sizes, channel capacities, etc), and resource usage of functional building blocks of the protocol. To gain statistical performance data by means of simulation, stimuli must also be taken into account in the early design models. We model stimuli as an aggregated background load on the transport infrastructure our protocol is acting upon, and the actual workload our protocol has to process. The conceptual model of our performance testing in the early design phases is illustrated in Figure 1.

According to this model, performance testing and evaluation of three design alternatives in a simple TCP-like protocol above channels of different quality were carried out by means of simulation. The new constructs of PerfSDL [9] and their use in modelling non-functional behaviour are introduced in Section 3. The simulation environment and the testing results are described in Section 4 and 5 respectively. The paper concludes and further directions are set in Section 6.

2.   STATE OF THE ART

Recently, there was a large discussion on performance integration within the SDL community. The study that was initiated by the ITU-T SG10 on support of performance engineering activities in the early stages was one of them. Tools are being developed nowadays to handle such activities. SPEET is one of those tools and was introduced in [12]. It uses the C code generated from the compilation of an SDL specification, which might include some probes added by the system developer in order to address interesting performance areas. Another tool is QUEST that was introduced in [3], which defines a new extension to SDL called QSDL. This is mainly achieved through introducing two new constructs in the SDL called load and machine. A similar tool is SPECS [2]. The tool that was introduced in [9] accomplishes the integration of functional and performance aspects by introducing an interface to a simulation environment used extensively for network performance evaluation and analysis. This is a major advantage in terms of less work to be done and in terms of speed capabilities. In the literature there are a number of papers giving surveys on the additional information needed for a performance evaluation of an SDL specification, e.g. [11].

3.   NEW SDL SEMANTICS IN PERFSDL

3.1   SIMULATION REQUIREMENTS

Real-time characteristics have to be added to SDL prior to any performance testing, e.g. the introduction of timed transitions, reinterpretation of concurrency, and finite input queues. In simulation-based approaches to performance testing this issue is of substantial importance. Moreover, aspects like scheduling and simulation time are closely related to those characteristics. Below we list those areas that need special treatment when simulating PerfSDL systems with performance aspects in mind.

Modeling Processing Time. Currently, in SDL, time advances when all queues of the system are empty [4]. This is generally interpreted this way

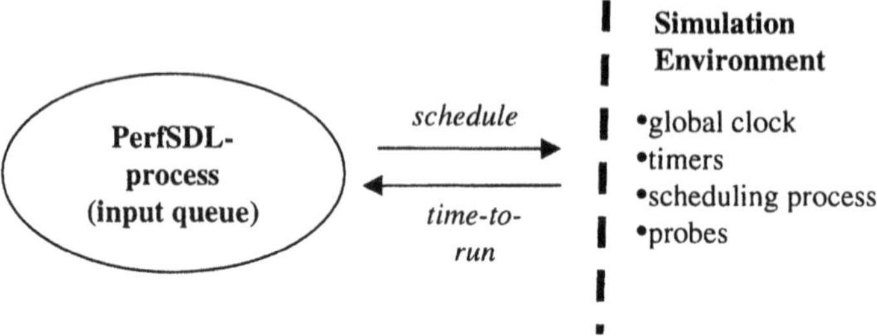

Figure 2   PerfSDL process interactions with the environment

to avoid a dramatic increase in the number of possible traces. In simulation, however, it is not adequate to advance time only when there is nothing to do, but also when there is any action to be taken.

Addition of Performance Sensors.   In order to gain quantitative measures on a system's performance, different counting events of time and recurrence of execution of transitions have to be applied. Time is crucial for measuring throughput, while recurrence of transitions reflects system bottleneck.

The Concept of Scheduling.   This concept is the most essential one during simulation. It specifies the time when the simulator is asked to act on certain change in the state of the model. This is not necessarily performed at the time when transitions start. When a process consumes an input signal, it should indicate to the simulator its reaction and when to handle it. Also the simulator has to consider the possible interactions that may happen after or during that reaction. In this context we consider the scheduling to be events that should occur after each progress in time, namely when consuming input or performing a transition.

Degree of parallelism.   Processes that run in a completely interleaving and independent manner can schedule their actions independently on the simulation time axis. Nevertheless, when processes share the same processor, as usually, a certain priority and scheduling strategy should be worked out between them. State transitions may consist of a number of actions, like decision and assignment, and therefore can not be handled as an atomic unit of scheduling. Instead of that, actions should be that unit.

3.2   NEW EXTENSIONS TO SDL

The formal interpretation of PerfSDL, as presented in [9], is based on PerfSDL-process, which can be characterised by the following:

- set of states, e.g. s_1, s_2, s_3;

- set of internal variables;

- set of initial states, e.g. s_0;

- transition function, e.g. tr_{12}.

This is called the underlying state machine, which requires the availability of a global clock in the system. The interaction with the environment, as in Figure 2, takes place by two types of controlling events:

- **schedule** and;

- **time-to-run**.

The environment acts on process scheduling requests, through **schedule** events, by assigning a time event for performing the task requested. Additionally, it activates relevant processes when there is some task to be handled by them through **time-to-run** events. This synchronisation is performed between each running process and the environment. PerfSDL processes can communicate with each other at block level. All levels higher than leaf block level are described in the simulation environment.

"Probes" in PerfSDL.   These have the ability to monitor, count and analyse the applied system. They are new elements that can be included during the design process but may be defined during testing. There are two kinds of probes - with or without time stamps. A counter counts the number of times a certain transition is visited, while stamps are used to record the time of specific events. The current status of some internal variables may be included in these stamps. In the following we will refer to probe events having time stamp requests by **stamp**(tr), where tr is the transition where we want to apply our analysis, e.g.: **stamp**(tr): **current state is** q_i **and input is** i_i

"Timed transitions" in PerfSDL.   This means taking time elapsing in account in each event that uses a processor's time. A trace, in PerfSDL, is a sequence of states and timed transitions:

$$trace : \theta = s_0(t_0)s1(t_1)s_2(t_2)s_3(t_3)....$$

$$transitions : tr_{01}(t_0)tr_{12}(t_1)tr_{23}(t_2).....$$

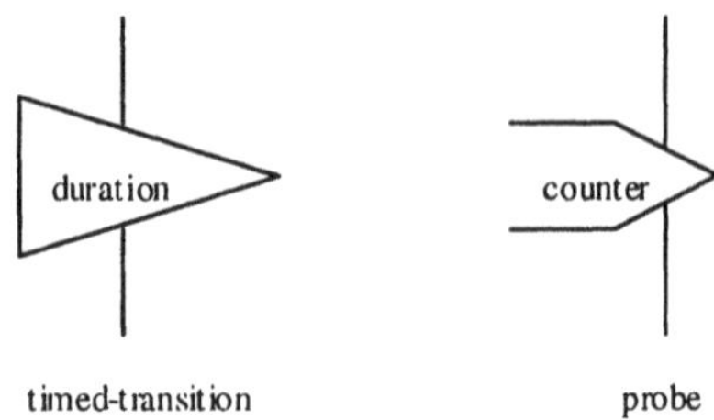

Figure 3   Introduction of new constructs for performance testing in PerfSDL

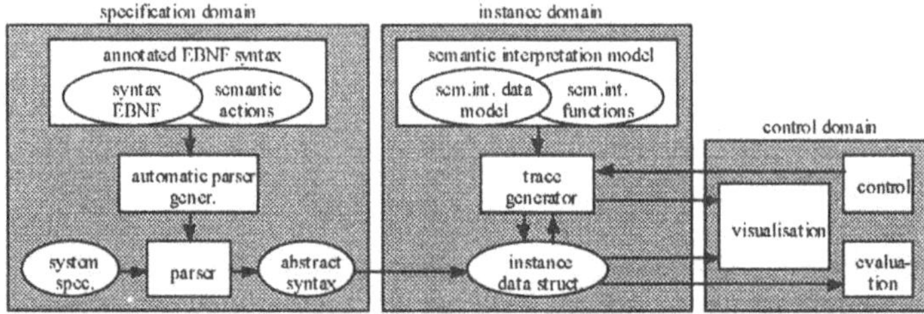

Figure 4   Behaviour interpretation of extended semantics in PerfSDL specifications

These times, e.g. t_1, include the consumption of inputs and execution of transitions. Figure 3 introduces the graphical representation of probes and timed transitions.

4.   SIMULATION ENVIRONMENT

For behaviour interpretation of system specifications using new constructs and new semantics of PerfSDL in our simulations, we built a prototyping framework. The main functions and data structures of the framework are illustrated on Figure 4. Because developing PerfSDL is also a subject of our research, our framework has to support prototyping PerfSDL specifications. Meanwhile its syntax and semantic interpretation is constantly evolving.

The framework includes automatic parser generation and abstract grammar definition from extended Backus-Naur Form (EBNF) descriptions, and an easily extensible object-oriented formal definition of the semantic interpretation model. Java has been used to implement an open framework of distributed visualisation, control and evaluation components.

The ITU-T Recommendation Z.100 defines the Specification and Description Language (SDL) with (1) its concrete grammars SDL/PR and SDL/GR,

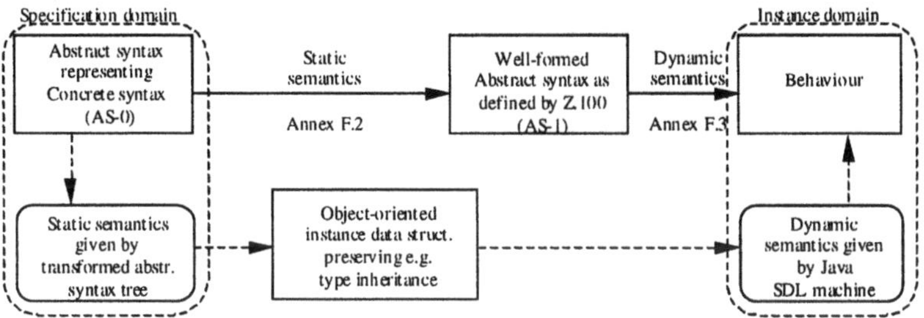

Figure 5   Alternative to Annex F derivation of PerfSDL behaviour

(2) an abstract syntax AS_1 equivalent to the concrete grammars, (3) rules to transform a system specification into abstract syntax, and (4) the SDL-machine semantics to interpret a specification given in terms of the abstract syntax [5]. Z.100 gives formal definition of SDL in Annex F2 and F3 describing the static and dynamic semantics respectively [6]. Their relation is illustrated by the top segment of Figure 5 along with the AS_0 and AS_1 syntaxes introduced by Z.100. AS_0 is not formally defined in the recommendation, but can be derived from the BNF syntax. AS_1 is given in Z.100 as the Abstract Grammar. However, with the F.2 transformation from the abstract syntax (representing concrete textual syntax) to AS_1 (input to the dynamic behaviour definition of SDL), specification domain information is lost. Such information would regard type inheritance hierarchies or virtuality. The bottom segment of Figure 5 illustrates our alternative way to derive dynamic behaviour of a PerfSDL specification with the representation forming the basis of behaviour interpretation that preserves such information, thus being closer to the specification domain.

Time and ordering of internal events play a key role in SDL's non-determinism. Due to the vague time semantics of SDL (time may or may not be advanced by signal transactions and transitions), tool builders and users of SDL choose legal, but different, interpretations of time semantics when simulating or interpreting SDL systems. A legal interpretation for example (most probably inspired by the way some leading tools handle advancing of time) is that time passes only if there are no enabled transitions (no valid input signals in queues) of processes. This interpretation is legal, reducing the set of possible traces and simplifying validation for example, but not appropriate for performance modelling [11]. A prototyping framework should support generation of possible traces of system behaviour based on different interpretation models. This is achieved by separation of different interpretation functions in a modular object-oriented model of formal semantics.

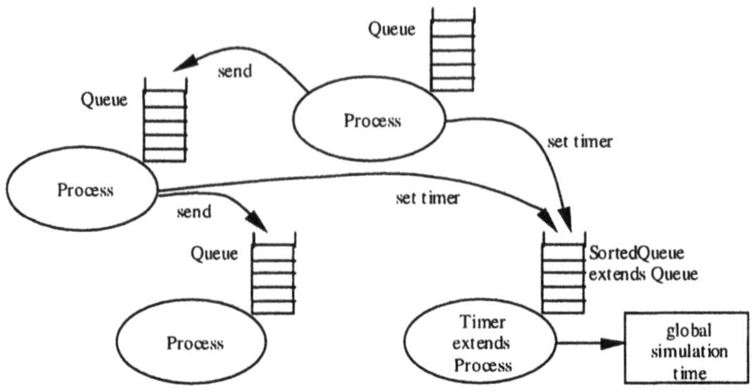

Figure 6   Time semantics interpreted with global timer process approach

Figure 6 illustrates an approach where the chosen interpretation of time semantics is orthogonal to the scheduler algorithm mentioned among the necessary implementation related aspects in Section 1. This solution is compatible with different interpretations of time semantics with respect to the generated set of possible traces. Whenever a process instance encounters setting of a timer, it is interpreted as sending a signal to a global timer process with parameters of signal name, sender id and expiration time. The timer process's queue is sorted by these expiration time values, and is handled by the scheduler the same way other processes are handled. Upon interpreting the reception of a signal by the timer process, it sets the global simulation time to the specified expiration time parameter, and sends a signal with the specified signal name parameter to the specified sender parameter. If the timer process is always scheduled last among all processes, the interpretation will generate a trace-set equivalent to the "time may not be advanced..." interpretation. If the scheduler randomly selects armed processes including this timer process, the resulting set of possible traces will be equivalent to the "may be advanced..." interpretation. Note that with this separation and orthogonal modelling of semantic interpretation functions, different concepts can be experimented with by prototyping in the modular object-oriented framework of semantic interpretation models.

5.   SIMULATION RESULTS

This section illustrates the applicability of the method introduced in this paper by inspecting some design aspects of the TCP transport protocol of the Internet. We limit our investigation to the achievement of high utilisation of network resources. We attempt to address the influence of design alternative choices on the overall performance of an implementation. Figure 7 shows

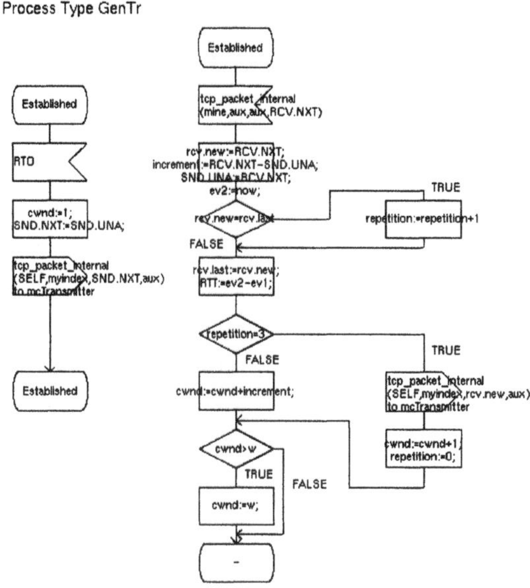

Figure 7   SDL specification with repetition set to 3 (third design alternative)

part of the SDL specification of a TCP transmitter entity. It contains the basic sliding-window aspects contained in TCP, with a lot of hidden details. Definitely, these details may have a substantial impact on performance issues in general, but they have no impact on the aspects we deal with. The model, containing transmitter, channel and receiver, gives a brief demonstration of the data transfer part of TCP with enough descriptiveness and soundness. In the diagram of Figure 7, the value of internal variable repetition is crucial, since it states when to re-inject some already sent packets. This specific part of the protocol shows how the overhead traffic will look when applied in a certain environment. There is an infinite number of design alternatives according to this aspect. They are design alternatives because they influence other parameters in the model, namely the decision of when to consider the occurance of a *Timeout* event. There are also some timers that depend on these settings. We choose three of them, in particular when repetition is equal to 1, 2 and 3. Our analysis of the performance, as mentioned earlier, will concentrate on the overhead traffic that is the major player in channel utilisation. By adding time stamps into the base specification we can obtain measurements of the repeatedly sent packets, and accordingly can compare design alternatives.

Figure 8 shows the log of four simulation measurements applied on the design alternative having repetition set to 3. These measurements are sketched to show how the choice of stamp events would help in obtaining different

Measurement1 Stamp (time,seq,ack)	Measurement2 Stamp (time,seq,0)	Measurement3 Stamp (time,0,ack)	Measurement4 Stamp (time,seq)
(0.0050,1,0)	(0.0050,1,0)	(0.0105,0,2)	(0.0155,2)
(0.0105,0,2)	(0.0155,2,0)	(0.0260,0,4)	(0.0205,3)
(0.0155,2,0)	(0.0205,3,0)	(0.0425,0,6)	(0.0310,4)
(0.0205,3,0)	(0.0310,4,0)	(0.0525,0,8)	(0.0360,5)
(0.0260,0,4)	(0.0360,5,0)	(0.0675,0,9)	(0.0410,6)
(0.0310,4,0)	(0.0410,6,0)	(0.0775,0,9)	(0.0460,7)
(0.0360,5,0)	(0.0460,7,0)	(0.0875,0,9)	(0.0510,8)
(0.0410,6,0)	(0.0510,8,0)	(0.6015,0,9)	(0.0560,9)
(0.0425,0,6)	(0.0560,9,0)	(0.6120,0,19)	(0.0610,10)
(0.0460,7,0)	(0.0610,10,0)	(0.6285,0,20)	(0.0660,11)
(0.0510,8,0)	(0.0660,11,0)	(0.6385,0,20)	(0.0710,12)
(0.0525,0,8)	(0.0710,12,0)	(0.6485,0,20)	(0.0760,13)
(0.0560,9,0)	(0.0760,13,0)	(0.6635,0,20)	(0.0810,14)
(0.0610,10,0)	(0.0810,14,0)	(0.6735,0,20)	(0.0860,15)
(0.0660,11,0)	(0.0860,15,0)	(0.6835,0,21)	(0.0910,16)
(0.0675,0,9)	(0.0910,16,0)	(0.6985,0,21)	(0.0960,17)
(0.0710,12,0)	(0.0960,17,0)	(0.7075,0,21)	(0.6065,18)
(0.0760,13,0)	(0.6015,9,0)	(0.7140,0,22)	(0.6170,19)

Figure 8   Measurements

statistics about system runs. They have been obtained by a simulation run of the PerfSDL specification of the TCP model. This model contains some timed-transitions and stamps applied in different places. Stamps are associated with certain transitions where inputs or outputs are in question. This way, recurrence of packets is counted and measurements related to throughput can be easily performed.

Measurement 1 for example uses two stamps for sending and receiving packets. It registers all sent packets even if they are repeated, like packet 9 at time 0.0675. Measurement 2 filters out the received packets or acknowledgement packets. That is why the ack parameter is replaced by 0. Meanwhile Measurement 3 does the opposite by showing the received packets only. Measurement 4 takes a further step and measures the packets sent with no repetition. These measurements were performed for each design alternative separately. The comparison of them is sketched in Figure 9. The comparison is done according to the amount of overhead traffic due to each design alternative. The diagrams describe the change in amount of traffic corresponding to repeated packets. On the horizontal axis there is the sequence number of packets, while on the vertical axis there is the number of repeated packets. The measurements were carried out using different models of the communication channel, and the design alternatives have different characteristics in these environments. In these comparisons we refer to the design alternatives having repetition values of 3, 2 and 1 as the 1st, 2nd and 3rd alternative respectively. The best choice is clearly the one with least overhead traffic. It is the 2nd, 3rd and 1st in the (a), (b) and (c) environments respectively.

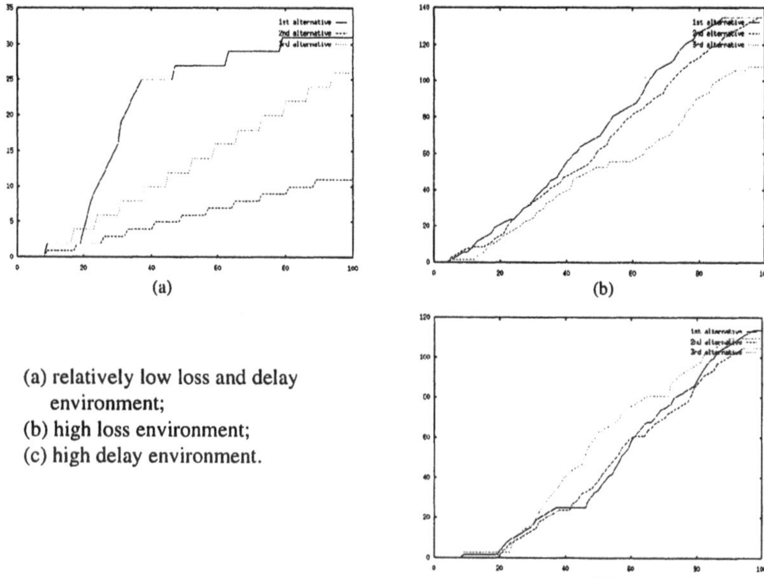

(a) relatively low loss and delay
   environment;
(b) high loss environment;
(c) high delay environment.

Figure 9   Amount of overhead traffic of design alternatives in different environments

6.   CONCLUSIONS

A model to integrate performance engineering into the early design phases of protocol engineering was presented. New constructs in PerfSDL and their use for expressing non-functional behaviour were introduced. Performance testing and evaluation of packet transport protocol alternatives were carried out by means of statistical data gathering in a proprietary simulation environment. Simulation results show the feasibility and value of the approach in improving the integration of non-functional design into the protocol engineering process. Further efforts are being put into the specification of a PerfSDL based performance evaluation and protocol design tool environment.

References

[1] Rolv Brak, Oystein Haugen, *Engineering Real Time Systems*, Prentice Hall International, 1993

[2] M. Butow, M. Mestern, C. Schapiro, P. Kritzinger. *Performance Modelling with the Formal Specification Language SDL.* FORTE/PSTV'96. Chapman & Hall 1996.

[3] M. Diefenbruch, E. Heck, Jorg Hintelmenn, B. Clostermann. *Performance*

Evaluation of SDL-Systems Adjunct by Queuing Models. Proc. 7th SDL forum. Elsevier 1995.

[4] Jan Ellsberger, Dieter Hogrefe, Amardeo Sarma, *SDL - Formal Object-oriented Language for Communicating Systems*, Prentice Hall Europe, 1997

[5] ITU-T Recommendation Z.100, CCITT *Specification and Description Language* (SDL) 03/1993

[6] ITU-T Recommendation Z.100, *Annex F. Static and Dynamic Semantics of SDL*, 03/1993

[7] ITU-T Recommendation Z.100, Appendices I and II: SDL Methodology Guidelines, 03/1993.

[8] ITU-T Recommendation Z.100, *Supplement 1: SDL+ methodology: Use of MSC and SDL (with ASN.1)*, 05/1997.

[9] Mazen Malek, *PerfSDL: an Interface to Protocol Performance Analysis by means of Simulation*, Accepted paper, 9th SDL Forum, Montreal, June 1999

[10] Andreas Mitschele-Thiel, *Results of the discussion*, Workshop on Performance and Time in SDL and MSC, Erlangen, February 1998

[11] Andreas Mitschele-Thiel, *Performance Evaluation of SDL Systems, SAM98*, 1st SDL and MSC workshop of the SDL Forum Society, Berlin, July 1998

[12] Martin Steppler and Matthias Lott. *SPEET - SDL Performance Evaluation Tool*. SDL'97: TIME FOR TESTING- SDL, MSC and Trends. Elsevier Science 1997.

VIII
TEST PRACTICE

21

DEVELOPMENT AND APPLICATION OF ATM PROTOCOL CONFORMANCE TEST SYSTEM

Sungwon Kang, Youngsoo Seo, Deukyoon Kang, Mijeong Hong,
Junhwan Yang, Ilkook Koh, Jaehwi Shin, *Sangjo Yoo, **Myungchul Kim
Korea Telecom Telecommunications Network Laboratory
Junmin-dong 463-1, Yusung-gu, Taejon, 305-390, Korea
* *Korea Advanced Institute of Science and Technology*
Gusung-dong 373-1, Yusung-gu, Taejon, 305-701, Korea
** *Information and Communication University*
Hwaam-dong 58-4, Yusung-gu, Taejon, 305-348, Korea
Email: { kangsw, sys01, dykang, lhdo, jhy, kok, jhshin } @sava.kotel.co.kr
*sangjo@sdivision.kaist.ac.kr
**mckim@icu.ac.kr

Abstract   This paper presents development and application of ATM Conformance Test System (ACTS), an automated test system for ATM protocols. ACTS is a test system that checks conformance of ATM terminal and network equipment implementing either ITU-T or ATM Forum user-network interface. This paper, after presenting the methodology and process used for developing ACTS, conducts case studies of its applications to real ATM equipment. By applying ACTS, we were able to detect numerous problems in protocol implementations of ATM equipment and analyze causes of the problems, thereby demonstrating the efficacy of ACTS as an efficient automated testing tool. Furthermore, by predicting the potential effects of the problems on interoperability, we show how ACTS can be used as a useful tool for ensuring interoperable ATM equipment.

Keywords: ATM, Communication Protocol, Conformance Testing

1.   INTRODUCTION

As the users demand faster and higher-volume multimedia telecommunication services, Asynchronous Transfer Mode (ATM) has been widely adopted as the base technology for Broadband-Integrated Services Digital Network (B-ISDN) services in many countries. Korea is not an exception and has started as a national project building a nation wide high-speed network based on ATM in 1993. A variety of equipment made by different vendors is expected to be deployed to construct the ATM network.

In order for communication equipment to interoperate, they should follow a common set of protocols. For this, ITU-T and ATM-Forum standardized ATM protocols. Therefore not only should ATM products follow the protocol standards but also it is essential to have efficient means to determine whether the protocol implementations indeed follow the standards correctly. The necessity of conformance test for ATM protocols as such means cannot be overemphasized because, in contrast with the case of PSTN, ATM networks rely on complicated protocols.

Most of the papers published so far about ATM protocols conformance test have focused on test generation methodologies [1, 2, 3]. On the other hand, only a few papers have been published on comprehensive ATM conformance test systems capable of testing for the complicated ATM protocol requirements of ITU-T and ATM Forum such as the ATM signaling protocol; e.g., [4] is about the design of an ATM conformance test system and [5] explains a commercial ATM test product. However, those systems generally could not keep up with the progress of standardization or meet the testing needs in a timely manner. Also conformance test system for the user-side ATM Forum UNI3.1 signaling protocol was not completely developed in those systems. Moreover, the published works on them were insufficient in showing the efficacy of their test systems by applying the systems and analyzing the application results for feedback.

In order to provide high-quality ATM protocol conformance tests without the limitations mentioned earlier, we started to develop ATM Protocol Conformance Test System (ACTS) in 1996. ACTS is a test system that determines conformance of ATM equipment against ATM UNI standards of either ITU-T or ATM Forum. For two years since then, ACTS has grown up so as to cover most of the major ATM UNI protocols including the user-side and network-side signaling protocol. In this paper, we discuss the development of ACTS and its application.

The rest of the paper is organized as follows; Section 2 surveys Signaling layer, SAAL layer, and ATM layer of ATM UNI protocol; test architectures for the layers are described in Section 3; in Section 4, we introduce the procedures and methods for developing Abstract Test Suite (ATS) and Executable Test Suite (ETS) and explain ACTS in detail; Section 5 conducts case studies of ACTS application and discusses expected interoperability problems based on the application results; finally, Section 6 concludes by explaining the future plan for ACTS development and further applications.

2.   ATM PROTOCOL STACK

ATM protocols are standardized both in ITU-T [6, 7, 8] and ATM Forum [9]. The ATM protocol stack can be functionally divided into the user plane, the control plane, and the management plane. ACTS focuses on the control plane shown as the shaded part of Figure 1.

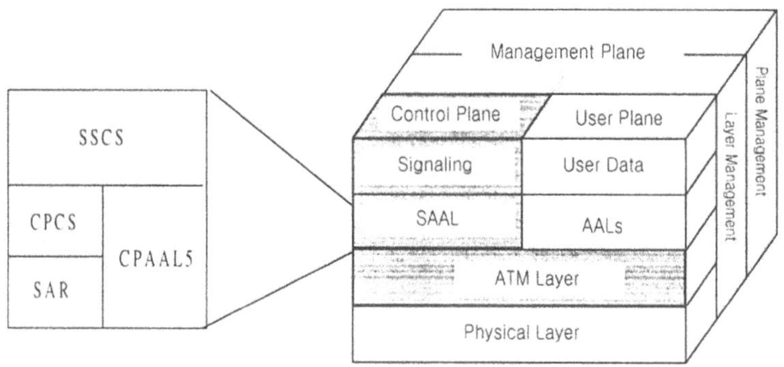

AAL: ATM Adaptation Layer

SSCS: Service Specific Convergence Sublayer

SAR: Segmentation And Reassembly

SAAL: Signaling AAL

CPCS: Common Part Convergence Sublayer

CPAAL5: Common Part AAL5

Figure 1   The structure of the ATM protocol stack

The signaling protocol specifies the formats of the signaling messages and the procedures to dynamically establish, maintain and release connections at UNI and to handle error cases.

The payload of ATM layer has a fixed size of 48 octets so that most of the higher layer services need to adapt variable length packets to the payload. This adaptation is provided by AAL. There are four AAL classes depending on applications: AAL1, AAL2, AAL3/4, and AAL5. The signaling protocol uses AAL5.

The ATM layer provides functions such as cell switching and cell multiplexing/demultiplexing between the AAL and the physical layer.

3.   DECIDING ON TEST ARCHITECTURES

ATS is a prerequisite of conformance test system development. In order to develop ATS we first have to decide on the test coverage, the test method and the test architecture. ISO9646 proposes the conformance test methods: the Local test method, the Distributed test method, the Coordinated test method, and the Remote test method [10]. The three test methods except the Remote

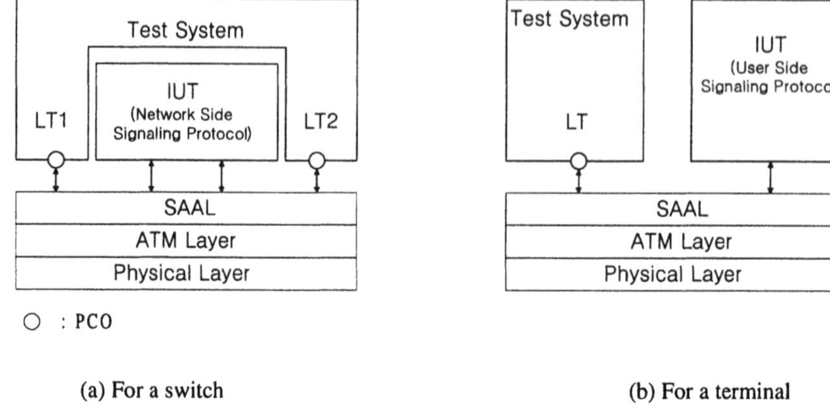

(a) For a switch   (b) For a terminal

Figure 2   Test architectures for the signaling protocol

test method need an Upper Tester (UT) and a Lower Tester (LT). In the Local test method, the test system and an Implementation Under Test (IUT) reside in the same system and, in the Distributed test method, the test coordination procedure is required between UT and LT. The coordinated test method uses Test Management Protocol (TMP) for test coordination. These three methods pose difficulties in *testing by the third party* because they need to access both the upper and the lower interfaces.

On the other hand, the Remote test method is suitable for the third party testing because it does not require UTs, although it tends to give lower test coverage than the others owing to absence of UTs. For these reasons, we

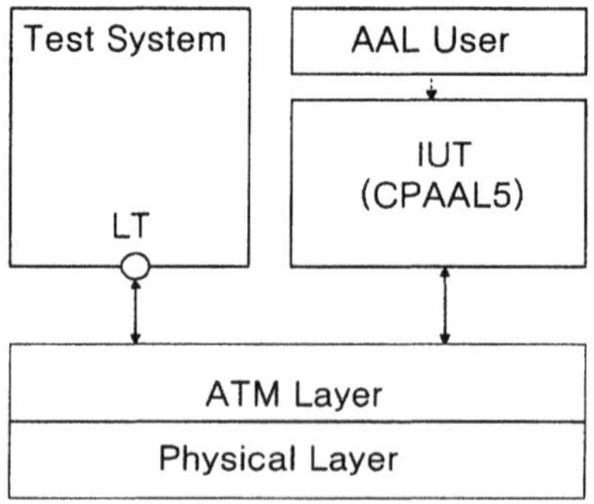

Figure 3   Test architecture for CPAAL5 protocol

adopted the Remote test method as the test method for each layer of the ATM protocols.

In order to test a switch or a terminal implementing the UNI signaling protocol, we selected the test architectures in Figure 2(a) and in Figure 2(b), respectively. In Figure 2(a), incoming and outgoing calls can be tested using two Points of Control and Observation (PCOs), LT1 and LT2.

SAAL is divided into two sublayers CPAAL5 and SSCS as shown in Figure 1. ACTS aims to test only CPAAL5, which is independent of the upper layer, and for that we choose the test architecture in Figure 3.

ATM layer implementations can be either in an Intermediate System (IS) or in an End System (ES). Figure 4(a) shows the test architecture for an IS and Figure 4(b) for an ES. To test for an IS, it takes two PCOs because it relays cells between two interfaces.

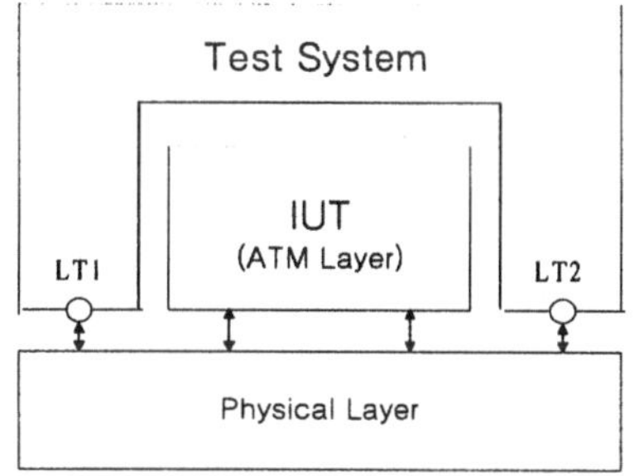

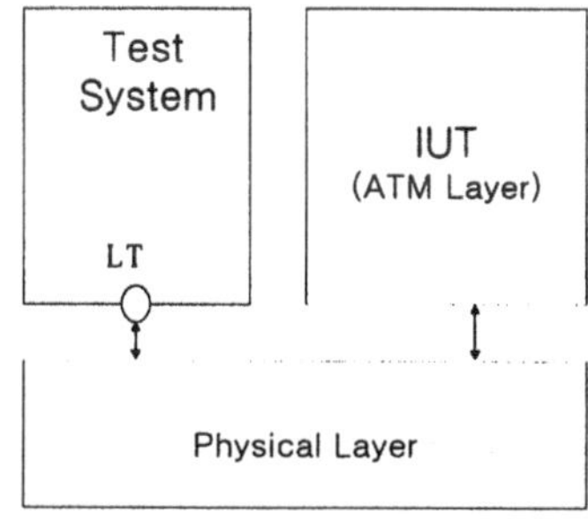

(a) For an intermediate system   (b) For an end system

Figure 4   Test architectures for the ATM layer protocol

4.   DEVELOPMENT OF THE TEST SYSTEM

In this section, we describe the test system development process carried out after the test architecture selection of Section 3 that reflects the characteristics of each layer. Section 4.1 explains the process of system development. Sections 4.2 and 4.3 describe respectively development of ATS and development of executable test software based on it.

4.1   METHOD AND PROCEDURE OF TEST SYSTEM DEVELOPMENT

In order to derive ATS from a protocol specification, we used *test matrix*, which is basically a tabular representation of a protocol Finite State Machine(FSM), as the intermediate step for deriving test cases. Test matrix utilizes both formal and informal description for each state of what message is produced and what action is taken by the protocol entity, from which all the test purposes are derived, which in turn are the bases for deriving individual test scenarios. Following such steps test system development proceeded as shown in Figure 5; 1) write ATS using a TTCN editor from test matrix and convert it to Machine Processible (MP) file, 2) generate executable test software in C program using a TTCN-to-C translator, 3) develop C program support code, 4) generate the platform-dependent executable test software by compiling and linking the results of steps 2) and 3).

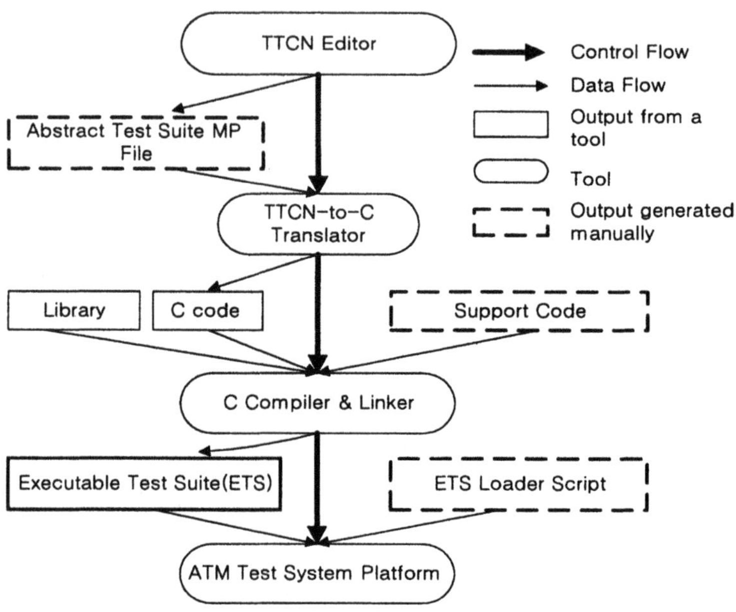

Figure 5   Flowchart for test system development procedure

4.2   DEVELOPMENT OF ATS

Protocol specification usually includes options and parameters in addition to mandatory requirements. Implementers may choose to implement or not to implement options and need to select values for parameters for the particular implementation. Questionnaire to identify options and parameters is called Protocol Implementation Conformance Statement (PICS). Questionnaire for the items not included in PICS but necessary for testing is called Protocol Implementation eXtra Information for Testing (PIXIT).

In this study, ATS was developed so that test cases can be automatically selected and parameter values can be conveniently substituted depending on the PICS and PIXIT contents for the particular implementations. Our ATSs can be easily converted by a TTCN-to-C translator to executable test software that can be used on any test platform when combined with suitable platform dependent programs.

Once test purposes and test methods are determined, the next step is to develop ATS, which describes a test scenario for each test purpose. One may consider a natural language for such description but we chose Tree and Tabular Combined Notation (TTCN) [10], a formal test description language standardized as ISO9646. The reasons are; 1) TTCN is an international standard language for describing communication protocol test scenarios, 2) By using a formal language, we can take advantage of automatic syntax checking using the TTCN tools and avoid possible vagueness and ambiguities in test scenario description, 3) Using automatic or semi- automatic program generation tools, efficient development of executable test systems is possible.

ATS for the UNI3.1 user-side signaling protocol was developed from scratch referring to ATM Forum ATS [11]. ATS for the network-side signaling protocol was developed, based on UNI 3.1 network-side signaling ATS [12], which is still under development in ATM Forum, by analyzing in detail the test purposes, the declarations, the constraints and the dynamic behaviors and correcting problems due to misinterpretation of the specification or errors in the test scenario descriptions. There was no ITU-T Q.2931 ATS known to us and it was developed by first comparing ITU-T Q.2931 and ATM Forum UNI 3.1 and then identifying differences between them.

ATS for CPAAL5 was developed in the situation that a similar or related ATS did not exist. CPCS was the main target in CPAAL5 testing and in order to access Cell Loss Priority and ATM User-to-User indication in ATM Cell Header Payload Type we directly looked into ATM cells instead of using the ATM primitives.[1]

For ATM layer ATS we used ATM Forum ATM Layer conformance test suite [14, 15] and since there is no ITU-T I.361 ATS we reused most of the test cases of ATM Forum ATSs and added test cases for Cell Rate Decoupling, Generic Flow Control, Operation and Maintenance and deleted or modified other test cases.

4.3   DEVELOPMENT OF EXECUTABLE TEST SOFTWARE

In order to derive ETS from ATS, in the case of the signaling protocol and the ATM layer we applied a TTCN-to-C translator to MP form ATS generated

---

[1] The ATS developed in this study was adopted by ATM Forum as the ATM Forum CPAAL5 Abstract Conformance Test Suite [13].

from a TTCN editor to obtain C program [5]. Sending and receiving PDUs depends on the lower layer and therefore cannot be automatically generated by a TTCN-to-C translator. The part not automatically generated should be manually programmed and we call such programs *support codes.*

4.3.1   Development of Executable Test Software for the Signaling Protocol. Sending and receiving PDUs of the signaling protocol was implemented using SAAL API supplied by ATM test platform [5]. The support code for PDU sending and receiving that we developed has the structure shown in Figure 6.

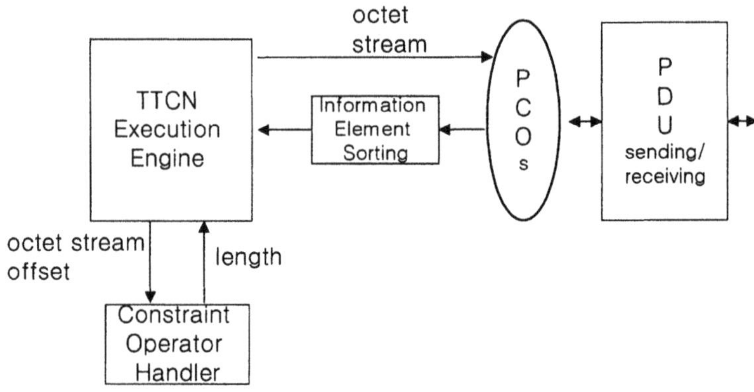

Figure 6   Structure of executable test software for the ATM signaling protocol

In Figure 6, the TTCN execution engine generated by the translator is essentially in charge of executing TTCN dynamic description. The TTCN execution engine sends and receives PDUs through PCOs, compares and calculates PDU constraints and received PDUs. When encountering constraint operators such as ANY and ANYOROMIT, it calls constraint operator handling routines with offset values indicating the location of the operators in the PDUs. Constraint operator handling routines analyze existence and the lengths of Information Elements (IEs) based on the octet string of PDU and offset values and return the results to the TTCN execution engine. Program codes for the operations ANY, ANYOROMIT and IF_PRESENT were realized so that they can be generally applied to IEs as units. When certain fields of IEs have such operations, special program codes were prepared for them.

PDUs of the signaling protocol are specified such that optional IEs can appear in any order. In order to make executable test software handle this, we included in support code a sorting routine that rearrange IEs in a predetermined order.

Most of the developed support code was commonly used both for the user-side and the network-side signaling test system and when discrepancy exists slight modifications sufficed.

4.3.2   Development of CPAAL5 Executable Test Software.   In the case of CPAAL5, we developed the whole test system by manual programming. CPAAL5 executable test software consists of 1) executable programs for test management module, 2) test library that supports easy programming of test cases described in TTCN and 3) executable test cases. Configuration of CPAAL5 test system is depicted in Figure 7.

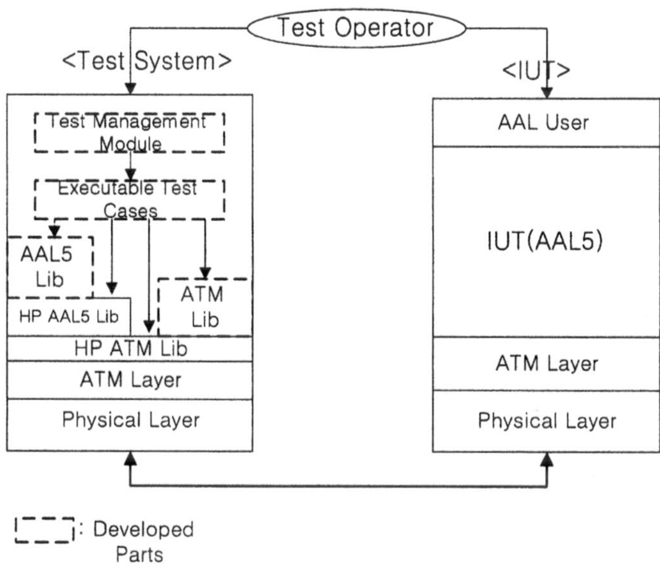

Figure 7   Configuration of CPAAL5 test system

In order to avoid dependency on test system platform, we developed test libraries for the functions : 1) initializing test modules, 2) setting of filter that only captures ATM cells with desired VPI/VCI, 3) to store ATM cells in capture buffer, 4) making up AAL5 PDUs by reassembling cells in the capture buffer, 5) reproducing the initial cell in the capture buffer and 6) reproducing cells sequentially. Test cases generally have the following functions: 1) setting of the capture filter, 2) setting of constraints, 3) receiving cells in a buffer, 4) reassembling cells into AAL5 PDUs, 5) comparing PDUs against predetermined constraints and 6) giving verdicts.

4.3.3   Development of Executable Test Software for the ATM Layer.
In order to load executable test software for the ATM layer on test system platform, there are certain prerequisites. For example, we must handle lengths and values of test suite parameters, information about the physical module and information about options. In general, the following cannot be generated by the translator and must be manually programmed: tester initialization, library for lower layer protocols and interface, initialization after execution of the individual test cases or the test suite and informally described test suite operations. We programmed support codes for them to get the final executable test software. The test software developed by the methods of Section 3 and 4 has been loaded on the ATM test system platform as shown in Table 1 and currently is in operation for conformance testing of ATM equipment.

Table 1   Specification of the ATM test system and system platform

	Protocol		Number of Test Cases	
			ITU-T	ATM Forum
Test System	Signaling Protocol	User-side	517	581
		Network-side	583	661
	CPAAL5		17	17
	ATM Layer	Intermediate System	49	47
		End System	40	37
System Platform	OS		HP UX 9.02	
	Line Interface		155Mb/s Optical (SONET/SDH) 2 Module	
	ATM Cell Processor		ATM Cell and AAL/UNI Signaling Protocol Processor 2 Module	

5.   APPLICATIONS OF ACTS

Generally, we follow the procedure described in Figure 8 in order to perform protocol conformance test. First, we decide if the IUT conforms to its specification by examining the IUT's PICS. This step is called *static conformance test*. If the IUT passes the test, a set of executable test cases is selected and their parameters are set to proper values according to the IUT's PICS and PIXIT. Then with the selected instantiated executable test cases, we test the IUT. This step is called *dynamic conformance test.*

In this section, we present case studies of dynamic conformance test by applying ACTS to ATM equipment, focusing mainly on the signaling protocol.

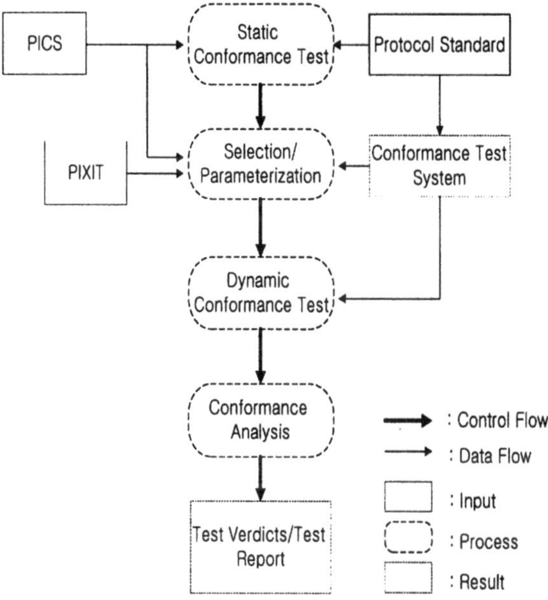

Figure 8   Procedure for protocol conformance testing

5.1   CASE STUDY I: CONFORMANCE TESTING OF ATM TERMINALS

In this conformance testing of two commercial ATM terminals *Ht* and *Ft* from vendors H and F, respectively, test cases for the user-side signaling protocol were selected in accordance with the individual features of the equipment. In both tests, there were fail and inconclusive verdicts.

The two terminals received fail verdicts for all the test cases of the incoming call test group. According to the specification [9], Broadband Low Layer Information (BLLI) and AAL Parameter (AALP) IEs are optional. Thus, the test cases of the incoming call test group were designed not to include those IEs in SETUP messages. However, the IUTs rejected SETUP by RELEASE COMPLETE. It was found out that *Ht* requires BLLI as mandatory and *Ft* requires both BLLI and AALP. Consequently, test cases of the call release test group using an incoming call received either fail or inconclusive verdicts.

5.2   CASE STUDY II: CONFORMANCE TESTING OF AN ATM SWITCH

In this conformance testing of the network-side signaling protocol, test cases were selected and applied to an ATM switch *Fs* from a vendor F. There were fail verdicts in the call release test group because of *state error, message sequence error, and IE content error.*

The *state error* occurs when the state of an IUT, i.e. the state of the signaling entity of a switch, is different from the expected state. According to [9], after sending RELEASE the state of the signaling entity should be *N12* but it became *N0.*

The *message sequence error* occurs when an IUT in a specific state sends a message different from what is expected. According to [9], after an IUT sends CALL PROCEEDING, if it cannot process a call for some reason, it should send RELEASE but *Fs* sent RELEASE COMPLETE.

The *IE content error* occurs when an IUT sends a message with incorrect IE based on the IEs of the received message. *Fs* relayed RELEASE with incorrect CAUSE IE based on the RELEASE message received from a calling party.

5.3   CASE STUDY III: CONFORMANCE TESTING OF AN ATM NT

In this conformance testing of an ATM Network Termination (NT) *En* from a vendor E, test cases of the network-side signaling protocol were selected and applied in accordance with its features. There were fail verdicts and their major causes are as follows; 1) *En* processed a call without inspecting the mandatory IEs of a received message; 2) *En* did not implement the restart procedure and its relevant messages; 3) *En* did not perform a proper VPI/VCI negotiation procedure and 4) there existed an error in the call release function upon receiving an incorrect signaling message.

5.4   EXPECTED INTEROPERABILITY PROBLEMS AMONG ATM EQUIPMENT

In this section, we discuss possible interoperability problems between a terminal and a switch and between two terminals based on conformance test results.

In this paper, interoperability means the capability of two or more systems that when they are interconnected they together perform given functions as desired. Note that even when each system conforms to a relevant specification individually, they might not be interoperable or vice versa because of incompleteness of specifications and/or conformance test. However, interoperability can be significantly improved by detecting possible interoperability problems and by fixing them in advance based on the results of conformance testing. Let us illustrate it with the following three cases:

1) Expected interoperability problems between *Ht* and *Fs*

Through conformance testing of *Fs*, we concluded that it had no conformance problem except some minor ones in the call release function. In other words, it successfully relayed BLLI and AALP IEs from the calling party to the called party as well as other IEs in the SETUP message. Thus, the incoming call may be established without problems as well as the outgoing call although there might exist minor turbulence in the call release.

2) Expected interoperability problems between *Ht* and *Ft*

As explained in Section 5.1, both *Ht* and *Ft* failed some test cases because they did not follow the specification in their selection of mandatory or optional information. For instance, they required BLLI as mandatory in an incoming SETUP message even though they are optional; this violation may result into severe interoperability problems with other ATM equipment.

Even though *Ht* and *Ft* required BLLI as mandatory, *Ht* additionally required AALP IE in an incoming SETUP message as mandatory. Thus calls originated from *Ft* could be rejected by *Ht*.

3) Expected interoperability problems between terminals and *En*

En has a problem with VPI/VCI negotiation; there is the possibility that two different VPI/VCIs are set up for a single call so that they might remain dangling even after the call release. Also *Ht* and *Ft* can request *En* to restart the indicated virtual channels or all virtual channels controlled by *En*. But *En* lacked this feature. Hence, a SETUP message from *En* included incorrect contents. Thus some application services on the terminal using signaling information may not work.

The above analyses are summarized in Table 2. In the table *A-B* means that *A* and *B* are directly connected with each other.

Table 2   Expected degrees of interoperability between IUTs

Configuration	Expected Interoperability	Reason
1. *Ht - Ft*	Low	A call initiated by *Ht* will be rejected by *Ft* if *Ht* does not send SETUP with AALP
2. *Ht - Fs*	High	No significant conformance problems discovered
3. *Ht - En*	Medium	Established calls as well as establishing calls may be interrupted or disconnected due to possible incorrect VPI/VCI negotiation of *En*
4. *Ft - En*	Medium	Same as the configuration 3. *Ht - En*

6.   CONCLUSION

In this paper, we have described the motivation behind developing ACTS, the development process and its applications. ACTS was developed so that ATM conformance tests are timely available as the testing needs arise and can keep up with change and progress of standardization. We applied ACTS to testing of various ATM equipment and analyzed the test results and discussed possible interoperability problems based on the results of conformance testing.

Through applications of ACTS, we observed that the ATM protocol implementations, especially those for the signaling protocol, had many problems. If we consider that these equipment probably have gone through many stages of testing during the development process before submitted for testing with ACTS, we may conclude that ACTS has a fairly good test coverage.

ACTS is used in Korea Telecom, the major carrier of Korea, for decision making related to the procurement of ATM equipment and can be used during the development stage as a test tool or during the operation stage as an analysis tool to track down problems. It can also be used to grant certification to vendors for standards conformance.

Currently, ACTS can perform conformance test for ATM layer, CPAAL5 layer, UNI 3.1 signaling protocol, and Q.2931. Now its scope is being extended to include signaling protocols of Network Node Interface (NNI). Since the ultimate goal of protocols is in achieving interoperation of communication services, the ability to verify protocol interoperability in a direct manner is an extremely important sort of testing. Recognizing such importance, we are developing an interoperability test system for Q.2931 and UNI 3.1 signaling protocols. Further augmented with a performance test system, ACTS is evolving into an integrated ATM test system which can perform conformance, interoperability, and performance test on a common system platform.

References

[1] Bechtold, R., Gattung, G., Henniger, O., Paule, C.,"Test case generation for ATM protocols using high-level Petri net models",IFIP TC6/WG6.1 The 9th International Workshop on Testing of Communication Systems (IWTCS'96), Darmstadt, Germany, 1996.

[2] Vandevenne,L., Sutter, D., Goossens, E., "Testing the broadband network: a quickly shifting demand for test equipment solutions", IFIP TC6/WG6.1 The 9th International Workshop on Testing of Communication Systems (IWTCS'96), Darmstadt, Germany, 1996.

[3] Witteman, M. F., Wuijtswinkel, R. C., "ATM braodband network testing using the ferry principle", IFIP TC6/WG6.1 The 9th International Workshop on Protocol Test Systems (IWPTS'93), Pau, France, 1993.

[4] Kim, K., Kim W., Hong, B., "Experiences with the design of B-ISDN Integrated System", IFIP TC6/WG6.1 The 8th International Workshop on Protocol Test Systems (IWPTS'95), Evry, France, September 1995.

[5] Hewlett Packard BSTS Manual, 1995-1996.

[6] ITU-T Recommendation I.361, "B-ISDN ATM Layer Specification", 1995.

[7] ITU-T Recommendation I.363, "B-ISDN ATM Adaptation Layer Specification", December 1993.

[8] ITU-T, "ITU-T draft Recommendation Q.2931: B-ISDN User-Network Interface Layer 3 Specification for Basic Call/Bearer Control", March 1994.

[9] ATM Forum, "ATM User-Network Interface Specification", Version 3.1, 1994.

[10] ISO/IEC, "ISO/IEC IS 9646 Part 1-7, Information Technology - Open Systems Interconnection - Conformance Testing Methodology and Framework", 1995.

[11] Collica, L., Yoo, S. J., Kang, S.,"96-0979: ATS for the UNI 3.1 Signaling -User-side", BTD- TEST-ATS-sig31user-01.00, 1996.7.

[12] ATM Forum, " Conformance Abstract Test Suite for Signalling (UNI 3.1) for the Network Side", af-test-0090, 1997.9.

[13] ATM Forum, "Conformance Abstract Test Suite for the ATM Adaptation Layer (AAL) Type 5 Common Part", af-test-0052, 1996.3.

[14] ATM-Forum, "ATM Layer Conformance Abstract Test Suite for Intermediate Systems", af- test-0030, 1995.4.

[15] ATM-Forum, "ATM Layer Conformance Abstract Test Suite for End Systems", af-test-0041, 1995.7.

[16] Burmeister, J., Rennoch, A., "Application of a LOTOS based Test Environment on AAL5", IFIP TC6/WG6.1 The 8th International Workshop on Protocol Test Systems (IWPTS'95), Evry, France, September 1995.

[17] Grabowski, J. Scheurer, R. Dai, Z. R., Hogrefe, D., "Applying SaMsTaG to the B-ISDN protocol SSCOP", IFIP TC6/WG6.1 The 10th International Workshop on Testing of Communication Systems (IWTCS'97), Cheju Island, Korea, September 1997.

Sungwon Kang received a B.A. from Seoul National University in Korea in 1982 and a M.S. and a Ph.D. in computer science from the University of Iowa in U.S.A in 1989 and 1992. In 1993, he joined Korea Telecom R&D Group and has been Head of Network Testing Department since 1997. From 1995 to 1996, he was a guest researcher at National Institute of Standards and Technology in U.S.A. In 1997, he served as the co-chair of the 10th International Workshop on Testing of Communicating Systems. His areas of interest include communication networks and communication protocol testing.

Youngsoo Seo received a B.S. and a M.S. in industrial engineering from Hanyang University in 1991 and 1993, respectively. Currently he is with Korea Telecom R&D Group as a member of technical staff. His research interests include protocol engineering and network engineering.

Deukyoon Kang has been working as a member of technical staff in Korea Telecom R&D Group since 1995. He has participated in projects such as ATM Signaling Protocol Validation and ATM Conformance Test S/W development. Currently, he is participating in the design of MPLS Test Architecture.

Mijeong Hong received a B.S. in electronic engineering from Kyungbuk University in 1990. Currently she is with Korea Telecom R&D Group as a member of technical staff.

Junhwan Yang received the B.S. and M.S. degrees in electronic engineering from Kon-kuk University in 1989 and 1991,respectively. Currently he is with the Korea Telecom R&D Group as a member of technical staff. His research interests include telecomunication/data network protocols, testing of telecommination/data network systems and developing test systems.

Ilkook Koh received B.S. in computer science from Chosun University in 1995. currentl he is with the Korea Telecom R&D Group as a member of technical staff. His research interests include protocol conformance testing on ATM/B-ISDN.

Jaehwi Shin received a B.S. and a M.S. in electronic engineering from Hanyang University and Seoul National University in 1991 and 1993, respectively. Currently he is with Korea Telecom R&D Group as a member of technical staff. His research interests include intelligent networks, communication protocol testing, digital signal processing and circuit theory.

Sangjo Yoo received a B.A. in electric communication engineering from Hanyang University in 1988 and M.S. in electrical engineering from the Korea Advanced Institute of Science and Technology in 1990. Currently he is with the Korea Telecom R&D Group as a member of technical staff.

Myungchul Kim received B.A. in Electronics Engineering from Ajou University in 1982, M.S. in Computer Science from the Korea Advanced Institute of Science and Technology in 1984, and Ph. D in Computer Science from the University of British Columbia, Vancouver, Canada, in 1993. Currently he is with the faculty of the Information and Communications University, Taejon, Korea. Before joining the university, he was a managing director in Korea Telecom R&D Group for 1984 - 1997 where he was in charge of research and development of protocol and QoS testing on ATM/B-ISDN, IN, PCS and Internet. He has also served as a member of Program Committees for IFIP International Workshop on Testing of Communicating Systems, IEEE International Conference on Distributed Computing Systems, and IFIP International Conference on Formal Description Technique / Protocol Specification, Testing and Verification, the chairman of Profile Test Specifications - Special Interest Group of Asia-Oceania Workshop (for 1994 - 1997), and co-chair of the 10th IWTCS'97. His research interests include Internet, protocol engineering, telecommunications, and mobile computing.

22

AUTOMATIC TEST CASE GENERATION FROM THE INDUSTRIAL POINT OF VIEW: CONFORMANCE TESTING IN ISKRATEL

Marjeta Frey-Pučko

IskraTEL, Telecommunications Systems, Ltd.

Ljubljanska cesta 24a, SI-4000 Kranj, Slovenia

pucko@iskratel.si

Monika Kapus-Kolar

Jožef Stefan Institute

Jamova 39, P.O.B. 3000, SI-1001 Ljubljana, Slovenia

monika.kapus-kolar@ijs.si

Roman Novak

Jožef Stefan Institute

Jamova 39, P.O.B. 3000, SI-1001 Ljubljana, Slovenia

roman.novak@ijs.si

Abstract   In this paper we present the tool iATS (integrated Automatic Test Sequence generator) for generation of conformance tests in development of digital switching systems SI2000. The tool generates test cases in TTCN form from an SDL specification of the service or protocol under test. Test cases are derived by implemented FSM-based methods. We also describe practical experience of using the tool, illustrated by some quantitative results.

Keywords: Practice of testing, test tools, conformance testing, automatic test case generation.

1.   INTRODUCTION

IskraTEL is a company specialised for production of telecommunications systems. Among other products, the company produces digital switching systems SI2000. One of the key issues for the success of the product is conformance of the implemented protocols and services with standards and requirements of specific market needs. Since the time consumption in manual test case generation and provision of adequate test coverage proved to be the most critical factors in conformance testing of the system SI2000, a need was detected for the automation of conformance test generation procedures.

In order to improve the software development process of SI2000 systems, a general decision to use formal languages was made already at the beginning of eighties. The first one actually used in the company was an SDL-like language called Specification Language One (SL1) [5]. We migrated to the SDL language after it reached its mature stage and its use was supported by commercially available tools [16]. Now SDL is successfully used in different phases and activities of the software development process. As a result of positive experience collected during the long-term use of formal languages in the company, the idea appeared in the middle of nineties to automate the generation of test cases for conformance testing by means of formal languages and methods. The goal was a tool for automatic generation of test cases in the standard TTCN (Tree and Tabular Combined Notation) form [9] from an SDL specification of the service or protocol to be tested.

Since no appropriate commercial tools were available on the market, the decision was made to develop our own tool adapted to the specific needs of conformance test generation for the system SI2000. The tool iATS (integrated Automatic Test Sequence generator) presented in this paper was then developed in the scope of the research and development project "Automation of test scenario generation for the system SI2000". The partners of the collaborative project were the Jožef Stefan Institute (academic partner) and IskraTEL (industrial partner).

The paper is organised as follows: section 2. presents the general testing framework in IskraTEL and exposes the main objectives in the development of the tool. Section 3. gives a detailed description of the tool. Section 4. contains basic guidelines for writing SDL specifications used as input of the tool. Section 5. summarises the results of practical use of the tool in test case generation for ISDN services and signalling protocols. Finally, section 6. discusses the advantages and shortcomings of the tool, and analyses perspectives for its improvement in the future.

2.   FRAMEWORK AND OBJECTIVES

Conformance testing is an important activity in the verification and validation phase in the development process of SI2000 systems. The aim of conformance testing is to prove the conformity of implemented protocols and services with standards or specific customer requirements. The testing architecture used corresponds to the ITU-T X.290 recommendation where the basic test configuration is represented by the upper and lower tester interacting with the IUT (Implementation Under Test). The interaction is performed by means of the execution of test suites and cases respectively. We use the terms "test case" and "test suite" as defined by ISO OSI terminology [9]. The definition of a similar term used in the following – test sequences – is adopted from [15]: test suites selected by test case generation methods are a (test) sequence of input-event and expected output-event pairs.

Test cases for conformance testing are coming from several sources. There is a set of conformance test cases collected in the past decade, which have been generated manually. Other test cases are either adopted from ETSI (after some minor corrections) or generated automatically by iATS. While the manually generated test suites are described in their abstract form (Abstract Test Suite – ATS [9]) mainly in MSC (Message Sequence Charts), the ETSI and the automatically generated test suites are described in TTCN. For actual execution on the system SI2000 they have to be translated into an executable form (Executable Test Suite – ETS). Afterwards they are executed on different testing platforms using the testing equipment of several providers, not allowing the execution of the same ETS on all platforms.

As the main objective of iATS development, the automatically generated ATSs should have been described in a unique form for all testing platforms possibly supported later by commercial tools for automatic ETS generation. Therefore the standardised TTCN form was selected. The second important objective was to integrate the form of the input formal specification into the existing formal specification environment adequately supported by existing commercial tools. For this reason, SDL was selected. Finally, the tool should have been designed also to derive benefit from previous testing results. In order to reuse the testing results for pre-tested components, the context test generation feature was defined for the cases where only the context should be tested in which the components are currently operating.

3.   THE IATS TOOL

At the very beginning of the tool development we decided to use methods for actual test case generation based on the FSM (Finite State Machine) model. The reason was the availability of many FSM-based methods with a well-defined mathematical background, which cover a wide spectrum of FSM properties. Another reason was that the model is very close to the EFSM (Extended Finite State Machine) model of SDL. Afterwards the remaining functionality of the tool was designed to support the selected FSM-based concept of test case generation. The tool differs from the test generation tools for conformance testing like TGV, TVEDA, TTCgeN or Autolink [4, 6, 17] basically in the approach because it is based on implicit test derivation methods without defining test purposes.

The development of the tool started in 1995. As a result of the pilot project, the first prototype was available in 1996. The functionality of the prototype included a simple compiler generating an FSM from a particular SDL specification of an ISDN service, and a test case generator implementing several FSM-based methods. It provided output in the form of test sequences described by transitions of the FSM, and in a TTCN-like notation. From this list of features it is evident that it had very limited applicability. Its purpose was only to prove the feasibility of the approach in the particular service example. We evaluated the prototype in real testing of the selected ISDN service and identified possibilities for improvement and generalisation of the tool to be applicable to a larger set of services and protocols. After that we started the second phase of the tool development which ended in 1998. The result was the tool with the functionality described herein.

Using iATS, test cases are generated in the next three steps:

- *Abstraction to FSM.* Each EFSM (i.e. process in the SDL specification) is first abstracted to a corresponding FSM. Afterwards, the FSMs for all processes involved are composed into a combined FSM modelling the complete SDL-specified behaviour.

- *Test sequence generation.* From the combined FSM, test sequences are generated using well-known UIO (Unique Input-Output) methods, or test generation methods for non-deterministic protocol machines. The selection of the method depends on the properties of the FSM.

- *Translation to TTCN.* The generated test sequences are automatically translated into TTCN test cases. While the behaviour part of the TTCN description is generated completely automatically, constraints in the declaration part have to be inserted manually.

The functionality of each step is covered by a corresponding tool component as follows:

1. *SDL-FSM compiler-simulator.* This component is used for abstracting each process involved in the given SDL specification to a FSM. Values of parameters are inserted by the user of the tool and afterwards considered as fixed.

2. *Tool for composition and construction of approximate machines.* From the FSMs constructed by abstraction from SDL processes, a composed FSM is generated. Construction of an approximate machine is possible if some constructs in the SDL specification have been declared by the

user as hidden, or if the SDL specification describes only the behaviour of the context of some pre-tested and correctly-working components.

3 *Test sequence generator and compiler into TTCN.* This tool generates test sequences from a given FSM on the basis of the selected test derivation method. The user may choose between the method suggested by the tool as default, and other implemented methods.

4 *Graphical user interface.* It integrates the first three components into a single tool in X-Windows environment, supports interaction with the user and provides a system of friendly help.

Although all the components have been developed specially for iATS, the first three components can also be used as standalone tools. The iATS tool has been developed for HP-UNIX and X-Windows environment. It is owned and used as an in-house tool by IskraTEL. We describe the functionality and theoretical foundations of the first three components from the list above more precisely in the following subsections. iATS is described in detail in [7].

As the input of the tool, any SDL specification in the textual form may be used, being created in correspondence with the methodology briefly described in section 4.. The tool generates two main outputs: a set of generated test sequences described by transitions of a FSM, and a TTCN description of test cases in the standardised form. As an auxiliary output, another description in style of TTCN is generated.

3.1   CONSTRUCTION OF FSM

We selected FSMs based on the Mealy machines. Since the set of all languages defined by FSMs is a subset of the set of all languages defined by SDL specifications based on the EFSM model, not all SDL specifications can be translated into FSMs. The language of an FSM [8] or an SDL specification is a set of all valid input/output sequences. An FSM and an SDL specification are considered behaviourally equivalent if their input/output behaviour is the same, i.e. the output sequences match for each possible sequence of input symbols.

The compiler consists of its front-end and back-end in the meaning defined in [1]. The front-end performs lexical, syntax and semantic analysis of the SDL specification, while the back-end actually translates the SDL specification into the FSM. The syntax analysis is adapted to the SDL-88 syntax definition of [19, 20]. The compiler's back end is actually a specially designed simulator of SDL processes.

The basic block of our translation is a single SDL process. Each process is translated separately and, at the end of the compilation, merged with the rest of the translated processes through FSM composition procedures. The translation of the input and output signals into the input and output symbols is almost straightforward. Signal parameters must be handled separately. In order to avoid state explosion problem, the set of allowed parameter values should be given in advance. No additional help, except the information about the type of the parameter, is given to the user in giving the right values by the system.

The mapping from SDL states to FSM states is not so clear since the state of the SDL process in execution can not be characterised only by the SDL state name. The state of the SDL process is determined by the state of all variables, signal parameters, timers, and by the content of the procedure call stack. In order to reduce the FSM only to the set of states, which actually appear during the SDL process execution, simulation is required. To a certain degree, we can avoid the state explosion problem and still get valuable test sequences by limiting the depth of the SDL process simulation. Still, all possible paths of execution need to be systematically simulated. The redundant states and transitions are removed from the final FSM. The output of the compiler is a set of FSMs where each FSM corresponds to one SDL process. The FSMs are composed in one FSM representing the input/output behaviour of the source SDL specification by sequential and parallel composition techniques.

A set of limitations is imposed on the SDL specification to make the translation into an FSM possible. Some limitations are introduced simply to reduce the complexity of the compilation process. These limitations, by our experience, do not strongly affect the expressive power of SDL when telecommunications services are in question. The input of the compiler may be an SDL specification in the textual form with the contents corresponding to the limitations described in the following.

The tool can handle the constructs "start", "state", "nextstate", "stop", "decision", "label" and "join" without any constraints. All transitions should have non-empty inputs except the first transition after the construct "start". The construct "output" is forbidden in the first transition following the construct "start", otherwise there are no restrictions. The resulting FSM may have empty output symbols on transitions. Multiple signals may be outputted on the same transition; they result in one new output symbol in the FSM output alphabet. Signal decomposition is forbidden. The construct "create" should be used without parameters. Timers should have no parameters. Variables can be of any predefined sort. User defined sorts are currently not supported. Supported are all predefined operators on variables of supported sorts. Procedures may have parameters called by value or by reference, and local variables. Nesting of procedures is forbidden. At the present stage, among frequently used constructs the constructs "save", "continuous signal", "import", "export", "viewed", "revealed", and "alternative" are forbidden. They are not supported for different reasons. In the case of "save", the translation into FSMs using an existing method [12] would be possible only for SDL processes with known contents of the input queue. To provide correct results, the same order of signals should be guaranteed during test case execution on a real object. This is in the example of the system SI2000 testing not possible. "continuous signal" would result in a transition with an empty (NULL) input symbol. Consequently, the generated FSM would be an unusable input for the test sequence generator. "import", "export", "viewed" and "revealed" would require knowledge of current values for variables used in different processes. This is not possible since the translation is based on simulation of one SDL process at a time. "alternative" was omitted from the list of supported constructs because it is practically never used in specifications relevant to conformance testing of SI2000.

In figure 1 an example of SDL to FSM conversion is given. User defines the parameter of the signal A to be 0. Only actual values of the variable x contribute to new states.

3.2   TEST SEQUENCE GENERATION

The test sequence generator takes as input a file containing the FSM for which the sequences are being generated, and a file with control information. The FSM file is a text file specifying the initial state of the FSM and listing for each transition its source, its destination, its input symbol and the resulting actions. An input symbol is either a sequence of ordinary input signals or a timer signal. The resulting actions specify the output symbol of the transition (a sequence of output signals) and an arbitrary number of settings and/or resettings of timers. If the FSM is non-deterministic, the use of timers is forbidden for the present, because they are difficult to handle during test generation.

Optionally, the FSM file can be edited to specify for each transition its cost, so that the generator can enhance the implemented test-sequence generation methods with additional cost minimisation. The cost of an individual transition can be an arbitrary non-negative number properly reflecting the difficulties associated with execution of the transition, for example the necessary time or resources. By editing the FSM file, one can also indicate which of the transitions have already been successfully tested. When building a test sequence as a composition of tests for individual transitions, the pre-tested transitions shall be ignored and the resulting composite sequence shorter.

The control information file can be generated with the help of a graphical interface that suggests which methods (and their parameters) are worth trying in the next run of the test sequence generator. The suggestions are based on the diagnostic information resulting from the previous runs. A possible suggestion might also be to make the FSM complete or to introduce the reliable reset capability. That can be done automatically, by introducing in each state the missing input/NULL loops in the first case, or a reset transition with user-defined input and output symbols in the second case.

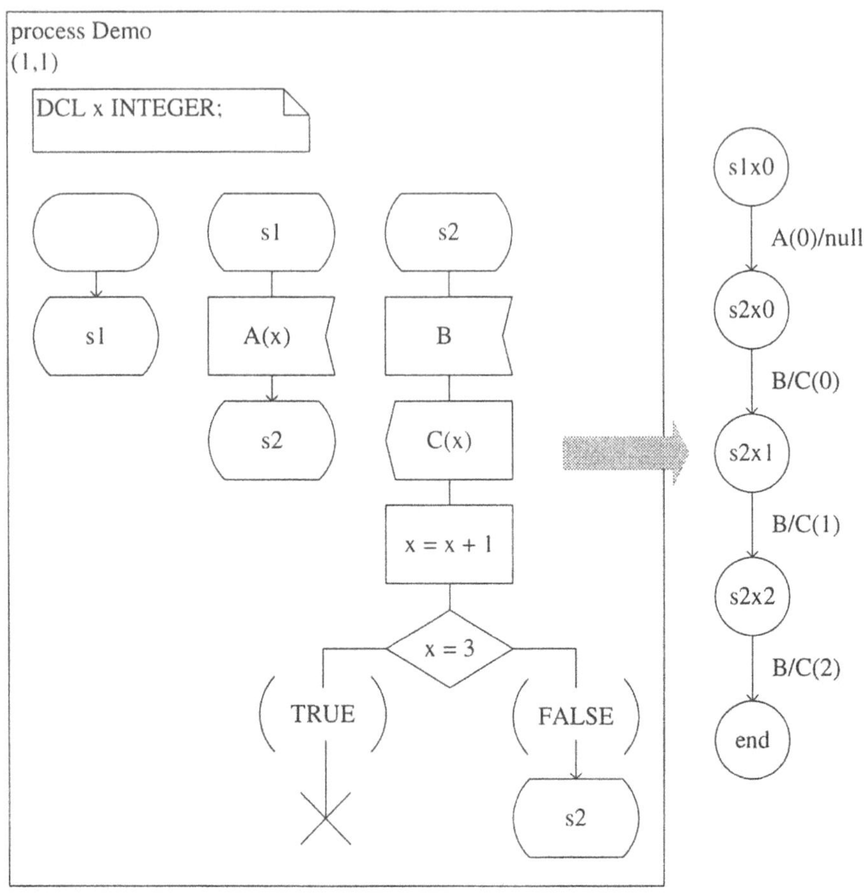

Figure 1   An example of SDL to FSM conversion

Based on the observations in [3, 15], we have elected to implement four existing test-sequence generation methods, for different purposes. For FSMs that are strongly connected, deterministic, with proper UIOs and with the correct number of states in the implementation, the methods proposed in [2, 11] have been implemented for transition testing. Method [2] is applicable to most practically interesting FSMs, while method [11] is seldom applicable, but generates extremely short sequences. If the reliable reset capability exists, both methods can be optionally preceded by testing of state and UIO implementation. In the absence of the reliable reset capability, the available method is [21], optionally without its state and UIO testing parts, for they might require catastrophically long test sequences. For a FSM that is non-deterministic (or even non-observable), only weakly connected, without proper UIOs or with an incorrect number of states in the implementation, the method proposed in [13] might help, if the reliable reset capability exists. To cope with states with very long characterisation sequences, the test sequences can be interactively optimised by gradually increasing the allowed length of the state-characterisation sequences as long as the so-called fuzziness degree [13] decreases.

The generated test sequences are basically represented by the corresponding sequences of input symbols of the FSM. For a deterministic FSM, that output file of the tool also specifies for each transition its source state, its output symbol and its cost. In addition, the cumulative cost is given for each test sequence and for their entire set. Another human-readable representation of the test cases is a tree in the TTCN [9] style, nicely divided into subtrees to fit into the designated page width. The representation is particularly convenient for a non-deterministic FSM, as each node of the tree is labelled with a list of the potentially corresponding states, and each output symbol with the cost of the most expensive among the transitions to which it potentially belongs.

For a deterministic FSM, the tool also represents the test cases in the standard machine-readable TTCN format. The file also specifies the necessary timer actions. Besides the timers specified in the given FSM, there is a special timer for each pair of a state and an input symbol. Expiration of the timer indicates that IUT reacted on the input symbol with a NULL output symbol. The duration of such a timer is a parameter of the test specification, so that it can be set to suit individual testing needs.

3.3   TESTING OF COMPONENTS

The black-box view of the IUT considered in the previous subsection is not appropriate when IUT is embedded within a complex system under test. In that case, grey-box testing methods are necessary where internal structure of the complete system under test is known in the sense of components structure. Test cases have to be generated for one particular component operating in the context of the remaining part of the system, which is assumed to be correctly implemented.

Basically we selected the model of a system with an embedded component given in [14]. We applied it as basis of test case generation for the context of pre-tested components with some minor simplifications and differences in interpretation. Firstly, the roles of the embedded component and context are changed: the embedded component actually to be tested is in our case the context represented by a context FSM (so-called component machine in [14]), and the rest of the system is a set of components represented by a single product FSM (in [14] called the context machine). Secondly, the set of input symbols (X) for the complete system under test equals to the set of input symbols of its components, i.e. no internal inputs are assumed. The same is assumed for the

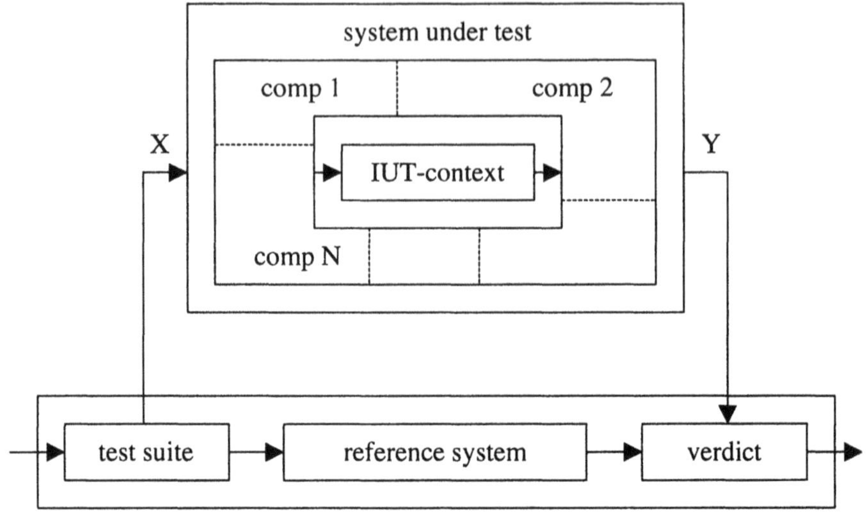

Figure 2   Architecture for testing of context

set of output symbols (Y). The applied architectural model is shown in figure
2. The described assumptions significantly reduce general applicability of the
model, but they adapt it for the use as a theoretical basis for test case generation
method implemented in iATS.

We generate test cases for the context of pre-tested components as follows.
As input, an SDL specification has to be made for each of the pre-tested components and the context to be tested. SDL specifications are translated into
FSMs. The transitions of pre-tested components are marked as already successfully tested (in the meaning described in the previous subsection), and the
FSMs of components (comp 1,...,comp N) are composed into a product FSM.
Afterwards an approximate machine for the context is constructed based on
composition of the product FSM and the context FSM (adaptation of method
of [14] in accordance with our assumptions about the model). For the constructed approximate machine, test sequences are generated using the generator
described in the previous section.

Besides limitations resulting from the assumptions described above, the
method has limited applicability also because of the combinatorial explosion
problem caused by use of FSM composition. The problem is not to construct
an approximate machine with considerable size, but to apply it as input of
the test sequence generator. Therefore it is strongly recommended to specify
each component by a single block containing one process and to restrict the
number of components. Another problem is that it is often difficult to pre-

test a component of a real system separately to prove its correct operation. However, the method can, to our experience, be successfully used for testing ISDN services in context, for example in testing interaction between services.

4.   WRITING SDL SPECIFICATIONS

The main purpose of an SDL specification used as an input of a test case generation tool is to describe the complete behaviour to be tested. To write an SDL specification we need a precisely defined input and output, methodological rules and an appropriate tool support. The input is adequate information on the functionality to be specified. For the case of ISDN services, information can be obtained from informal service specifications available in standards or internal documents of the system SI2000. The output is an SDL specification of the functionality meeting some requirements about its form and contents. Methodological rules define how from the input an output should be generated. Among software tools supporting the specification activity, an editor (preferably graphical) and a syntax checker are the most essential. A semantic checker proves to be also a very valuable tool to check the specification against logical errors, such as deadlocks for example. We used Verilog GEODE Editor and Simulator.

An SDL specification used as an input of the iATS tool can give prospective results in test scenario generation if the abstraction level and precision in the specification of the functionality to be tested is appropriate for the derivation of conformance test cases, and if the structure of the specification and the properties of the contained EFSM assure optimal results regarding to the implemented test derivation methods.

The requirements belonging to different specification problem domains imply different methodological rules. A very important question is how to reach the appropriate abstraction level and which details of behaviour to present in the specification. For the generation of conformance test cases of ISDN services, for example, we need to specify the behaviour of the system under test communicating with a user where all the observable communication should be described. The system actions are therefore observed only at the user-system interface but actually they are the result of sequences of actions executed in different parts of the system: switching system, fixed network, related mobile network etc.

Our methodology is based on a set of rules. Due to different specification problem domains the rules are divided into five categories:

1 *Abstraction rules* define at which abstraction level the SDL specification describes the given functionality.

2 *Naming rules* specify how names of all elements of the SDL specification are defined. The rules impose restrictions on the structure and contents of

names of the complete specification, blocks, processes, signals, channels and signal routes.

3 *Structure rules* define how the specification has to be structured into blocks and processes and how communication paths between them have to be specified. They also recommend how the specified functionality should be structured into components in means of test scenario generation for testing of single components.

4 *Adaptation rules* define how the form and contents of the specification have to be adapted to the properties of the test scenario generation tool. While use of a subset of those rules is mandatory to provide an acceptable input for the iATS tool, other rules may be used optionally to provide input for iATS giving more optimal results in test scenario generation. The mandatory rules impose restrictions particularly on use of some SDL constructs and their combinations. The optional rules deal more with the number of SDL processes and properties of the EFSM inside the SDL specification that are expected to give the best result in the sense of test suite length and test coverage.

5 *Mixed rules* concern more than one specification domain. The intersected domains are adaptation, structure and naming.

Defining the rules we considered two existing methodologies: ITU basic methodological guidelines for use of SDL [10] and the IskraTEL SDL methodology [16]. The reason for consideration of the latter was the need to tailor iATS to the previously existing SDL specifications, which have been developed in accordance with that methodology. Although we adopted some of its naming and adaptation rules, the most of the rules of our methodology were newly defined.

5.   FROM SDL TO ETS

In the last year the tool was used for generation of test cases for seven ISDN services and for some parts of the SS7 and DSS1 signalling protocols. Our experience is described as follows:

Writing SDL specifications. SDL specifications were already available for all services and the selected protocol parts from the design phase. They were not usable for test case generation for two reasons: firstly the specifications contained all the details of internal system behaviour and secondly they served as formal basis for automatic code generation in the implementation phase. Therefore we created new SDL specifications describing the observable service behaviour at a much higher abstraction level (the user-system interface). The length and complexity of the specifications were dependent on the complexity of the service functionality and the number of users involved. The

Table 1   Some quantitative results for the services CLIP and 3PTY

	service CLIP	*service 3PTY*
SDL specification length	349 lines	995 lines
number of FSM states	8	36
effort of writing SDL spec.	0.25 man/month	0.4 man/month
time for test suite generation by iATS – TTCN, no constraints	8 seconds	0.3 hour
total time of test suite generation – TTCN with constraints	1 hour	2.3 hours
time for automatic ETS generation	30 seconds	45 seconds

specified behaviour was described by a single SDL block containing a single SDL process. Some data on the length of the specification and effort required for the services CLIP and Three-Party Service (3PTY) can be found in table 1.

TTCN test case generation. For the created SDL specifications we generated test cases using iATS. As the common usable method for test sequence generation from automatically generated FSMs proved the method of [13]. The actual fault coverage was as defined for the method of [13]. The method of [11] was applicable for none of the automatically generated FSMs from the set of the created SDL specifications. Since the functionality was described in the specifications by a single SDL process, the FSM was small enough to generate practically usable test cases. Time for generation depended completely on the properties of the FSM and, consequently, the selection test sequence generation method. The TTCN constraints were manually inserted into the automatically generated test cases. In table 1 we present some quantitative results on time required for generation for the services CLIP and 3PTY.

Translation into the executable code. The way of translation of test cases in TTCN into the executable code depended on the testing platform and equipment. While the translation for the test cases later executed on the Tekelec Chameleon 32 equipment was performed automatically using the Expert TTCN-C compiler and an in-house tool, the form for the Alcatel 8610 equipment required quite a lot of manual work. Since there is no tool for automatic translation into the Alcatel 8610 form available for the present, iATS has not yet become the main tool for designing tests within IskraTEL.

Time spent and time saved. Generally, we saved by automatic generation of test cases in TTCN (ATS) between 20% and 50% of the total time needed for complete manual test case generation. For the cases where also the ETS was generated automatically, we saved additional 20% of the total time.

6.   CONCLUDING REMARKS

We have presented the tool iATS for automatic generation of test cases for conformance testing. The tool has been developed to generate conformance test cases for services and protocols in the development of digital switching systems SI2000. Its main advantage is that it generates a TTCN test suite from the given SDL specification automatically except the selection of (fixed) signal parameter values and the definition of TTCN constraints. These are currently also the main disadvantages of the tool. Practical use of the tool has shown that the first disadvantage decreases the general applicability of the tool, while the second one results only in time consumption for completion of a generated test suite. Developing the tool, we evaluated several existing methods for test sequence derivation and developed a methodology for writing SDL specifications used as the input of the tool.

Our main goal for the future is to improve the iATS tool by removing the limitation of fixed parameter values in test sequences derivation. In this way iATS is currently being enhanced with an additional test case generation method based on the EFSM model [18]. Here some theoretical work is also being done, trying to make the method work for more than one SDL process. We also plan to build into the tool the possibility of generating the TTCN constraints more automatically using ADTs.

To our experience, formal description techniques, especially SDL and TTCN, are successfully used and becoming well accepted by the industry. The most important contributing factors for their success in the industrial use are the appropriate tool support in all steps from writing specifications to generation of the executable code, and the systematic training of system developers possibly already at the undergraduate level.

Acknowledgments

We wish to thank the Ministry of Science and Technology of the Republic of Slovenia, and IskraTEL for financing this project. Next we wish to thank A. Ciglič, D. Kodrič, N. Maloku and M. Stojsavljevič from IskraTEL for constructive comments and provision of a platform for test scenario execution on the real system. Finally we would like to express our gratitude to the project participants V. Avbelj, B. Močnik, B. Slivnik and R. Verlič, who contributed to the iATS tool in an essential way.

References

[1] Aho, A.V., Sethi, R., and J.D. Ullman. (1986). *Compilers: Principles, Techniques, and Tools*, Addison-Wesley Series in Computer Science, Addison-Wesley Publishing Company, Bell Telephone Laboratories.

[2] Aho, A.V., Dabhura, A.T., Lee, D., and M.U. Uyar. (1991). An optimiza-

tion technique for protocol conformance test generation based on UIO sequences and rural Chinese postman tours, *IEEE Transactions on Communications*, vol. 39, no. 11, pp. 1604-1615.

[3] Anido, R., and A.R. Cavalli. (1995). Guaranteeing full fault coverage for UIO-based testing methods, in *Proceedings of the 8th Int. Workshop on Protocol Test Systems*, Evry, France, pp. 221-236.

[4] Doldi, L., et al. (1996). Assessment of automatic generation methods of conformance test suites in an industrial context, in *Testing of Communicating Systems*, Chapman & Hall, pp. 347-361.

[5] Exel, M., Popovič, B., and F. Prijatelj. (1982). SL1 language - A specification and design tool for switching systems software development, *IEEE Transactions on Communications* (Special Issue on Comm. Software).

[6] Fernandez, J.-C., Jard, C., Jeron, T., and C. Viho. (1997). An experiment in automatic generation of test suites for protocols with verification technology, *Science of Computer Programming*, vol. 29, no. 1-2, pp. 123-145.

[7] Frey-Pučko, M., Kapus-Kolar, M., Novak, R., Verlič, R., Močnik, B., Slivnik, B., and V. Avbelj. (1998). *Automatic Test Scenario Generation for the System SI2000*, Final report, IskraTEL (in Slovene).

[8] Hopcroft, J.E., and J.D. Ullman. (1979). *Introduction to Automata Theory, Languages and Computation*, Addison-Wesley Publishing Company.

[9] ISO. (1997). *ISO/IEC 9646-3, Tree and Tabular Combined Notation (TTCN)*, Second Edition.

[10] ITU-T. (1993). *Z.100*, Appendix I, SDL Methodology Guidelines.

[11] Jiren, L., and D. Jun. (1994). A new approach to protocol conformance test generation based upon UIO sequences, *Chinese Journal of Advanced Software Research*, vol. 1, no. 4, pp. 373-381.

[12] Luo, G., Das, A., and G. von Bochmann. (1994). Software testing based on SDL specifications with Save, *IEEE Transactions on Software Engineering*, vol. 20, no. 1, pp. 72-87.

[13] Luo, A., Petrenko, A., and G. von Bochmann. (1994). *Selecting Test Sequences for Partially-Specified Nondeterministic Finite State Machines*, Technical Report, Departement d'IRO, Univ. de Montreal, Canada.

[14] Petrenko. A., Yevtuschenko, N., and G. von Bochmann. (1996). Fault models for testing in context, in *Formal Description Techniques IX*, Chapman & Hall, pp. 163-178.

[15] Ramalingam, T., Das, A., and K. Thulasiraman. (1995) Fault detection and diagnosis capabilities of test sequence selection methods based on the FSM model, *Computer Communications*, vol. 18, no. 2, pp. 113-122.

[16] Robnik, A. (1995). Experiences of using SDL collected in IskraTEL SDL methodology, in *Formal Description Techniques VIII*, North-Holland Elsevier, pp. 221-236.

[17] Schmitt, M., Ek., A., Grabowski, J., Hogrefe, D., and B. Koch. (1998) Autolink – Putting SDL-based test generation into practice, *Testing of Communicating Systems*, Kluwer Academic Publishers, pp. 227-243.

[18] Ural, H., and A. Williams. (1994). Test generation by exposing control and data dependencies within specifications in SDL, in *Formal Description Techniques VII*, North-Holland Elsevier, pp. 335-350.

[19] Verilog. (1993). *GEODE Product Documentation*, Appendix C: SDL Concrete Syntax.

[20] Verilog. (1993). *GEODE Product Documentation*, Chapter 5: SDL Extensions/Restrictions.

[21] Yao, M., Petrenko, A., and G. von Bochmann. (1993). Conformance testing of protocol machines without reset, in *Protocol Specification, Testing, and Verification XIII*, Elsevier Science Publishers, pp. 241-256.

Marjeta Frey-Pučko received the B.S., M.S. and Ph.D. degrees in computer science from the University of Ljubljana, Slovenia, in 1988, 1991 and 1995, respectively. In 1988 she joined the Department of Digital Communications and Networks at the Jožef Stefan Institute in Ljubljana. In the years 1995-1998 she was the manager of the project "Automation of test scenario generation for system SI2000". Since 1998 she has been with IskraTEL where she is responsible for process improvement in the development of the system SI2000. She is also with the Jožef Stefan Institute as a part-time researcher. Her main research interests concern communications systems engineering, verification and validation techniques, and use of formal methods.

Monika Kapus-Kolar received the B.S. degree in electrical engineering from the University of Maribor, Slovenia, in 1981, and the M.S. and Ph.D. degrees in computer science from the University of Ljubljana, Slovenia, in 1984 and 1989, respectively. Since 1981 she has been with the Jožef Stefan Institute, Ljubljana, where she is currently a researcher at the Department of Digital Communications and Networks. She is also a part-time lecturer at the University of Maribor. Her current research interests include formal specification techniques and methods for development of distributed systems and computer networks.

Roman Novak received his B.S., M.S. and Ph.D. in computer science from the University of Ljubljana in 1992, 1995 and 1998, respectively. He works as a researcher at the Department of Digital Communications and Networks at the Jožef Stefan Institute in Ljubljana, Slovenia. Since 1997 he has been also a part-time lecturer at the Faculty of Computer and Information Science at the University of Ljubljana in the fields of programming techniques, algorithms and data structures. His current research interests include distributed systems, communication protocols and security.

23

EXTERNAL CONFORMANCE REQUIREMENTS: CONCEPTS, METHODS AND TOOLS

Rafał Artych
Krzysztof M. Brzeziński
Institute of Telecommunications
Warsaw University of Technology
{ rartych,kb } @tele.pw.edu.pl

Abstract   In this paper we propose to re-use and extend the elements of standard conformance testing methodology, in order to cover particular needs of a telecommunications Network Operator. In this way we intend to contribute to the "industrial relevance" of formal techniques. We present the idea and formal notation of external requirements imposed by a Network Operator on protocol implementations. We also report on the support tool ORB.

Keywords: Validation, protocols, requirements, conformance testing, PICS, ORB

1.   INTRODUCTION

In order to achieve (preserve, prove) correctness in particular phases of system's life-cycle, various theories, methods and processes (applications of methods) have been researched and applied. Some of them, like conformance testing, have become state-of-the-art. However, the applicability of individual methods and their acceptable level of formality depend on the context. What is obvious in a university environment, may be rejected as unacceptable by a commercial network operator.

This problem has been recognized and analyzed, mostly in relation to the testing phase of the system life-cycle [1, 2]. Accordingly, after the initial emphasis on fundamental theories (such as test generation algorithms, testing architectures, defining and formalizing the TTCN language), the interest of researchers has visibly turned to making the available methodology more industrially relevant: acceptable and actually used. This is also the general framework of our work, which concentrates on the role of techniques related to protocol conformance testing in the activities of a Network Operator.

We are not aiming at making any particular subset of conformance testing techniques more industrially relevant *per se*. Instead, we propose to apply a "leverage strategy", which can be summarized as follows:

- identify an informal element within the system life-cycle: a weak link in the "formality chain" that makes other, well developed formal techniques relatively unattractive to the industrial user;

- formalize this element of the life-cycle, and develop the appropriate tools that would support and automate its execution;

- respect, as far as possible, the well established procedures of the user and their need for flexibility (which also pertains to the formal, or at least organized treatment of informal elements).

Our conjecture is that, when a weak link is formalized, the user will see sense in actively using other, already developed formal techniques for other parts of the process, because their use will then become natural.

We concentrate on a particular example of this general strategy. We consider the system life-cycle as seen from the perspective of a commercial Operator of a public telecommunications network, pursuing its own commercial goals on a deregulated, competitive market. This life-cycle differs from the usually considered design and implementation-oriented process [3], because the Operator does not design or implement the system and does not place the equipment on the market for sale. Formal techniques are available for stating the abstract properties of a telecommunications protocol and for establishing conformance of a given implementation to a standardized definition of this protocol. However, to demand "conformance with ETSI 300xxx" is grossly insufficient for the planning and construction of a particular telecommunications network. For contract procedures it is necessary to specify which protocols are to be implemented, which protocol standards are to be followed and which options are to be chosen in a particular piece of equipment, according to its planned role and in line with the Operator's strategy of growth and service deployment.

The common practice of the Operator has been either to refrain from stating their own requirements ("conformance with protocol standards and national requirements"), or to produce lengthy, verbose textual documents under the common name of Technical Requirements, which were in an obscure relation to international standards and to each other. Both solutions were obviously inadequate. Therefore, we will treat the equipment procurement phase as a "weak point" mentioned before. Any mistakes or misunderstandings in this phase can have serious consequences for interoperability of equipment and the overall quality of service.

In this paper we make the following contributions: we introduce the notion of *Operator Requirements* (ORs), identify their relation to requirements inherent in a specification of a protocol, formalize both the notation of ORs and the process of their design and use, describe the support tool ORB, and report on the initial experience with introducing the formalized approach into the activities of a real Network Operator.

We draw from the generic techniques related to conformance testing, as described in ISO 9646 [4] ETS300406, ETR212 and [5]. Our technique of expressing the ORs is similar to a protocol profile. Note that we are not dealing with conformance testing itself; in particular, we are not interested in whether conformance tests are actually conducted or how the test suite has been derived.

The need to formalize and automate other parts of industrial processes related to testing has been identified and addressed previously, e.g. in the context of ATS management [6]. To our knowledge, the particular set of concepts presented in this paper has not been considered before, apart from our own preliminary attempt [7].

The rest of this paper is organized as follows. In section 2 we identify two different classes of requirements: integral and external, and discuss the particular needs of a Network Operator. In section 3 we introduce the proposed structure and notation of Operator Requirements (ORs). Section 4 illustrates how the structure of ORs can be applied to generic problems. In section 5 we briefly discuss the design cycle of ORs. Section 6 is devoted to the presentation of the ORB tool, developed to support and automate this design cycle. Section 7 concludes the paper.

2.   TYPES OF REQUIREMENTS

Two notions (models) of conformance prevail: more theoretical [8, 9], based on implementation relations, and more practical (pragmatic) [4], based on a collection of conformance requirements that will have to be satisfied by an implementation. We adopt the latter model, which is the basis of "industrial" conformance testing, as described in ISO 9646. Within a system life-cycle an inherent discontinuity point can be identified, in which the system under development changes its nature (fig. 1): from abstract object (S – specification) to real equipment (I – implementation). Due to this discontinuity point, the organized empirical observation (testing) is considered as the only relevant method of revealing whether the implemented system is "correct". However, "the notion of validity of an implementation w.r.t. a specification is not expressed in the specification itself" [9]. Similarly, one could say that requirements (which define the link between S and I) are fundamentally different from the specification (which is abstract and should not refer to the implementation).

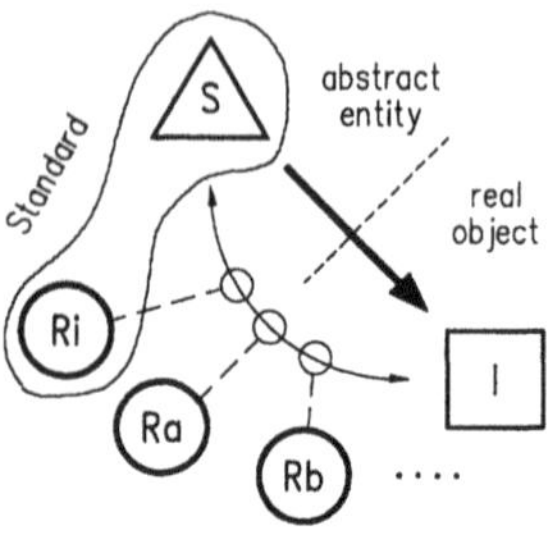

Figure 1   Requirements in system design cycle

Let us introduce two kinds of requirements: *integral* (R_i in fig. 1) and *external* (R_a, R_b, ...). Integral requirements are inherently related to the specification of a protocol. These requirements are considered in the methodology of conformance testing. Two types of integral requirements are distinguished: *static* and *dynamic* (fig. 2). A Static Conformance Requirement is defined as "one of the requirements that specify the limitations on the combinations of implemented capabilities... "[8]. Static requirements are reflected in an ICS proforma (Implementation Conformance Statement), which is a part of a standard. Technically, static requirements are modal qualifiers that assign a Status to every capability/feature of a protocol (M–Mandatory to be implemented, O–Optional, X–prohibited from being implemented, etc.). To state the implemented capabilities, the Supplier has to fill in the Support column in the proforma (this is the prerequisite for test case selection and subsequent testing). Dynamic conformance requirements specify the permitted observable behaviour within the implemented capabilities, and cannot be easily extracted (separated) from the specification.

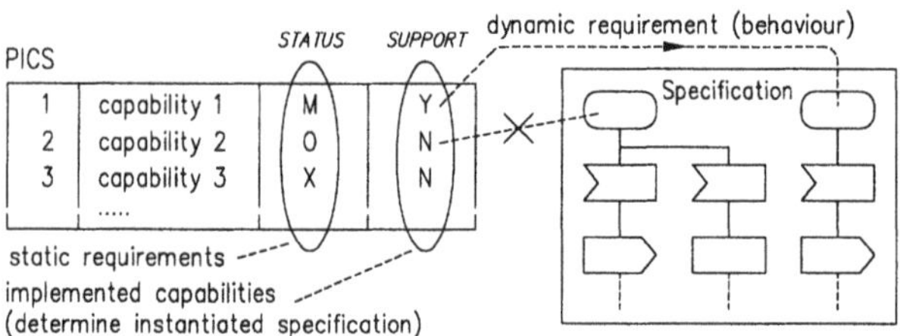

Figure 2   Relations between requirements and specification

Standard specifications, with their associated integral requirements (e.g. ETSI standards), do not contain any self-reference, or obligation that they should be applied in a given context. The implementation options allowed by static requirements are fixed for a given protocol, irrespective of whether a particular choice makes sense or is desired in a given piece of equipment.

To address the *need* to conform, external requirements are also necessary. External requirements are disjoint from the integral ones and cannot be deduced from a specification. They may be imposed by different organizations and for various reasons (e.g. international or national normative requirements, private user requirements). Let us consider two examples of external requirements.

Normative (regulatory) international requirements w.r.t. the terminal equipment, and terminal equipment only (TBR documents), contain clauses such as: "We do not require that integral requirement X be fulfilled (in fact we don't care), because it is not essential from our point of view". Such statement does not mean that the integral requirement in question becomes in any way invalid, but does indicate that it is inappropriate for regulatory use and will not be further investigated [5].

The most concise contents of private requirements of a Network Operator could state: "It is mandatory for system X, intended for our network, to implement the protocol standard Y (i.e. to fulfil all its integral requirements), AND to fulfil all normative requirements". However, such briefness will almost certainly not match the real needs of the Operator.

In general, external requirements may be less restrictive or more restrictive than integral requirements, or may pertain to other properties of the system, not covered by integral requirements (e.g. when protocol X, and not Y, is required to be implemented). It is also generally true that the behaviour of a protocol within its individual capabilities (i.e. the dynamic requirements) will not be subject to any intentional changes. Therefore external requirements are essentially static requirements.

In the sequel we will concentrate on the external requirements imposed by a Network Operator (Operator Requirements – ORs). The discussed concepts may also be applicable to other kinds of external requirements. The crucial observation is that a Network Operator has the right to establish its private requirements that pertain to devices installed in its network, according to its own commercial and technical policy.

The general aims of Operator Requirements are: to impose consistency and interoperability within a particular network and at its borders, to document the technical capabilities of a network as a whole and individual systems within this network, and as a template for technical clauses in contracts (in order to make them more effective). The Operator is bound by legal provisions and will not install in its network any devices that are contrary to normative regulations. This also usually means that ORs cannot contradict the integral requirements. Therefore, the main question is: how to ensure that resulting ORs are both internally consistent and compatible with other valid requirements (integral and regulatory). Obviously, it seems expedient to take advantage of as many elements of the standardized methodology as possible, such as: specifying (or rather reflecting) a set of conformance requirements in an ICS-like document, formulating a profile by introducing Requirement Lists (RL) that change the status of protocol features in ICS documents, identifying and concentrating on essential external requirements [5], and using the methodology of conformance testing: existing test suites, existing test instruments and existing facilities at test houses (the ORs will simply result in a particular parameterization of a test suite).

The main idea of ORs is to change the status of chosen protocol capabilities (features), while leaving the status of other features as defined in a base specification. The change is done in reference to the standard ICS documents (PICS) that are part of the standard. For example, it will usually be necessary to state that it is mandatory for the implementation to support both the incoming call and the outgoing call capability, even if this does not follow from any standard (fig. 3). The technique that we intend to use is not new and basically

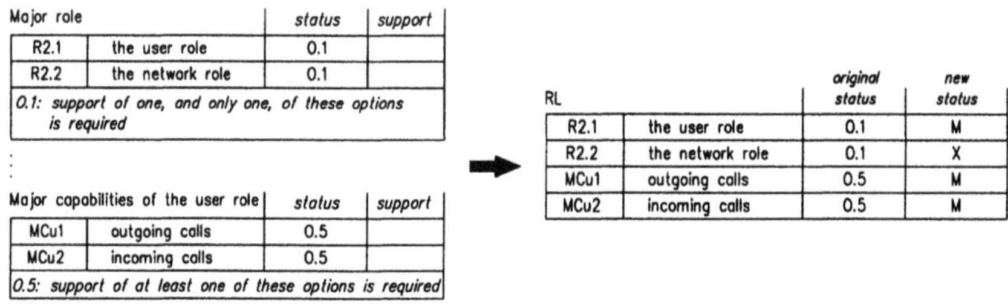

Figure 3   Example of external requirements imposed upon integral ones

amounts to specifying a *profile*. However, the notion of a profile will have to be extended. For example, the Operator will need to state that given services shall be supported as well. Also note that any implementation status of any feature is meaningful only if a protocol itself is required by the Operator to be implemented. In this way the ORs can be very concise: they do not repeat, rephrase or interpret the contents of a standard protocol specification (unlike formerly used, narrative "Technical Requirements"), but directly identify the differences.

3.   STRUCTURE AND NOTATION

ORs are constructed from a set of tables of three types: DP (Declaration of Parameters), RL (Requirements List) and CS (Conformance Statement). The structure of CS and RL tables had originally been defined elsewhere [4], but for the purpose of ORs was suitably modified. DP tables are used to specify the status of items defined by the Operator, and to document the support of these items (or commitment to these items) by the Operator. CS tables serve to specify the status of items defined by the Operator, and to capture the support answers given by an external party (such as an equipment supplier). RL tables are used to change the original status of an item defined in a protocol specification to another status, indicated by the Operator.

The entries in the tables are linked to form a hierarchical structure. The structure itself is defined for each application (instance) of ORs. The rules of the correct linking of objects within the tables are referred to as *Requirements Calculus*. The structure of tables is illustrated below. Note that the structure of DP and CS tables is identical (ICS type), but the use of these tables differs.

ICS(DP/CS):

⟨header⟩					
Item	⟨Item description⟩	Reference	Conditions	Operator's status	Status

RL:

⟨header⟩						
No.	Item	⟨Item description⟩	Status	Conditions	Operator's status	Ref.

Before going into more details, we present a tiny sample of real ORs:

Table CS28. HOLD - general					
References to [141-1], unless stated otherwise					
Item	General service aspects	Reference	Conditions for status	Op's status	Status
1	not sending notifications to the remote user	[141-1]9.2.1	DP1/2 else	x o	

Table RL22. Requirements on items used in HOLD PICS [141-2]						
No.	Item	Feature with modified status	Status	Conditions for status	Op's status	Ref.
5	A1/R3.2	support requirements for interworking with private ISDNs	o	not CS28/1 and DP1/4 else	m o	

The values of Status fields belong to a set
$$STAT = \{i, m, o, oe.i, om.i, x, na, nr, c\}, \text{ where:}$$

i irrelevant (out of scope)

m mandatory

o optional

$oe.i$ optional: exactly one item needs to be implemented

$om.i$ optional: one or more items need to be implemented

x excluded (implementation prohibited)

na not applicable

nr not a requirement

c (only in RL tables) any previous status (e.g. conditional)

Support fields in DP and CS tables may assume values from a set SUPP={Yes,No,$-$,$\emptyset$}. Value '$-$' denotes "no support answer required". Value "$\emptyset$" (empty field) is assumed before any support answer is given. This value, if used in condition expressions, is treated in a special way (note that the Requirements Calculus involves multi-valued logic).

The actual status is calculated by evaluating predicates and conditions. A predicate is associated with a table. If a predicate evaluates to FALSE, then all the status values in the table assume value 'i' (in case of RL tables 'i' is interpreted as "ignore any change in the original status"). A single item (entry) in a table may have multiple status values, associated with multiple condition expressions in a multi-line arrangement (see the mini-example of real ORs). Subsequent expressions are evaluated from top to bottom (TTCN-like), and the status associated with a TRUE expression becomes the current status. If all expressions evaluate to FALSE, the default 'na' status is applied. Both predicates and conditions are logical expressions built of support answers (Yes=TRUE, No=FALSE). The support answers are identified by reference ($ref \in REF$) to a support field in one of the existing tables within the ORs (reference by $\langle$ table_identifier$\rangle$/$\langle$ item_number $\rangle$, e.g. DP1/2) or to a support field in an external PICS table (e.g. [092-2]A1/R1 refers to a specific support field in ETS300092-2). To find a support value, the function $supp : REF \rightarrow SUPP$ is applied.

Function $stat_check : STAT \times STAT \rightarrow \{$TRUE,FALSE$\}$ controls the correctness (validity) of status changes specified in RL tables. The value $stat_check(s1, s2) =$TRUE indicates that a change from the original status $s1$ to the new status $s2$ is allowed w.r.t. integrity and consistency of ORs. The current status of an item determines the correct values of a support answer, according to function $supp\_check$: $STAT \times SUPP \rightarrow \{$TRUE,FALSE$\}$. The answer is acceptable only if $supp\_check$ evaluates to TRUE.

We use the following definition of both functions (interesting results can be obtained by tuning this definition):

	stat_check						*supp_check*		
	Operators's status:						support answer:		
base status	na	nr	m	o	x	i	Y	N	—
na	T	F	F	F	F	F	F	F	T
nr	T	T	F	T	T	T	T	T	T
m	F	F	T	F	F	F	T	F	F
o	F	F	T	T	T	F	T	T	F
x	F	F	F	F	T	F	F	T	F
c	T	F	T	T	T	F	-	-	-
i	T	T	T	T	T	T	T	T	T

4.   USE PATTERNS

DP and CS tables.   ORs may be used for two purposes:

(a) as a general declaration (specification) of technical and service-related properties of a network;

(b) as a specification of technical requirements imposed on a particular piece of equipment.

For each type of application the DP tables, which store the parameters of ORs, are used in a different way. DP tables, which are always filled by the Operator, can be fully completed (to indicate the current state of the network and its global properties) or partially completed, with the Support column left blank — this column will be completed (still by the Operator) in order to characterize the required type of equipment.

In application (a), it is assumed that the blank Support fields in DP tables have, or may have, value 'Y'. This means that the reader of ORs may expect a particular feature/capability to appear somewhere in the Operator's network, but not necessarily in every piece of equipment. Application (b) corresponds to contract procedures: the Operator has to fully identify the type of equipment required. In this application status 'o' in a DP table means that the Operator has freedom to point to a particular capability as required or not required, while the support value 'Y' means that the Operator does consider the capability as required.

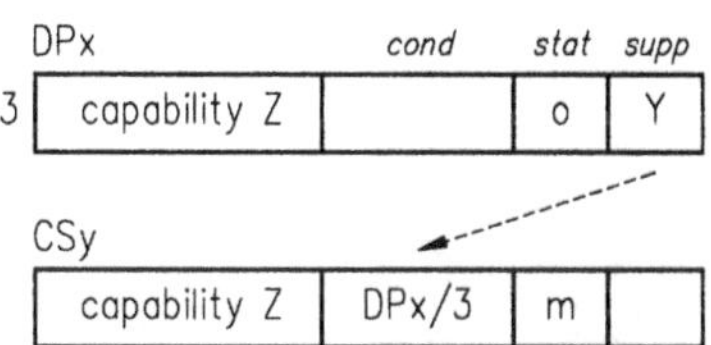

Figure 4   Indirect parameterization

In this use pattern, each DP table is accompanied by a similar CS table, as shown in fig. 4. The Operator could state its requirements directly in the Status field of a CS table. However, the use pattern described above allows the Operator to benefit from the same semantic support mechanisms that also serve the Supplier, who fills the Support fields of CS tables.

Phased introduction of a protocol.   Another use pattern addresses the tricky problem of the phased introduction of capabilities into the network. Consider the following fragment of ORs:

DP:	condition	Op.status	Support
phase_1	—	i	Y
phase_2	—	i	N

CS:	condition	Op.status	Support
feature_a	phase_2	M	
	phase_1	O	

RL:	status	condition	Op.status
feature_b	M	phase_1	NR
feature_c	O	phase_2	M

This example shows how informal, non-protocol items (such as the identification of a phase of network development) can be formally accommodated in ORs. Feature_a, defined by the Operator, becomes mandatory to be implemented in phase 2. In the earlier phase it remains optional (note the order in which condition expressions are evaluated). Feature_b, which is mandatory according to integral requirements, in the initial phase of network development becomes "not required". Feature_c will become mandatory later.

5.   ORS AS A PROCESS

ORs, as an entity, must undergo a full design cycle under the control of the Operator. This design cycle is orthogonal to the design cycle of the system (compare fig. 1). The OR design cycle has its own peculiarities, but there is no reason why it should not follow the well known general principles. The simplified, waterfall-like model of this life-cycle is outlined in fig. 5.

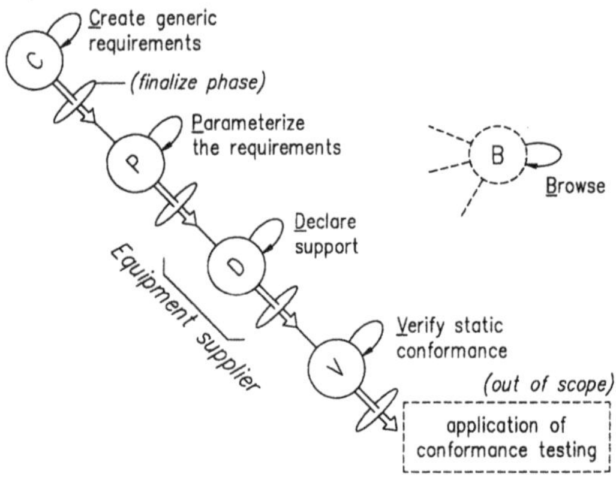

Figure 5   The life-cycle of ORs

The outcome of the OR design cycle can be directly used to parameterize the test suite in the process of conformance testing w.r.t. the ORs, which is a non-standard application of the standard methodology.

6.   SUPPORT TOOL — THE ORB SYSTEM

Our experience shows that in most practical cases ORs are complex. Control over their correctness is a nontrivial task, which calls for a computer support tool. Therefore, the concepts presented above have been embodied in the ORB (Operator Requirements Backbone) support system. Technically, ORB is a database application implemented in Borland Delphi with BDE (Borland Database Engine), running under MS Windows 95/98. This simple execution platform is well suited for the job, and is readily available to the Operator.

Static structure of ORs.   According to the declaration made in the Introduction, ORs handled by ORB should closely follow the metaphor of a familiar technical document. A single OR document handled by ORB, called a project, is a sequence of chapters. Each chapter consists of its title, a sequence of textual notes, and an arbitrary number of DP, CS and RL tables. Chapters form a tree (i.e. a chapter may contain an arbitrary number of subchapters, which are chapters themselves):

> *ORB-project*:= { *Chapter* }+
>> *Chapter*:= *Title* { *Chapter-element* }*
>> *Chapter_element*:= *Text_note* | *Table* | *Chapter*
>> *Title*:=text
>> *Text_note*:=text
>> *Table* := *DP_table* | *CS_table* | RL_table
>> *DP_table*:=ICS_table
>> *CS_table*:=ICS_table
>
> ('+' means: one or more, '*' means: zero or more)

Altogether, a project is equivalent to a technical annex of a contract, an internal document of the Operator, etc. To exploit the familiar metaphor, documents produced by ORB are formatted exactly as expected by the technical personnel of the Operator. The document is also produced in a portable format (HTML), in which it can be made available for corporate-wide browsing, as well as in a format compatible with the MS Word text editor.

ORB Functionality.   The waterfall model presented in fig. 5 pertained to a single project and a single thread of activity. ORB supports the realistic version of the life-cycle (fig. 6), which takes into account the multitude of projects and their versions under development. Project management functions are an important part of system functionality. Individual phases of the OR life-cycle are handled by separate modules — Actors, with the following functionality:

- CREATOR (C): create a generic OR document for a particular signalling protocol (create and modify the structure of ORs; edit the contents of tables; perform automatic semantic checks)

- PARAMETERIZER (P): prepare a requirements document for a given contract (parameterize the ORs by filling in the remaining Support fields of DP tables; check correctness)

- DECLARATOR (D): declare the implementation of capabilities (fill in the blank Support fields in CS tables; assign the values from standard PICS documents to "external variables" of ORB; calculate condition values from the actual Support answers)

- VERIFIER (V): verify the support answers (perform full Static Conformance Review; generate the Conformance Report)

- BROWSER (B): browse the project in any phase of its life-cycle.

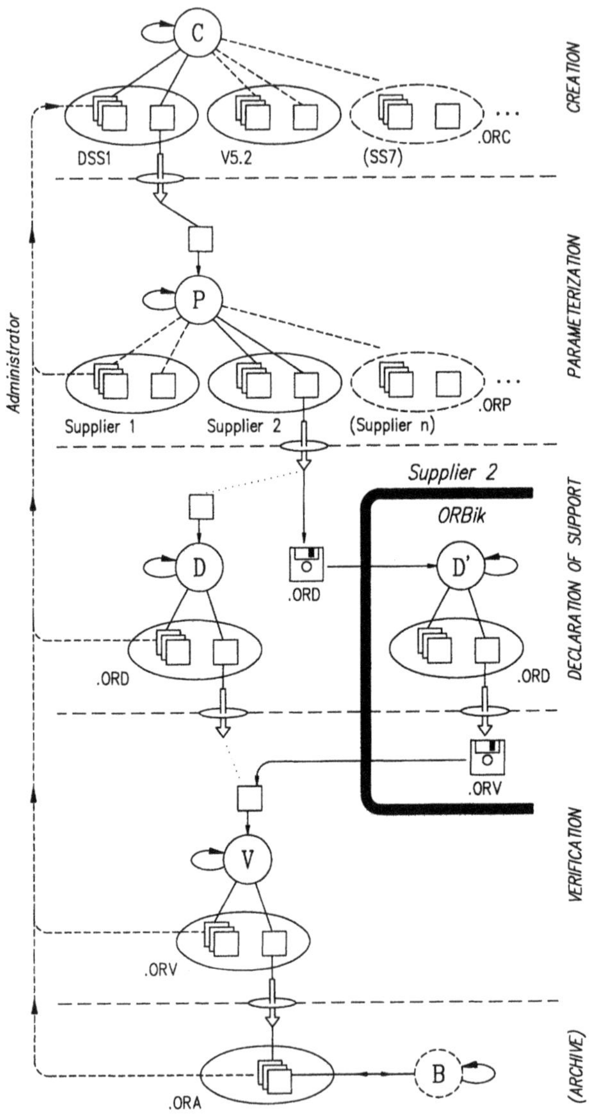

Figure 6   The OR life-cycle supported by ORB

ORB in use.   The ORB toolset consists in fact of two separate systems: the full ORB system (which contains all the Actors) and the ORBik system with a single actor (DECLARATOR only), intended primarily for the Equipment Supplier.

a) A designer, using the CREATOR, creates generic Operator Requirements for a given signalling protocol (or rather for a profile of signalling protocols to be used in a given signalling relation). The final version of ORs in this phase reflects a current state of the network, but is not parameterized (some Support fields in DP tables are empty). The project is then finalized (changes its phase) and can no longer be modified by its designer.

b) When a need arises to procure a piece of network equipment of a given type, a "contract officer" of the Operator prepares a parameterized version of Requirements (fills in the remaining free Support fields in the DP tables), using the PARAMETERIZER. The same project can be used to prepare multiple, differently parameterized versions (e.g. for different Suppliers). The finalized parameterized version is saved on diskette and is handed over to a Supplier, together with the ORBik system.

c) The Supplier, using the DECLARATOR module of ORBik, declares the implemented capabilities in the Support fields of the CS tables. Apart from the ORs themselves, the Supplier also fills in the standard PICS proformas, identified by the ORs. The DECLARATOR points to particular support answers in these PICS documents that have to be copied into the electronic document handled by ORBik. The finalized version of ORs is saved on diskette and handed over to the Operator, together with standard PICS documents.

d) In parallel, or as part of earlier design activities, the designer may use the DECLARATOR built into the ORB system to perform "what-if" simulations.

e) Using the VERIFIER, the "contract officer" checks the correctness of support answers. Static Conformance Review [4] is performed automatically, and its results are included in a Conformance Report.

f) The personnel of the Operator can browse any of the OR documents, using the BROWSER. The available functions are hypertext-like, which is invaluable in a large document of a technical nature.

g) To allow for real-life events, the Administrator (a user with special rights) can recover the original version of the document at any development phase.

Semantic support.   ORB implements the Requirements Calculus, which allows for extensive automatic semantic checks on the structure of ORs and values given by the User. The creation of ORs is the crucial stage. For example, internal consistency of ORs requires that there shall be no circular dependencies involving condition expressions, for items in RL tables the *stat_check* function shall evaluate to TRUE for any modified value of the Status field, and for each declared parameter the *supp_check* function shall evaluate to TRUE. The example of inconsistency in ORs is shown in fig. 7.

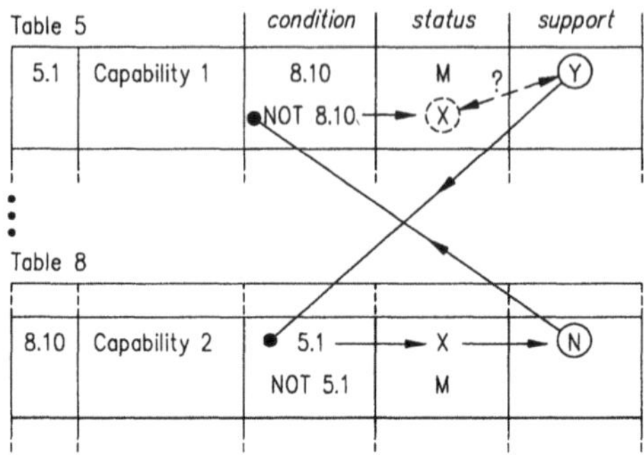

Figure 7   Example of circular reference

In real ORs, "multi-hop" dependencies are commonplace, and manual discovery of possible loops is a hopeless task. For this single reason, the ORB tool appears to be the enabling component of the whole approach. In other stages, the *supp_check* function is extensively used. The Static Conformance Review is successful if for each item:

$supp_check(status(item), support(item)) =$TRUE.

7.   CONCLUSION

In this paper we proposed a particular "leverage strategy" for making formal methods more relevant to the industry. The application of this strategy to conformance testing in the context of a Network Operator leads to the following effects:

- the elements of conformance testing methodology used to formally state the ORs, are introduced for well understood commercial reasons;

- conformance testing w.r.t. the ORs (e.g. for internal acceptance purposes) becomes a task for a test laboratory run by the Operator;

- consequently, the Operator is compelled to acquire the methodology of conformance testing;

- similarly, the formal aspect of protocol standards (SDL diagrams, PICS proformas), which is the prerequisite of conformance testing methodology, becomes less obscure.

Introducing the formalized Operator Requirements into procedures and structures of the Operator was a methodical, step-wise, and sometimes painful process. The proposed structure of ORs was first applied "manually", to DSS1 and INAP protocols. After a period of confusion, the improvement over the previously used informal documents was acknowledged, but the inconvenience of the paper form of ORs became evident. The introduction of ORB answered this criticism. The first real application of ORB was to handle (transcribe) the ORs for DSS1 protocols. The ORs themselves had been developed "manually", and were available on paper. The transcription process went effortlessly. ORB almost immediately found a few logical errors (inconsistencies) in what was believed to be the final version of ORs and was about to be approved. This demonstrates non-triviality of the problem and usefulness of the tool. The Operator is currently making arrangements for field trials of the ORB system in its R&D department. In section 6 we reported on features that appear to be minor technical details (such as the printout format). However, it has become clear to us that this kind of minor details, rather than the complexity of semantic mechanisms built into the tool, is essential for winning acceptance with traditionally-minded personnel. Neither the methodology of formulating and expressing the ORs, nor the support toolset are complete. We are currently working on the Requirements Calculus, and the ORB tool is being linked with the Means of Testing (MOT), which will allow for the semi-automatic test selection.

Acknowledgments

The ORB system was designed and implemented in LTiV (Laboratory of Testing and Verification, Warsaw University of Technology) on contract from the TP S.A., the leading Polish network operator. The authors wish to thank the TP S.A. for the kind permission to publish information about the system, and anonymous referees for helpful remarks.

References

[1] Lai, R., Leung W.(1995) Academic and industrial protocol testing - the gap and the means of convergence. Computer Networks and ISDN Systems, 27 (1995), pp.537-547

[2] Lai, R. (1996) How could research on testing of communicating systems become more industrially relevant? In Proc. IWTCS'96, pp.3-13, Darmstadt, Germany

[3] Making Better Standards; A guide from Technical Committee MTS. ETSI, 1996

[4] ITU-T/ISO 9646; Conformance Testing Methodology and Framework

[5] ETSI TCR-TR 017 MTS; Technical Basis for Regulation (TBR) specification methodology

[6] Danet, P.-Y., Desecures, E. (1995) Management and maintenance of TTCN Abstract Test Suites. In Proc. IWPTS'95, pp.367-375, Evry, France

[7] Brzeziński, K.M. (1998) Towards Formality of Technical Requirements for Protocols. In Proc. EWDC'98, pp.71-74, Gdańsk, Poland

[8] ITU-T Z.500/ISO Framework on Formal Methods in Conformance Testing

[9] Leduc, G. (1992) A framework based on implementation relations for implementing LOTOS specifications. Computer Networks and ISDN Systems, 25 (1992), pp.23-41

24

TIME SIMULATION METHODS FOR TESTING PROTOCOL SOFTWARE EMBEDDED IN COMMUNICATING SYSTEMS

J. Latvakoski, H. Honka

VTT Electronics, Kaitoväylä 1, P.O.Box 1100, FIN-90571, Oulu, Finland. Tel. +358 8 551 2111, Fax. +358 8 551 2320, E-mail: juhani.latvakoski@ele.vtt.fi, hannu.honka@ele.vtt.fi

Abstract   This paper deals with time simulation methods that are applicable for testing protocol software embedded in communicating systems. The novelty of the presented approach is that simulation time is used to control the level of non-determinism caused by the environment of the software under test. The test system is controlled in such a manner that external non-determinism is allowed, reduced, or eliminated according to the instructions given by the tester. The described methods have been evaluated in and applied to testing protocols for several standards. A TETRA terminal test system has been outlined as an example, demonstrating how time simulation can be integrated with a TTCN tester and target implementation generated from the SDL model.

Keywords: Embedded software, protocol testing, time simulation, SDL, TTCN

1.   INTRODUCTION

The protocol software testing of products consisting embedded communicating systems can be carried out in product environment, laboratory environment, and on workstations. The focus of this paper is on workstation testing, which is here referred to as simulation-based testing of protocol software. This kind of software validation involves simulating the environment of the component or subsystem being assessed. Thus, the target operating system is emulated, while the I/O interfaces, missing subsystems, and the communication network are simulated. Through the employment of this method in protocol integration testing, we have found out that it is the repeatability of tests that causes the most significant problems. The repeatability issue can be divided into such problem areas as *controllability of testing, observability of testing and test re-*

sult analysis. On closer examination, these subproblems reveal rather complex interdependencies in terms of time-dependent behaviour and non-determinism.

Controllability denotes the ease of producing a specific output from a specific input. Lack of controllability, on the other hand, is viewed as an inability to control the global state of a distributed system, caused by uncontrolled local clocks and by the use of message queues. Asynchronous local clocks of separate nodes make debugging difficult, the stopping of one node will not affect the other nodes. This is likely to cause serious problems to the debugging control. *Observability* implies the ease of determining, whether specified inputs are affecting the outputs. Adequate observability is essential for monitoring the dynamic behaviour of the entire test system (outputs and states) and for analysing the test results. Some of the changes in time-dependent behaviour may be ascribed to the instrumentation (probe effect) used for observing the system behaviour during testing. Furthermore, there may not be enough time to produce traces on the lower layers, which limits the possibilities of testing. The *analysis of test results* can be based on the test case verdicts (ISO/IEC 9646 1991) and/or trace analysis (Sarikaya 1988, Bochmann & Bellal 1990, Wvong 1991, Bochmann *et al.* 1991). When applying test case verdicts, non-determinism may lead to an explosion of alternatives in TTCN behaviour trees. Thus, the order of observed events and their timing information may vary in an unpredictable way, which, however, may still be acceptable.

In this paper, we use the term *internal non-determinism* referring to the non-deterministic implementation of the software under test (Weiss 1988, Schütz 1990, Carver & Tai 1991, Tai *et al.* 1991). *External non-determinism*, in turn, is triggered by the concurrent (and reactive) environment of the software (Kim *et al.* 1995). The internal behaviour of the software being tested can be serialised, so as to work out the deterministic variant of implementation (Tai 1985, Weiss 1988, Auer & Korhonen 1995). Another approach involves using deterministic scheduling and synchronisation in the operating system implementation (Schütz 1990). Both of these approaches may require changes in the implementation in order to initially remove non-determinism. In this case, however, a set of serialisation variants of the original implementation is tested, instead of testing the original concurrent program.

The *simulation time* can be handled using local clocks (de la Puente *et al.* 1993, de la Puente *et al.* 1994), and relying on the Lamport (1978) synchronising algorithm, for example. It is also possible to manage simulation time based on the local clocks using conservative (Candy & Misra 1979) or optimistic (Jefferson 1985) approaches (Fujimoto 1995). A single global simulation clock can also be applied (Billoir 1996). In addition, the advance of simulation time can be continuous or discrete (Taha 1988, Seila 1995). The instrumentation can be kept active in a real product so that its behaviour remains consistent during testing (Thome 1993, Liband 1995, Dotseth & Kuzara 1993,

Billoir 1996). Additional hardware and tools can be used for implementing non-intrusive monitoring. According to another, different approach, null virtual simulation time could be used in the execution of the instrumentation code parts.

We are proposing that the test system is controlled in such a manner that external non-determinism is allowed, reduced or eliminated according to the instructions given by the tester. *The novelty of this approach is that the simulation time is used for controlling the level of non-determinism caused by the environment of the software under test.*

2.   SIMULATION CONCEPT

The software being tested consists of application software and system software, i.e. the operating system. In our approach, the target operating system has been emulated on top of the host operating system (Figure 1).

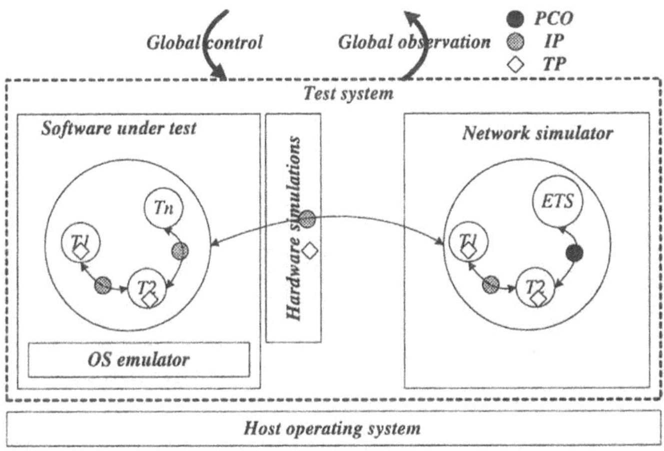

Figure 1   Test system.

Hardware simulations are required for simulating I/O, interrupts, and the behaviour of external hardware, such as processors and ASICs. The purpose of the network simulator is to implement a service provider for test cases. For example, when integrating the Ln, Ln-1, Ln-2, Ln-3, . . . , layers of a protocol stack, Ln-1, Ln-2, Ln-3, . . . , the layers of the peer protocol need to be simulated. Test cases are designed to fulfil a specific test requirement (test purpose), and their implementation is shown as an executable test suite (ETS) in Figure 1.

The global control and observation mechanism applies the PCO concept of ISO. In addition, the interaction point (IP) has been designed to serve as a message passing interface between any modules or any real-time tasks of the test system, or any I/O points between the software under test and its environment. At the test point (TP), some data items are changed and/or traced for testing purposes. The data may include an internal signal of a module or an I/O signal between the software under test and its environment. A signal may consist of a block of data which does not include any message identifier, or alternatively, of a single piece of data or, finally, of no data at all. A message is a block of data characterised by a message identifier.

3.   TIME SIMULATION

In this context, simulation time is equivalent to the behavioural time simulated for testing purposes. *Tick*, a simulated timer interrupt, serves here as a base unit, each tick increasing the time value by one. In our concept, the simulation time can be selected to be *local or global* (Figure 2). Simulation time is of the local type, if each node of the distributed test system has its own independent time reference. Accordingly, simulation time is global, if all the nodes share a common time reference.

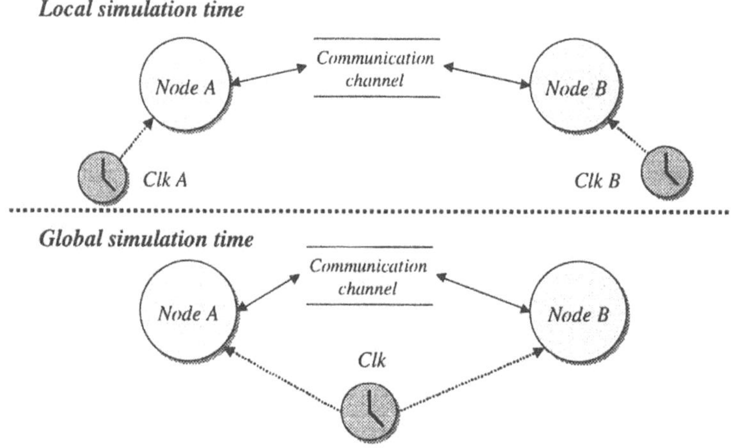

Figure 2   Local and global simulation time.

3.1   BEHAVIOURAL TIME VS. REAL-TIME

Behavioural time is based on timer services provided by the operating system. The smallest time unit of these timer services is usually a timer interrupt generated by the hardware or by a simulator. Real time, on the contrary, is based on the continuous advance of a real-time reference provided by an accurate hardware clock. The behaviour of the software under test is, naturally, dependent on real-time factors, including real-time delays of task execution times (Te), real-time delays caused by the instrumentation, i.e., the probe effect (Tp), message passing delays in mailboxes $(Tmbx)$, and message passing delays in the communication channels (*Tcc*) of the distributed simulation environment. Delays are also caused by the operating system due to the scheduling and synchronisation of real-time tasks (*Tos*), which, however, are not shown in Figure 3. In our approach, the real-time delays, such as *Te, Tp, Tmbx, Tos and Tcc*, are included in the testing of performance constraints. The testing is carried out in the target hardware environment. However, these delays can be totally eliminated in the testing of the behavioural time-dependent features in a host by means of time simulation. When behavioural time is simulated, it can advance only when simulated timer interrupts occur. Task execution time (*Te*), instrumentation (*Tp*), message passing in mailboxes (*Tmbx*), as well as the scheduling and synchronisation of real-time tasks (*Tos*) are assumed to be zero all the time. This is achieved by changing the real-time delay between two timer interrupts dynamically in such a way that there is just enough time for the software to carry out all the required operations before the next timer interrupt occurs. Similarly, the messages in the communication channels (*Tcc*) of the distributed simulation environment are transferred in null behavioural time. This is accomplished by keeping the order of timer interrupts and messages in the communication channels constant.

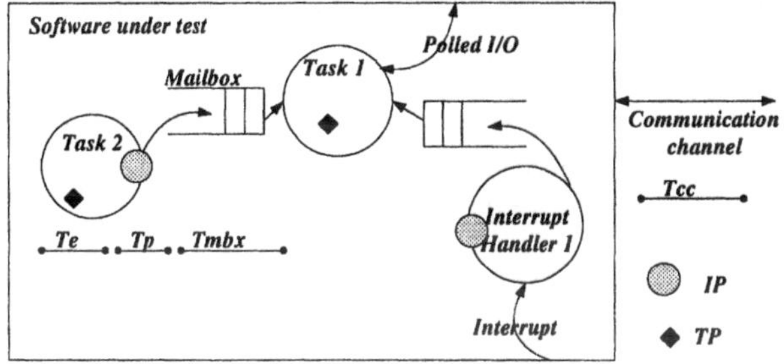

Figure 3   Time-dependent behaviour of a test system.

3.2   CONTINUOUS VS. DISCRETE SIMULATION

3.2.1   Continuous Simulation.   The time reference (local or global) can be based on the clock of the host. In this case, the timer delays have to be scaled to allow the execution of the test, as shown in the following (fixed scaled timing):

$$ST = S * N * Tick, \text{ where} \tag{1}$$
$$ST = \text{simulated timer delay}$$
$$S = \text{scaling factor}$$
$$N * Tick = \text{timer delay in the target HW environment}$$

In practice, the scaling factor depends on the load (L) and on the performance (P) of the host. Therefore, (1) can be presented as follows:

$$ST = S(L, P) * N * Tick \qquad (2)$$

The equation (2) can be demonstrated quite easily, since the execution of a test case with a constant scaling factor (S) for timer (T), the test results produced by separate test runs are usually different. Sometimes, scaling requires changing during a test session, because different test cases are likely to have different timing requirements. At other times, a previously passed test case may not be passed, due to someone having started some heavy operation after logging onto the same host. Furthermore, new code in the tested software (such as new traces, instrumentation for measuring test coverage, or new features) may change the timing and cause the tests to fail due to the timers not having been adjusted to compensate for the added code. In this case, premature timeouts may occur.

It is very difficult to design a function capable of predicting the load and the performance in a reliable way. One possible solution is, instead of adjusting the scaling factor, letting it remain one, to scale the length of the simulated tick ($STick$) dynamically, while keeping the tick multiplier (N) at the target environment value. Thus, (1) can be written as follows:

$$
\begin{aligned}
(1)\ ST &= S * N * Tick;\ Let(S = 1)\ and\ (Tick = STick) \\
\Rightarrow 1 * T &= N * Stick \\
\Rightarrow T &= N * Tick(L, P) \qquad (3)
\end{aligned}
$$

This allows the length of the tick to be scaled dynamically according to the load and the performance. This kind of timing is called dynamically scaled timing, involving dynamic adjustment of the length of the *simulated tick* during execution, in response to changes in the load and performance of the host (Figure 4). Since each tick is processed separately, this method can also be called *continuous simulation*. The fixed scaled timing involves scaling of the length of the target timer interrupt by a fixed multiplier, and therefore, the provided ticks are based on the clock of the host (i.e., the wall clock).

3.2.2   Discrete Simulation.   The purpose of continuous simulation is to eliminate the effects of load and performance by scaling individual tick delays dynamically. However, in real systems, timer delays are usually longer than a single tick. This means that there is not necessarily anything happening in the test system during the individual ticks. Therefore, it can be estimated that improved performance may be achieved through a discrete handling of simulation time. This calls for an ability to relate the time advance to discrete steps (simulation cycles) instead of individual ticks.

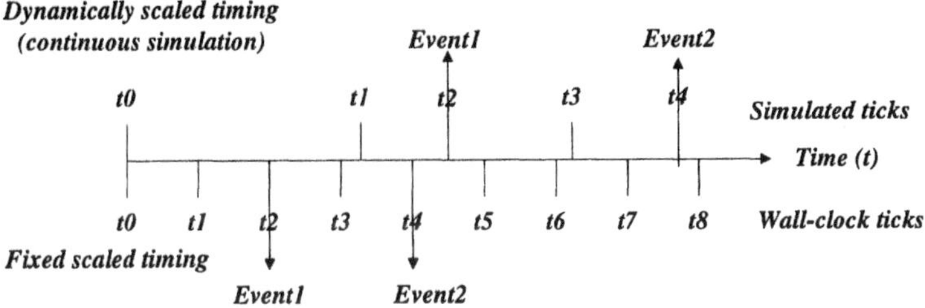

Figure 4   Continuous simulation and fixed scaled timing.

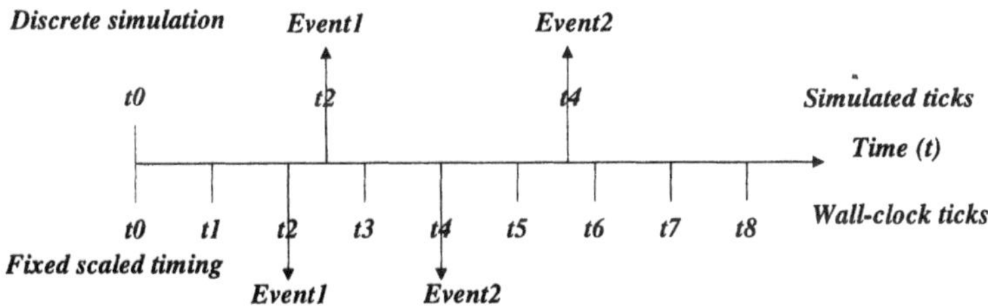

Figure 5   Discrete simulation and fixed scaled timing.

Discrete time simulation requires that each component of the test system has knowledge of the next expiring timer; this information can be used in defining the next step of simulation time. The tick multiplier (N) in equation (3) can be utilised in defining the next simulation cycle. The discrete simulation time principle involves a continual adjustment of the length of each simulation cycle during execution, according to the next expiring timer of the test system. Therefore, simulation time advances in discrete steps first from *t0* to *t2*, and then from *t2* to *t4*, and so on (Figure 5).

3.3   CONTROL OF SIMULATION TIME

The control of simulation time is carried out by means of mode selection (Figure 6). The mode of simulation time can be local fixed scaled timing, global fixed scaled timing, sequential continuous timing, parallel continuous timing, deterministic continuous timing, or deterministic discrete timing. The features of the modes are described briefly in the following:

Local or global fixed scaled timing: Timing interrupts in each component in the test system are generated by one or more simulation clocks. Communication and timing signals between the test system components are allowed to form without restrictions, which means that external non-determinism is permitted.

Sequential continuous timing: A global simulation clock generates timing interrupts in each component belonging to the test system in a predefined sequential order; ensuring that each component has processed the tick before it is sent to the next component. Thus, external non-determinism is reduced by means of a specific sequential order being predefined for the start of the event procedure(s), which are triggered by simulation clock ticks. However, some degree of external non-determinism is allowed, because the communication between test system components occurs without restrictions.

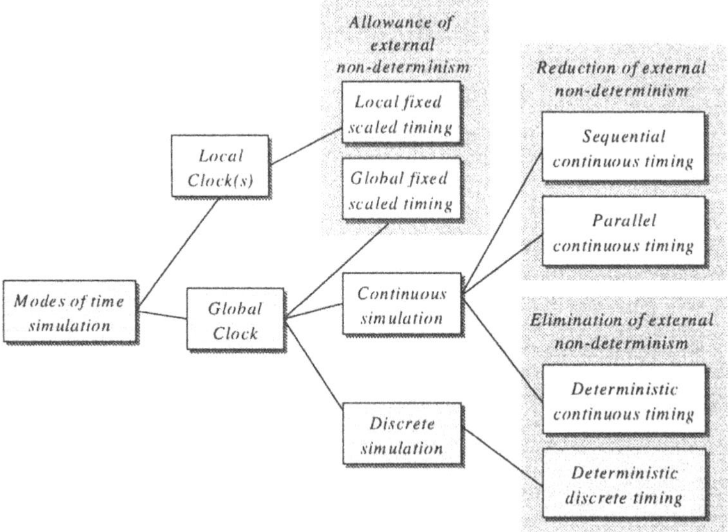

Figure 6   Control modes of simulation time.

Parallel continuous timing: Global simulation clock generates timing interrupts in parallel with all the components belonging to the test system. It also ensures that all the components have processed the tick before the next tick is sent. Thus, external non-determinism is reduced by forcing time to advance according to simulation clock ticks. However, timeouts may still occur in a non-deterministic order in the test system nodes, due to the fact that the ticks are delivered in parallel. In addition, communication between test system components is allowed to take place without restrictions.

Deterministic continuous timing: Global simulation clock works as a simulation scheduler, i.e., it delivers an execution token to each test system component at a given time and allows the component to accumulate its time by one tick. If any event procedures are triggered by the time increment, the system has to become stable before the time can advance. This allows eliminating external non-determinism and forcing concurrent behaviour to become sequential.

Deterministic discrete timing: Global simulation clock works as a simulation scheduler and delivers an execution token for each test system component at a time. It allows the component to accumulate its time by a specific amount of ticks (≥ 1). If any event procedures are triggered by the time increment, the system has to become stable before the time can advance. This allows eliminating external non-determinism and forcing concurrent behaviour to become sequential.

In summary, in the elimination of external non-determinism, simulation scheduling serialises the communication between the test environment and the software under test. Then, only a partial order of events is tested. In reducing external non-determinism, simulation time is used to enhance the controllability of testing while allowing several possible orders for the events. By allowing external non-determinism, simulation time is used to scale timing in such a way that the tests can be executed in a host.

3.4   ALGORITHM

The basic idea of deterministic discrete timing is that time is not allowed to advance before all the events of the current simulation cycle have been completed and the system has reached a stable state.

```
IF not token owner THEN
    IF received event is token-release THEN
        IF node has sent a message to another node
        AND not pending token-request related to it THEN
            set hold-token
        ELSE call SCHEDULE ENDIF
    END
    ELSE IF received event is token-request THEN
        Set pending token-request
    END
END
ELSEIF hold-token THEN
    IF received event is token-request THEN
        Set pending token-request
        call SCHEDULE
    END
ENDIF

PROCEDURE SCHEDULE
    IF pending token-request THEN
        Send Token to the required node
    ELSE
        go to the next simulation cycle
        Send Token to the next node
    ENDIF
END
```

Figure 7   Deterministic discrete timing algorithm.

A token is handed over to each component of the test system one at a time. In addition, a token may indicate timing advance, or just present an execution order to the component. The high-level control procedure of deterministic discrete timing is presented in Figure 7.

4.   CASE STUDY: TETRA TEST SYSTEM

This chapter outlines a real-life system for testing the TETRA mobile terminal software on a HP-UX workstation, simulating the target system. Here, we are demonstrating how time simulation can be applied in conjunction with TTCN (ISO/IEC 9646-3) test cases and an SDL-based (ITU Z.100) implementation of the MCU software for the TETRA terminal.

The simulation of time and target system relies on MOSIM, which is a simulation-based tool and platform, designed for testing embedded software (Honka 1992, Latvakoski 1997). The time simulation concept and algorithms presented in this paper were implemented as parts of MOSIM libraries. The MOSIM libraries embody time simulation probes, offering a generic interface for the time simulation and scheduling algorithms. The following description focuses on depicting how the time simulation interfaces with TTCN and SDL.

4.1   ARCHITECTURE

The architecture of the TETRA test system is illustrated in Figure 8. The system comprises four main components and invisible write servers running as separate UNIX processes: TETRA Mobile MCU (software to be tested, IUT) and DSP Simulator, TETRA Base Station L2 Simulator, L3 TTCN Tester, and MOSIM Test Controller.

In Figure 8, the grey areas embrace the modules and functionality that enable the inter-working of the components. Communication takes place via the MOSIM protocol. The actual implementation of the protocol is based on asynchronous UNIX Domain or TCP/IP socket communication, as well as on the write server concept. The motivation of the write server concept is to guarantee deadlock-free communication and fair scheduling of the components. In addition to the reliable delivery of the application messages, MOSIM protocol takes care of the delivery of global timing, proper event ordering, well-controlled start-up and shut-down, and synchronous debugging support.

The behavioural part of ETS is produced by ITEX CCG (Telelogic 1998), which generates C code from TTCN ATS. The purpose of the TTCN Run-time System (RTS) is to interface ETS with the test system. RTS contains three interfaces: timer management, snapshot (PCO) queues, and sending/receiving events via PCOs. The C code generated by ITEX CCG provides hooks for the implementation of these interfaces. The majority of the L2 Simulator is generated by SDT CCG (Telelogic 1998) from the SDL system modelling

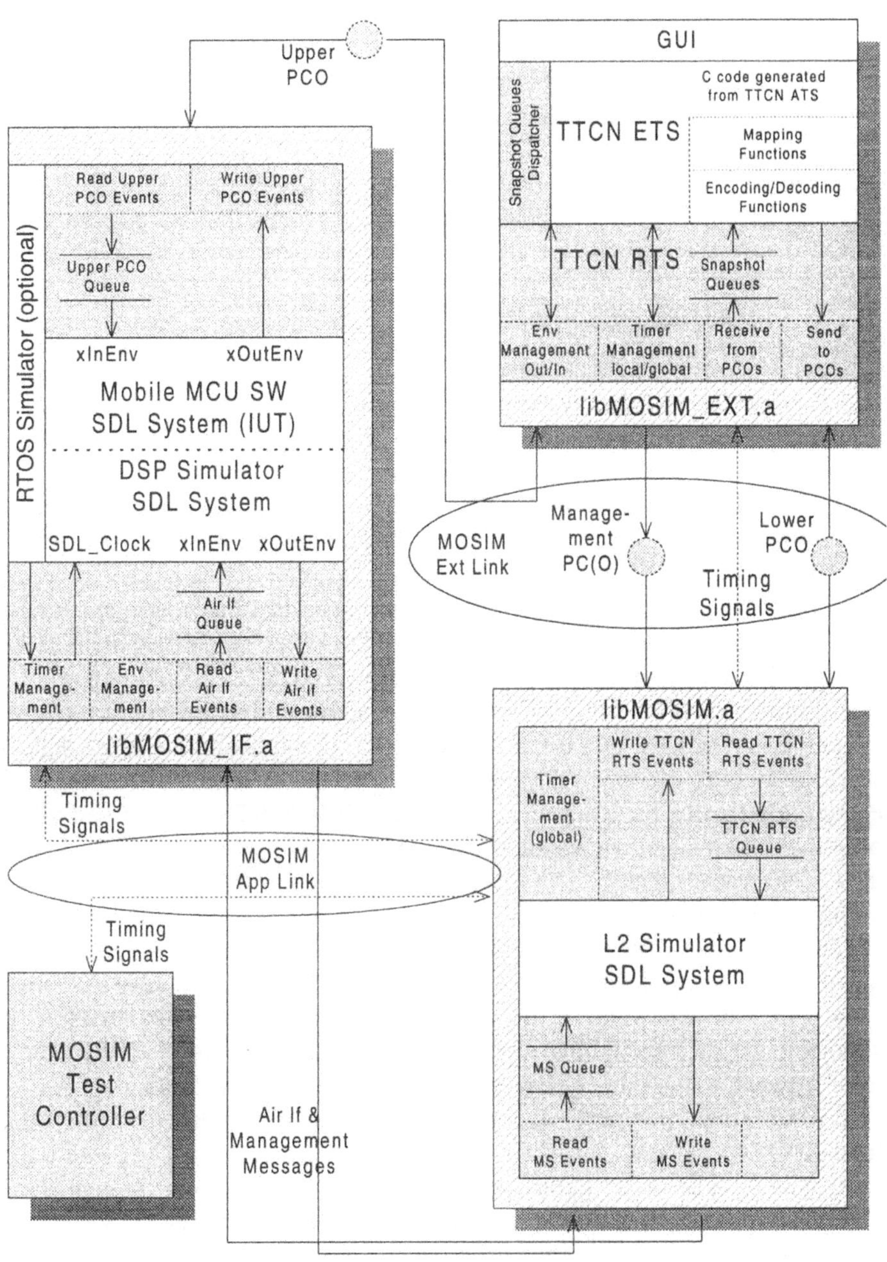

Figure 8   TETRA Test System Architecture.

MAC, LLC and MM layers. SDT Simulator UI is optionally available for facilitating the debugging at the SDL level. The Mobile SDL System can be executed under the SDT Simulator, which implies that an SDT scheduler is applied. However, there is a disadvantage: the SDT scheduling policy is significantly different from that of RTOS, which implies that testing under SDT Simulator does not cover all the behaviour combinations that are possible on RTOS.

4.2   TIMING MANAGEMENT

Figure 9 depicts a timing scenario that may take place in the TETRA test system. This scenario and timing management are explained in the following, starting from the TTCN Tester interface and completing with the Mobile interface.

TTCN Tester Interface: TTCN RTS timer management is capable of working in both local and global timing modes. It contains two interface functions called by libMOSIM_EXT.a: one for switching between timing modes and the other for catching timing signals. In the local mode, the timer management is implemented as a delta list based on the use of the UNIX setitimer system service. In the global mode, the timing signals arrive from MOSIM. The arrival of a new timing signal is notified by invoking the interface function that instigates the appropriate actions with timer management, i.e., updating the timer delta list and launching the timer(s) expired. As soon as all actions belonging to the current time tick have been completed (i.e., TTCN RTS goes into an idle state), the timing signal is receipted by sending the idle signal to MOSIM. In practice, this means that the idle signal is sent as soon as the TTCN RTS goes on stand-by in the snapshot queue.

In the continuous timing algorithm, each timing signal has to be receipted separately. This may lead to a remarkable overhead when a massive amount of test cases are executed. This overhead can be decisively decreased by employing the discrete timing algorithm in which the moment of the next time-depended action (i.e., TTCN timer) is estimated and indicated.

L2 Simulator Interface: Actual timing signals are hidden from the L2 Simulator SDL System, because libMOSIM.a assumes control of handling those signals. The effect upon timing becomes implicitly true through the MOSIM simulation clock that is a part of libMOSIM.a. The current value of the simulated time is periodically inquired by the function SDL_Clock. The time passed to the SDL system is the simulated time, either local or global, depending on the MOSIM setting. The switching between different timing modes is internally managed by libMOSIM.a. Therefore, whether the timing mode is local or global, is not explicitly known by the L2 Simulator.

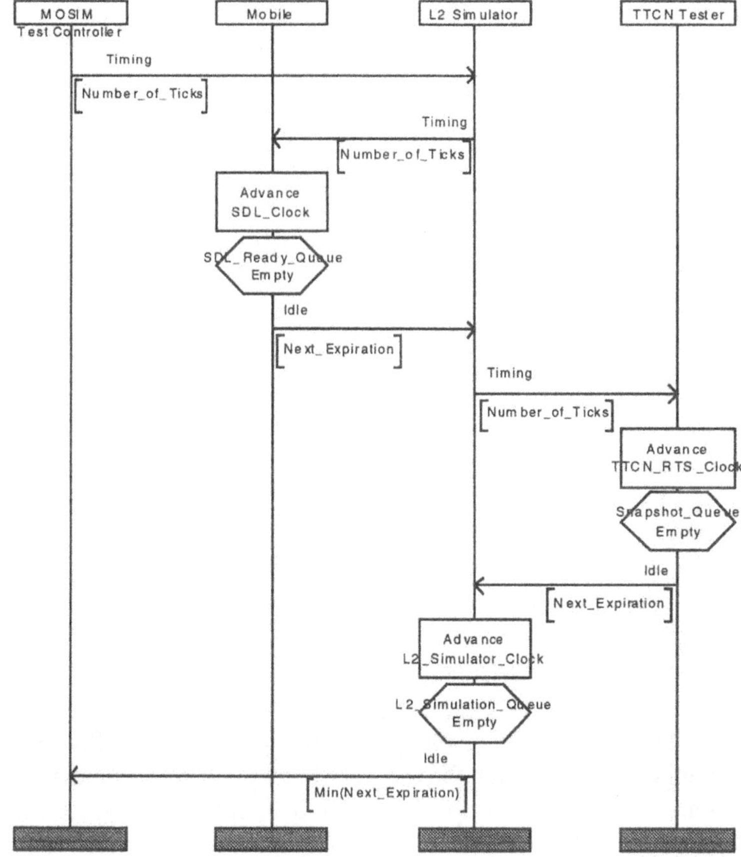

Figure 9   A timing scenario example.

Mobile SDL System Inteface: The timing modes available are, similarly
to those of the TTCN Tester, local and global. In the local timing mode,
the standard SDT implementation (i.e., high resolution time) of the function
SDL_Clock is utilised (i.e., the SDL System is polling time by calling this
function periodically from xMainLoop). In the global timing mode, the time
value returned by SDL_Clock is based on the global clock provided by MOSIM.
The switching between the timing modes is controlled by libMOSIM_IF.a
and implemented in the functions of SDT_enable_timer_ticks (local time) and
SDT_disable_timer_ticks (global time). According to the timing algorithm, an
idle signal is sent to MOSIM by the SDL System as soon as all the activities
scheduled for the current tick period have been completed. This is deduced
from the status of the SDL Process ready queue by polling SDL_Transition_Prio
in xInEnv. The idle signal is sent as soon as the ready queue is empty. The

overhead arising from continuous timing signals can be decreased by employing the discrete timing algorithm in which the estimated moment of the next time-depended action (i.e., SDL timer) is returned to MOSIM as an idle signal parameter.

In the case of the RTOS Simulator, a standard MOSIM-RTOS timing interface is employed. In the local timing mode, the internal RTOS timer ticks are controlling the RTOS scheduler and timers. In the global timing mode, timer ticks are fed by MOSIM. Accordingly, an idle signal is sent by the RTOS Simulator as soon as all the activities scheduled for the tick period(s) have been completed.

5.   CONCLUDING REMARKS

The time simulation methods described in this paper have been successfully employed in the integration testing of protocols for several standards, such as GSM and TETRA. The results indicate that the TTCN test sessions executed on a workstation simulating the target system are now repeatable and that tests can be executed automatically overnight. However, this requires a careful selection of the order of test execution. If an error is detected during a test session, the session will probably have to be aborted and restarted in order to prevent the emergence of potential side effects in the next test sets. The use of a global simulation clock has also facilitated the high-level controllability of debugging in distributed test systems by enabling the stopping of the time flow at exactly the increment of time, at which the software breakpoint occurs in the debugger (synchronous debugging). Furthermore, the demands arising from execution time no more restrict the capacity of observing time-critical lower protocol layers.

The developed system has turned out to be highly flexible, thanks to it allowing the selection of the time simulation mode according to the testing requirements. The main criteria for selection can be based on the behaviour of the target system, and, in particular, on whether the target system behaves in a non-deterministic way or not. However, the controllability of testing, the repeatability of tests, and the automating of test result analysis may require the use of a single global clock instead of a set of local clocks, so as to reduce non-determinism initially. The main benefit of the deterministic discrete timing algorithm can be found in the improved performance of the test system. For example, when using this algorithm, the execution time of an average test run of a GSM Phase 2 mobile terminal test system has been reduced to 68% of the time achieved with sequential continuous timing.

Acknowledgements: We would like to thank Mr. Juha Lehtikangas, Mr. Matti Sangi and Mr. Tero Manninen for their kind co-operation.

References

Auer, A. & Korhonen, J. 1995. State testing of embedded software. In: Conference Papers of 3rd European International Conference on Software Testing Analysis & Review. EuroSTAR'95. Jacksonville, FL: Software Quality Engineering. 13 p.

Billoir, T. 1996. Methods and tools for developing distributed real-time systems. Real-Time Magazine. No. 2. Pp.73-78.

Bochmann, G. & Bellal, O. B. 1990. Test result analysis with respect to formal specifications. In: de Meer, J., Mackert, L. & Effelsberg, W (ed.) Protocol Test Systems II. The Netherlands: Elsevier Science Publishers B.V. Pp. 103-117.

Bochmann, G., Desbiens, D., Dubuc, M., Ouimet, D. & Saba, F. 1991. Test result analysis and validation of test verdicts. In: Davidson, I. & Litwack, D. W. (ed.) Protocol Test Systems III. The Netherlands: Elsevier Science Publishers B.V. Pp. 263-274.

Candy, K. M. & Misra, J. 1979. Distributed simulation: A case study in design and verification of distributed programs. IEEE Transactions on Software Engineering. Vol. 5, no. 5. Pp 440-452.

Carver, R. H. & Tai, K. C. 1991. Replay and testing for concurrent programs. IEEE Software. Vol. 8, no. 2. Pp. 66-74.

Dotseth, M. & Kuzara, E. 1993. Real-time debugging of embedded operating systems. In: Proceedings of the Sixth Annual Embedded Systems Conference. Volume 1. San Francisco, CA: Miller Freeman Inc. Pp. 217-226.

Fujimoto, R. M. 1995. Parallel and distributed simulation. In: Alexopoulos, C., Kang, K., Lilegdon, W. & Goldsman, D. (ed.) Winter Simulation Conference Proceedings. San Diego, CA: WSC'95. Pp. 118-125.

ISO/IEC 9646 (1-5), 1991. Information technology - open systems interconnection - conformance testing methodology and framework. Ed 1. Geneva, Switzerland: IEC. 271 p. ITU Z.100 1994. Specification and description language (SDL). Ed 1. Geneva, Switzerland: ITU. 237 p.

Jefferson, D. R. 1985. Virtual time. ACM transactions on programming languages. Vol. 7, no. 3. 1985. Pp. 404-425.

Honka, H. 1992. A simulation-based approach to testing embedded software. Espoo: Technical Research Centre of Finland. 118 p. VTT Publications 124

Kim, M., Chanson, S.T. & Yoo, S. 1995. Design for testability of protocols based on formal specifications. In: Cavalli, A. & Budkowski, S. (ed.) Protocol Test Systems VIII. Proceedings of the IFIP WG6.1 TC6 8th International Workshop on Protocol Test Systems. Great Britain: Chapman & Hall. Pp. 252-263.

Lamport, L. 1978. Time, clocks, and the ordering of events in a distributed system. Communications of the ACM. Vol. 21, no. 7. Pp 558-564.

Latvakoski, J. 1997. Integration test automation of embedded communication software. Espoo: Technical Research Centre of Finland. 97 p. (VTT Publications 318)

Liband, J. 1995. Techniques for real-time debugging. Embedded Systems Programming. Vol. 8, no. 4. Pp. 34-56.

de la Puente, J., Alonso, A., Leon, G. & Duenas, J. C. 1993. Distributed execution of specifications. In: Real-Time Systems. Vol. 5, no 2-3. Pp. 213-234.

de la Puente, J., Leon, G., Alonso, A., Viana, J. C., Ruz, M. A. & Sink, E. W. 1994. Analysis of real-time object communication. IPTES Project EP5570 report IPTES-UPM-3-V2.6. Madrid, Spain: UPM. 57 p.

Sarikaya, B. 1988, Protocol test generation, trace analysis and verification techniques. In: Proceedings of the Second Workshop on Software Testing, Verification, and Analysis. Washington D.C.: IEEE Computer Society Press. Pp. 123-130.

Schütz, W. 1990. A test strategy for the distributed real-time system MARS. In: Proceedings of the COMPEURO '90 Conference. Washington D.C.: IEEE Computer Society Press. Pp. 20-27.

Seila, A. F. 1995. Introduction to simulation. In: Alexopoulos, C., Kang, K., Lilegdon, W. & Goldsman, D. (ed.) Winter Simulation Conference Proceedings. San Diego, CA: WSC'95. Pp. 7-14.

Tai, K. C. 1985. On testing concurrent programs. Proceedings of COMPSAC '85. Chicago, IL: IEEE Computer Society Press. Pp. 310-317.

Tai, K. C., Carver, R. H. & Obaid, E. E. 1991. Debugging concurrent Ada programs by deterministic execution. IEEE Transactions on Software Engineering. Vol. 17, no. 1. Pp. 45-63.

Taha, H. A., 1988. Simulation modeling and SIMNET. New Jersey: Prentice-Hall Inc. 397p.

Telelogic 1998. Telelogic Tau 3.4 User's Manual.

Thome, B., 1993. Systems engineering: principles and practice of computer-based systems engineering. England: John Wiley & Sons. 394 p.

Weiss, S. N. 1988. A formal framework for the study of concurrent program testing. In: Proceedings of the Second Workshop on Software Testing, Verification, and Analysis. Washington D.C.: IEEE Computer Society Press. Pp. 106-113.

Wvong, R. 1991. LAPB conformance testing using trace analysis. In: Jonsson, B., Parrow, J. & Pehrson B. (ed.) Participants Proceedings of the 11th International IFIP WG6.1 Symposium on Protocol Specification, Testing, and Verification. Stockholm. Sweden. Pp. 248-261.

Author Index

Keyword Index

GPSR Compliance
The European Union's (EU) General Product Safety Regulation (GPSR) is a set
of rules that requires consumer products to be safe and our obligations to
ensure this.

If you have any concerns about our products, you can contact us on

ProductSafety@springernature.com

In case Publisher is established outside the EU, the EU authorized
representative is:

Springer Nature Customer Service Center GmbH
Europaplatz 3
69115 Heidelberg, Germany

www.ingramcontent.com/pod-product-compliance
Ingram Content Group UK Ltd.
Pitfield, Milton Keynes, MK11 3LW, UK
UKHW020918130726
13719UKWH00012B/175